John Gould

Handbook to the Birds of Australia

John Gould

Handbook to the Birds of Australia

ISBN/EAN: 9783741182488

Manufactured in Europe, USA, Canada, Australia, Japa

Cover: Foto ©Andreas Hilbeck / pixelio.de

Manufactured and distributed by brebook publishing software (www.brebook.com)

John Gould

Handbook to the Birds of Australia

HANDBOOK

TO THE

BIRDS OF AUSTRALIA.

HANDBOOK

TO THE

BIRDS OF AUSTRALIA.

BY

JOHN GOULD, F.R.S., ETC.

AUTHOR OF THE 'BIRDS OF AUSTRALIA,' 'MAMMALS OF
AUSTRALIA,' 'BIRDS OF EUROPE,' 'BIRDS OF ASIA,'
MONOGRAPHS OF THE TROCHILIDÆ,
RAMPHASTIDÆ, TROGONIDÆ,
ODONTOPHORINÆ, ETC.

IN TWO VOLUMES.

VOL. I.

LONDON:
PUBLISHED BY THE AUTHOR,
20, CHARLOTTE STREET BEDFORD SQUARE.
1865.

PRINTED BY TAYLOR AND FRANCIS, RED LION COURT, FLEET STREET.

PREFACE.

Nearly twenty years have elapsed since my folio work on the Birds of Australia was completed. During that period many new species have been discovered, and much additional information acquired respecting those comprised therein; consequently it appeared to me that a careful *résumé* of the entire subject would be acceptable to the possessors of the former edition, as well as to the many persons in Australia who are now turning their attention to the ornithology of the country in which they are resident. Indeed I have been assured that such a work is greatly needed to enable the explorer during his journeyings, or the student in his quiet home, to identify the species that may come under his notice, and as a means by which the curators of the museums now established in the various colonies may arrange and name the collections intrusted to their charge. With these views the present Handbook has been prepared.

In order to facilitate a reference to the larger work whenever it may be required, the name under which each species appears therein, together with the number of the volume, and of the plate on which it is figured, is indicated at the head of the respective descriptions, in a bold and conspicuous type.

Modern research having ascertained that many of the species believed at the time I wrote to be new had been previously described by Latham and others, the specific names assigned to them by those authors have, in obedience to the law of priority, been restored; and the generic terms formerly employed have been changed wherever, in accordance with the advanced state of ornithological science, I have deemed it necessary.

Should any of my Australian readers consider that too many divisions have been made, and too many generic terms employed, I would direct their attention to the works recently published on the birds of other countries, and to the divisions and genera which their authors have deemed necessary, and they will at once perceive that I have not gone further in this direction than my ornithological compeers; I would also remind them that a true judgment on the necessity of so many separations cannot be formed from the perusal of a work on the birds of any single country, but can only be clearly understood when all the known birds of the world are brought under review. As an illustration of my meaning, I may mention that, in by far the greater number of cases, when only a single species of a genus is to be found in one part of the globe, others are to be met with elsewhere: this is the case with the genera *Garrulus*, *Pica*, *Nucifraga*, and *Pyrrhocorax*, of each of which only a single species occurs in Britain, while there are others in India or Africa. So it is with the Australian birds, many of the forms, of which only a single species is found in that country, being also existent in New Guinea and the neighbouring islands. In many instances the remarks on the various genera contained in the Handbook will be found to comprise references to these additional species.

The frequent repetition of the late Mr. Gilbert's name renders it necessary for me to state that he accompanied me to Australia, and diligently sought materials and information in behalf of my work, in those parts of the country which I was unable to visit myself, until I lost an able coadjutor, and science a devoted follower, by his premature death at the hands of the treacherous natives during Leichardt's expedition to Port Essington—a fate which also befell Strange, whose name likewise frequently occurs.

I would wish to remind the reader that many of the passages detailing the occurrence of certain species have reference to the date of my visit to the country. It may be possible—and, indeed it is most likely—that flocks of Parrakeets no longer fly over the houses and chase each other in the streets of Hobart Town and Adelaide, that no longer does the noble Bustard stalk over the flats of the Upper Hunter, nor the Emus feed and breed on the Liverpool plains, as they did at that time; and if this be so, surely the Australians should at once bestir themselves to render protection to these and many other native birds: otherwise very many of them, like the fine Parrot (*Nestor productus*) of Norfolk Island, will soon become extinct.

The Australian birds at present known amount, in round numbers, to about six hundred and sixty species.

HANDBOOK
TO THE
BIRDS OF AUSTRALIA.

INTRODUCTION.

IT will be evident to all who may turn their attention to the fauna of the great southern land called Australia, that it comprises peculiarities unexampled in any other portion of the globe. The total absence of the *Quadrumana*, the *Carnivora*, and the *Ruminantia* from its mammals, and of the *Vulturidæ* and *Picidæ* from (and the feeble representation of the *Gallinaceæ* among) its birds, are facts too striking to be overlooked. On the other hand, it possesses almost exclusively two orders of mammals, the *Marsupialia* and *Monotremata*, comprising many singular forms and a great number of species—among them being the large family of Kangaroos, of which only two or three abnormal species are known to exist elsewhere. In like manner the *Peramelidæ* and *Phascogalæ*, the Wombats and the Koalas, are almost as strictly confined to it: those singular forms the *Ornithorhynchus* and *Echidna* are only found there; and, supposing the Dingo to have been introduced by human agency, it is probable that no more highly organized placental land-animal than a *Vespertilio* or a *Mus* occurs in the whole of its Mammalia. Although Australia is destitute of many of the great groups of birds inhabiting India and its islands, she possesses many other equally singular forms especially adapted to find their existence among her very remarkable flora and her equally remarkable insects. Of these, her sixty species of Parrots, scarcely one of which is found beyond her limits, are unrivalled for size and beauty of plumage; conspicuous also are the extensive group of Honey-eaters forming the family *Meliphagidæ*, the elegant *Maluri*, the forest-loving Bower-birds, and the graceful Menuræ: all these, combined, furnish abundant illustration of the remark with which I commenced. The absence of such birds as Hornbills from Australia is evidently due to the circumstance of her flora not comprising any of the numerous large fruit-bearing trees which occur in India and Africa, and which are so essential to the existence of these birds; in like manner she is destitute of Woodpeckers, because the bark of her trees is not adapted for the shelter of the insects upon which they love to feed; neither do her few berry-bearing shrubs offer attractions to the *Eurylaimi* or the omnivorous Barbets and Trogons. No true Wagtail (*Motacilla*) trips over her hard-baked soil in pursuit of Aphides and other minute insects, as in Britain; no *Saxicola* enlivens with its sprightly actions her sterile wastes; and feebly indeed, among her birds, are represented the melodious notes which are freely poured forth by many of the species inhabiting countries north of the equator, and which render the spring such a joyous period in England. No Mavis has she to usher in the morning, and terminate the summer day, with its vigorous note; no Philomel to break the stillness of night with its joyous song: quietude, as regards the voice of birds, reigns supreme; or if there be any exception to this rule, it is the noisy screams of her Parrots, the monotonous though liquid notes of some of her Honey-eaters, the loud call of the Menura, or the warbling of the Reed-birds. Some parts of the avifauna of Australia are, however, very similar to that of other countries: Eagles, Hawks, Harriers, and Owls play their accustomed parts;

while Swifts, Swallows, Martins, and Flycatchers perform the same offices as with us; the nocturnal group of *Caprimulgidæ* are not wanting where *Phasmidæ* and *Cicadæ* abound; Petrels, Gulls, Terns, and Penguins frequent her seas; her rock-bound shores have their Cormorants; the sandy spits of her rivers their Pelicans, Sandpipers, and Plovers; and her swamps and morasses are tenanted by Ducks, Grebes, Gallinules, and Rails of the same types as those inhabiting her antipodes. But these, in nearly every case, are distinct species: she possesses no true *Anser*; but every one knows that she has that "rara avis in terris," the Black Swan. It is, however, in the interior of the country (adorned with the universally spread *Eucalypti*, extensive belts of *Banksiæ*, forests of *Xanthorrhææ*, *Melaluceæ*, &c.), and in the heated brushes which clothe its south-eastern portions (where stately palms spread their leaves over the eccalcobion or hatching-mound of the *Talegalla* and the theatreion or playing-bower of the *Ptilonorhynchus*), that we find an avifauna different from that of all other countries, and which presents us with such genera as *Ægotheles*, *Falcunculus*, *Collaricincla*, *Grallina*, *Cinclosoma*, *Menura*, *Psophodes*, *Malurus*, *Sericornis*, *Ephthianura*, *Pardalotus*, *Chlamydera*, *Struthidea*, *Licmetis*, *Calyptorhynchus*, *Platycercus*, *Euphema*, *Climacteris*, *Scythrops*, *Myzantha*, *Talegalla*, *Leipoa*, *Pedionomus*, *Cladorhynchus*, *Tribonyx*, *Cereopsis*, *Anseranas*, and *Biziura*.

Some few other genera, such as *Graucalus*, *Artamus*, and *Halcyon*, are represented in the Indian Islands, on the peninsula of India, and even in some portions of the continent of Asia; and many more genera, and in some instances the same species, extend to New Guinea. The productions of this latter country are, in fact, so similar to those of Australia, that, zoologically speaking, they cannot be separated. In writing thus I, of course, include the southern country of Tasmania, but not New Zealand or its satellites, Lord Howe's, Norfolk, and Phillip Islands, and other adjacent specks in the ocean, the culminating points probably of some great sunken continent, where bird-life reigned supreme; for we have no evidence that any mammals, either placental or marsupial, except Bats, ever roamed over its surface—mighty birds taking the place of Mammalia, as is evidenced by the remains of the great *Struthiones* which are almost daily exhumed from the morasses, alluvial beds, and caves of New Zealand.

Australia, lying as it does between the 10th and 45th degrees of south latitude, is subject to many varieties of climate. The northern portions are visited by monsoons; while the southern have seasons similar to those which occur in countries lying under the same degree of latitude north of the line; in a word, summer and winter are as in England, but of course at reversed periods of the year; as a natural consequence, when the sun retires for a period from any portion of the land, vegetation sleeps and insect-life becomes inert. Bird-life follows the law of nature, as seen in the northern hemisphere, and is much more rife at one season than at another. The Swallow and its congeners come and go as regularly in the southern portion of Australia as in England; and so do the Cuckoos, of which there are several species, and not only a single one as with us. Besides these, there are many other birds that are thus influenced; but the extent of their journeying has not been clearly ascertained further than that they generally proceed north when the sun retires, and return when he approaches: that they do not cross the equator is certain, for we should then find these peculiar species northward of the line, which we never do. There are also some non-migratory species which appear to perform a kind of exodus, and entirely forsake the part of the country in which they have been accustomed to dwell, and to betake themselves to some distant region, where they remain for five or ten years, or even for a longer period, and whence they as suddenly disappear as they had arrived. Some remarkable instances of this kind came under my own observation; for instance, the beautiful little warbling Grass Parrakeet (*Melopsittacus undulatus*), which prior to 1838 was so rare in the southern parts of Australia that only a single example had been sent to Europe, arrived in that year in such countless multitudes on the Liverpool Plains, that I could have procured any number of specimens, and more than once their delicate bodies formed an excellent article of food for myself and party. The *Calopsitta Novæ Hollandiæ* forms another case in point, and the beautiful Harlequin Bronze-winged Pigeon (*Phaps histrionica*) a third; this latter bird occurred in such numbers on the plains near the Namoi in 1839, that eight fell to a single discharge of my gun; both the settlers and natives assured me they had suddenly arrived, and had never before been observed in that part of the country. The aborigines who were with me, and of whom I must speak in the highest praise, from the readiness with which they rendered me their assistance, affirmed, upon learning the nature of my pursuits, that they had come to meet me! The *Tribonyx ventralis* may be cited as another species whose movements are influenced in the same way. This bird visited the colony of Swan River in 1833, and that of South Australia in 1840, in such countless myriads, that whole fields of corn were trodden down and destroyed in a single night; and even the streets and gardens of Adelaide were, according to Captain Sturt, alive with them. Similarly to what occurs in America and on other great masses of land, we find in Australia the law of representation markedly carried out, as it mostly is where the same conditions exist. For instance, the beautiful frill-necked Bower-bird of the scrubby plains of New South Wales is represented in north-western Australia by a nearly allied species, which makes its elegant bower in similar situations. The *Podargus humeralis*, which inhabits the *Angophora*-flats of New South Wales, is in like manner represented by the *P. brachypterus* in Western Australia, which presents a similar character of country; and so it is with many other species, both of mammals and birds.

These remarks might have been greatly prolonged; but I think sufficient has been said in the way of introduction to the subject which has to follow, a short succinct account of the Birds of Australia (a kind of handbook to my large illustrated work), in which I shall confine my remarks to the birds of the Australian continent, Tasmania, and those islands of the Great Barrier Reef which properly belong to Australia.

The history of the various species will be given in nearly the same order as in the folio edition, and will commence with the Raptores or birds of prey, to which will succeed the Insessores, or perching-birds; these will be followed by the Rasores, or Pigeons, the feebly represented Gallinaceæ, Quails, Partridges, &c.; then the Grallatores, or Plovers, Sandpipers, Ibises, Herons, &c.; and, lastly, the Natatores, which comprise all the various water-birds. This, however, is not to be regarded as a natural arrangement, but as one which offers great facilities for the study of the avifauna of a single country.

Order RAPTORES.

This Order, the members of which perform such important offices in the great scheme of nature, is but feebly represented in Australia as compared with Europe, Asia, and the other portions of the globe of similar extent; for in India alone, according to Mr. Jerdon, there are fifty-nine diurnal and twenty-two nocturnal birds of prey, while in Australia the number of the former is under thirty, and of the latter about ten. The absence of any great Equine, Bovine, Cervine, or Antilopine quadrupeds in Australia is doubtless the reason why her fauna contains no Vultures to act as scavengers, as they do in India and Africa when those huge beasts fall a prey to the large Carnivora. One typical Eagle and one equally typical Osprey play their accustomed parts in Australia, as do also the two or three species of true Falcons. Of Buzzards she has but one; but Kites and Harriers abound, as might be expected, in a country where reptiles are abundant, and which is visited at irregular periods by caterpillars to such an extent that the entire face of the country would be ravaged and rendered a desert were it not for these useful birds. Species of the Goshawk and Sparrowhawk type too are not wanting, to keep within bounds the smaller Mammalia and birds. Of the nocturnal Rapaces, the members of the genus *Strix* play the most important part; of these there are four very well-defined species which, in their structure and in the character of their plumage, assimilate most closely with the *Strix javanica* of India and the *Strix flammea* of Europe. These true nocturnes prey to a great extent upon the numerous species of small rodents which literally swarm in some parts of the country; while the huge yellow-eyed Owls (forming the genus *Hieracoglaux*) of the brushes feed upon birds, youthful Koalas, the night-loving Phalangistas, and Bandicoots.

Family FALCONIDÆ.

Genus AQUILA, *Briss.*

Sp. 1. AQUILA AUDAX.

Wedge-tailed Eagle.

Mountain Eagle of New South Wales, Collins, New South Wales, vol. ii. pl. in p. 288.
Vultur audax, Lath. Ind. Orn., supp. p. 2.
Falco fucosus, Cuv. Règn. Anim., 1st edit. pl. 3. f. 1.
Aquila fucosa, Cuv. Règn. Anim., 2nd edit. pl. 8. f. 1.
—— *albirostris*, Vieill. 2nde édit. du Nouv. Dict. d'Hist. Nat., tom. i. p. 229.
—— *audax*, Gray, Ann. Nat. Hist., vol. xi. p. 189.
—— *cuneicaudata*, Brehm, Isis, 1845, p. 356.
—— (*Uroaëtus*) *audax*, Kaup, Classif. der Säug. und Vög., p. 12.
Wol-dja, Aborigines of the mountain and lowland districts of Western Australia (Gilbert).
Eagle Hawk, Colonists of New South Wales.

Aquila fucosa, Gould, Birds of Australia, fol., vol. i. pl. 1.

This noble bird is so universally spread over the southern portion of Australia, that it is quite unnecessary for me to enter more minutely into detail respecting the extent of its range, than to say that it is equally distributed over the whole of the country from Swan River on the west to Moreton Bay on the east; it is also as numerous in Tasmania and on all the larger islands in Bass's Straits, being of course more plentiful in such districts as are suited to its habits, and where the character of the country is congenial to the animals upon which it subsists. I have not yet seen it, in any collection, either from the northern portion of Australia or any other country. In all probability it will hereafter be found that its range extends over corresponding latitudes in the southern hemisphere to those over which the Golden Eagle (*Aquila chrysaëtos*) does in the northern: the two birds are, in fact, beautiful analogues of each other in their respective habitats, and doubtless perform similar offices in the great scheme of creation.

Since the above passage was written, Mr. Gilbert has observed the bird on the Cobourg Peninsula; he did not, however, obtain a specimen; but the fact of its having been seen there by so careful an observer proves that its range extends to the extreme northern portion of the country; still I believe that it becomes more and more scarce as we advance northwards from the south coast.

All that has been said by previous writers respecting the courage, power, and rapacity of the Golden Eagle applies with equal force to the *Aquila audax*: in size they are also nearly alike; but the lengthened and wedge-shaped form of its tail gives to the Australian bird a far more pleasing and elegant contour.

I find by my notes that one of those I killed weighed nine pounds, and measured six feet eight inches from tip to tip of the opposite pinions; but far larger individuals than this have, I should say, come under my notice. The natural disposition of the Wedge-tailed Eagle leads it to frequent the interior portion of the country rather than the shores or the neighbourhood of the sea. It preys indiscriminately on all the smaller species of Kangaroo which tenant the plains and the open crowns of the hills, and whose retreats, from the wonderful acuteness of its vision, it descries while soaring and performing those graceful evolutions and circles in the air so frequently seen by the residents of the countries it inhabits: neither is the noble Bustard, whose weight is twice that of its enemy, and who finds a more secure asylum on the extensive plains of the interior than most animals, safe from its attacks; its tremendous stoop and powerful grasp, in fact, carry inevitable destruction to its victim, be it ever so large and formidable. The breeders of sheep find in this bird an enemy which commits extensive ravages among their lambs; and consequently in its turn it is persecuted unrelentingly by the shepherds of the stock-owners, who employ every artifice in their power to effect its extirpation; and in Tasmania considerable rewards are offered for the accomplishment of the same end. The tracts of untrodden ground and the vastness of the impenetrable forests will, however, for a long series of years to come afford it an asylum, secure from the inroads of the destroying hand of man; still, with every one waging war upon it, its numbers must necessarily be considerably diminished. For the sake of the refuse thrown away by the Kangaroo-hunters it will often follow them for many miles, and even for days together. I clearly ascertained that, although it mostly feeds upon living prey, it does not refuse to devour carrion or animals almost in a state of putridity. During one of my journeys into the interior to the northward of the Liverpool Plains, I saw no less than thirty or forty assembled together around the carcass of a dead bullock, some gorged to the full, perched upon the neighbouring trees, the rest still in the enjoyment of the feast.

Those nests that I had opportunities of observing were placed on the most inaccessible trees, and were of a very large size, nearly flat, and built of sticks and boughs. Although, during the months of August and September, I repeatedly shot the birds from their eyrie, in which there were eggs, I was quite unable to obtain them—no one but the aborigines (of which none remain in Tasmania) being capable of ascending such trees, many of which rise to more than a hundred feet before giving off a branch. But during the year 1864 a fine egg was presented to me by George French Angas, Esq.; and my son Charles, now engaged upon a geological survey of Tasmania, informs me that he has obtained others on that island. The egg is very similar in form and size to those of the Golden Eagle of Europe. It is clouded with large blotches of pale purple, and small specks and dashes of yellowish umber-brown on a stone-coloured ground, and is three inches in length by two and a half in breadth.

The adults have the head, throat, and all the upper and under surface blackish brown, stained on the edges and extremities of many of the feathers, particularly the wing- and upper tail-coverts, with pale brown; back and sides of the neck rusty red; irides hazel; cere and space round the eye yellowish white; bill yellowish horn-colour, passing into black at the tip; feet light yellow.

The young have the head and back of the neck deep fawn-colour, striated with lighter; all the feathers of the upper surface largely tipped and stained with fawn and rusty red; tail indistinctly barred near the extremity; throat and breast blackish brown, each feather largely tipped with rufous; the abdomen blackish brown.

Modern systematists consider that this bird was first characterized by Latham under the specific appellation of *audax*; I have therefore retained that name in lieu of *fucosa*, by which the bird has hitherto been known.

Genus HIERAËTUS, *Kaup.*

Sp. 2.   HIERAËTUS MORPHNOÏDES.

LITTLE EAGLE.

Aquila morphnoïdes, Gould, in Proc. of Zool. Soc., part viii. p. 161.
Butaëtus morphnoïdes, Blyth, Journ. of Asiat. Soc. Beng., vol. xiv. p. 546.
Hieraëtus morphnoïdes, Kaup, Classif. der Säug. und Vög.

Aquila morphnoïdes, Gould, Birds of Australia, fol., vol. i. pl. 2.

Since my discovery of this species on the Upper Hunter, in 1839, but few examples have been obtained, and it is still an exceedingly rare bird in the collections of Europe. My original specimen is now at Philadelphia, while, in England, a second is in the possession of T. C. Eyton, Esq., and I believe a third is comprised in the fine series of Raptorial Birds formed by J. H. Gurney, Esq., in the Museum at Norwich. At present no example is to be found in the National Collection. Captain Sturt obtained it at the Dépôt in South Australia; and Mr. White, of the Reed Beds, near Adelaide, has sent me a drawing of another obtained by him in the same country.

The Little Australian Eagle, which is about the size of the Common and Rough-legged Buzzards (*Buteo vulgaris* and *Archibuteo lagopus*), forms a beautiful representative of the *Hieraëtus pennatus* of Europe, its specific distinctions from which are its larger size, the total absence of the white mark on the shoulder, and the cere and feet being of a lead-colour instead of yellowish-olive.

The specimen obtained by myself was shot at Yarrundi on the river Hunter. I was led to the discovery of the bird by finding its nest, containing a single egg, upon which it had been sitting for some time. I regret to add that, although I several times visited the place after killing the bird, all my attempts at procuring the other sex were entirely unsuccessful. The nest was of a large size, and was placed close to the bole, about one-fourth of the height from the top, of one of the highest gum-trees; the egg was bluish white with very faint traces of brown blotchings, two inches and two lines long by one inch and nine lines broad.

Face, crown of the head, and throat blackish brown, tinged with rufous, giving it a striated appearance, and bounded in front above the nostrils with whitish; feathers at the back of the head, which are lengthened into a short occipital crest, back of the head, back, and sides of the neck, all the under surface, thighs, and under tail-coverts rufous, all but the thighs and under tail-coverts with a stripe of black down the centre of each feather; back, rump, and wings brown, the centre of the wing lighter; primaries brownish black, becoming darker at the tip, and barred throughout with greyish buff, which is conspicuous on the under surface, but scarcely perceptible on the upper, except at the base of the inner webs; under surface of the wing mottled with reddish brown and black; tail mottled greyish brown, crossed by seven or eight distinct bars of blackish brown, the tips being lighter; cere and bill lead-colour, passing into black at the tip; eye reddish hazel, surrounded by a blackish-brown eyelash; feet lead-colour.

Total length 21½ in.; bill 1¾; wing 15; tail 9½; tarsi 2¾.

Very recently I have received a second specimen of the egg of this species, presented to me by Mr. S. White of Adelaide, who, I believe, obtained it in the interior of South Australia, and which, although very similar to the one above described, differs in being more extensively blotched with yellowish brown and pale purple, the latter hue appearing as if beneath the surface of the shell.

Genus POLIOAËTUS, *Kaup.*

Sp. 3. POLIOAËTUS LEUCOGASTER.

WHITE-BELLIED SEA-EAGLE.

Falco leucogaster, Lath. Ind. Orn., vol. i. p. 13.
White-bellied Eagle, Lath. Gen. Syn., vol. i. p. 83.
Haliæetus leucogaster, Gould, Syn. Birds of Australia, part iii.
—— *sphenurus*, Gould, in Proc. of Zool. Soc., 1837, part v. p. 138, young.
—— (*Pontoëtus*) *leucogaster*, Kaup, Classif. der Säug. und Vög., p. 122.
Cuncuma leucogaster, List of Birds in Brit. Mus. Coll., part i. 2nd edit. p. 24.

Ichthyiaetus leucogaster, Gould, Birds of Australia, fol., vol. i. pl. 3.

I have little doubt that this noble species of Sea-eagle will be found to extend its range over all those portions of the Australian continent that present situations suitable to its peculiar habits. It has been observed along the whole southern coast, from Moreton Bay on the east to Swan River on the west, including Tasmania and all the small islands in Bass's Straits. It has neither the boldness nor the courage of the Wedge-tailed Eagle, *Aquila audax*, whose quarry is frequently the Kangaroo and the Bustard; and although, at first sight, its appearance would warrant the supposition that it pursues the same means for obtaining living prey as the true *Pandion*, by the act of submersion, yet I can affirm that this is not the case, and that it never plunges beneath the surface of the water, but depends almost entirely for its subsistence upon the dead Cetacea, fish, etc., that may be thrown up by the sea and left on the shore by the receding waves; to which, in all probability, are added living mollusks and other lower marine animals: its peculiar province is consequently the sea-shore; and it especially delights to take up its abode on the borders of small bays and inlets of the sea, and rivers as high as they are influenced by the tide; nevertheless it is to be met with, though more rarely, on the borders of lakes and inland streams, but never in the forests or sterile plains of the interior. As it is almost invariably seen in pairs, it would appear to be permanently mated; each pair inhabiting a particular bay or inlet, to the exclusion of others of the same species. Unless disturbed or harassed, the White-bellied Sea-eagle does not shun the abode of man, but becomes fearless and familiar. Among the numerous places in which I observed it in 1839 was the Cove of Sydney, where one or two were daily seen performing their aërial gyrations above the shipping and over the tops of the houses: if I mistake not, they were the same pair of birds that found a safe retreat in Elizabeth Bay, skirting the property of Alexander Macleay, Esq., where they might be frequently seen perched on the bare limb of a tree by the water's edge, forming an interesting and ornamental addition to the scene. In Tasmania it is especially abundant in D'Entrecasteaux Channel, and along the banks of the Derwent and the Tamar; and there was scarcely one of the little islets in Bass's Straits but was inhabited by a pair of these birds, which, in these cases, subsisted in a great measure on the Petrels and Penguins which resort there in great numbers to breed and which are very easily captured.

With regard to the nidification of the White-bellied Sea-eagle, I could not fail to remark how readily the birds accommodate themselves to the different circumstances in which they are placed; for while on the mainland they invariably construct their large flat nest on a fork of the most lofty trees, on the islands, where not a tree is to be found, it is placed on the flat surface of a large stone, the materials of which it is formed being twigs and branches of the Barilla, a low shrub which is there plentiful. While traversing the woods in Recherche Bay, I observed a nest of this species near the top of a noble stringy-bark tree (*Eucalyptus*), the bole of which measured forty-one feet round, and was certainly upwards of 200 feet high; this had probably been the site of a nest for many years, being secure even from the attacks of the natives, expert as they were at climbing. On a small island, of about forty acres in extent, opposite the settlement of Flinders, I shot a fully fledged young bird, which was perched upon the cone of a rock; and I then, for the first time, discovered my error in characterizing, in the 'Proceedings of the Zoological Society of London,' and in my 'Synopsis,' the bird in this state as a different species, under the name of *Haliæetus sphenurus*, an error which I take this opportunity to correct. The eggs are almost invariably two in number, of a dull white, faintly stained with reddish brown, two inches and nine lines long, by two inches and three lines broad.

This Sea-Eagle may be frequently seen floating about in the air above its hunting-ground, in circles, with the tips of its motionless wings turned upwards; the great breadth and roundness of the pinions, and the shortness of the neck and tail, giving it no unapt resemblance to a large butterfly.

The sexes are alike in plumage, but the female is considerably larger than her mate.

Adults have the head, neck, all the under surface, and the terminal third of the tail-feathers white; primaries and base of the tail blackish brown, the remainder of the plumage grey; irides dark brown; bill bluish horn-colour, with the tip black; cere, lores, and horny space over the eye bluish lead-colour slightly tinged with green; legs and feet yellowish white; nails black.

The young have the head, back of the neck, and throat light buff; all the upper surface and wings light chocolate-brown, each feather tipped with buffy white; tail light buffy white at the base, passing into deep brown towards the tip, which is white; chest brown, each feather margined with buff; abdomen mingled buff and brown, the latter colour occupying the margins of the feathers; under tail-coverts, and the under surface of the tail-feathers white.

Since the above was written, I have received an example from the north coast; and I have no doubt that its range extends over the islands of the eastern archipelago and the peninsula of India generally; indeed Mr. Jerdon states that it inhabits the sea-shores of the latter country, and occasionally ascends its larger rivers.

Whether as a result of the progress of civilization and the destructive hand of man this fine bird has been extirpated from the precincts of the great city of Sydney and similarly populous places is for the present race of Australians to say; in all probability, this to a certain extent has been the case; still the bird will hold its own in other parts of the colony for a long time to come; yet (and it is pitiable to comtemplate such a contingency) a period will doubtless arrive when the bays and inlets of the southern coast of Australia will no longer be adorned by the presence of this elegant species.

An opinion has been expressed that the enormous nests observed by Captains Cook and Flinders had been constructed by some species of *Dinornis*; but it is quite evident from the account given by Flinders that they must have been formed by a bird of the Raptorial order, and I have no doubt that they were the nests of the present bird.

"Near Point Possession," says Flinders, "were found two nests of extraordinary magnitude. They were built upon the ground, from which they rose above two feet; and were of vast circumference and great interior capacity, the branches of trees and other matter, of which each nest was composed, being enough to fill a small cart. Captain Cook found one of these enormous nests upon Eagle Island, on the east coast." Subsequently Flinders found another of these nests, in which were "several masses resembling those which contain the hair and bones of mice, and are disgorged by the Owls after the flesh is digested. These masses were larger, and consisted of the hair of seals and of land animals, of the scaly feathers of Penguins, and the bones of birds and small quadrupeds."—*Flinders's Voyage*, vol. i. pp. 64 and 81.

Genus HALIASTUR, Selby.

Sp. 4. HALIASTUR LEUCOSTERNUS, Gould.

WHITE-BREASTED SEA-EAGLE.

White-breasted Rufous Eagle, Lath. Gen. Hist., vol. i. p. 218.
Haliaëtus leucosternus, Gould, in Proc. of Zool. Soc., part v. p. 138.
Falco pondicerianus, Shaw, Nat. Misc., pl. 389.
Haliaëtus (Ictinoaëtus) leucosternon, Kaup, Isis, 1847, p. 270.
Gwerara, Aborigines of New South Wales.
Me-ne-u-roo, Aborigines of the Cobourg Peninsula.

Haliastur leucosternus, Gould, Birds of Australia, fol., vol. i. pl. iv.

In size and in the general markings of its plumage, this beautiful species is closely allied to the *Haliastur intermedius* of Java, and the *H. Indus* of India; but the total absence of the coloured stripe down the centre of the white feathers which clothe the head, neck, and breast of the Australian bird, at once distinguishes it from its Indian and Javan allies.

A more beautiful instance of analogy than that which these three birds offer to our notice can scarcely be imagined, and I feel assured that in their habits they are equally similar.

The White-breasted Sea Eagle is very common on the northern and eastern portions of Australia, where it takes up its abode in the most secluded and retired parts of bays and inlets of the sea. Upon one occasion only did I meet with it within the colony of New South Wales, but I have several times received specimens from Moreton Bay: the individual alluded to above was observed soaring over the brushes of the Lower Hunter. The chief food of this species is fish and crustaceans, which it captures either by plunging down or by dexterously throwing out its foot while flying close to the water, such fish as swim near the surface being of course the only ones that become a prey to it: sometimes the captured fish is borne off to the bird's favourite perch, which is generally a branch overhanging the water, while at others, particularly if the bird be disturbed, it is borne aloft in circles over the head of the intruder and devoured while the bird is on the wing, with apparent ease. Its flight is slow and heavy near the ground, but at a considerable elevation it is easy and buoyant.

In speaking of *H. Indus*, Colonel Sykes says, "It is seen constantly passing up and down rivers at a considerable height, but prepared to fall at an instant on its prey; usually it seizes when on the wing, but occasionally dips entirely under water, appearing to rise again with difficulty; the stomach of many specimens examined all contained fish and flesh, except one, in which a crab was found."

"This species," says Gilbert in his notes from Port Essington, "is pretty generally spread throughout the peninsula and the neighbouring islands, and may be said to be tolerably abundant. It breeds from the beginning of July to the end of August. I succeeded in finding two nests, each of which contained two eggs; but I am told that three are sometimes found. The nest is formed of sticks, with fine twigs or coarse grass as a lining; it is about two feet in diameter, and built in a strong fork of the dead part of a tree: both of those I found were about thirty feet from the ground and about two hundred yards from the beach. The eggs, which are two inches and two lines in length by one inch and eight lines in breadth, are of a dirty white, having the surface spread over with numerous hair-like streaks and very minute dots of reddish brown, the former prevailing and assuming the form of hieroglyphics—these singular markings being most numerous at one end, sometimes at the larger, at others at the smaller, the difference even occurring in the two eggs of the same nest."

The sexes are alike in colour; the young, on the other hand, differ considerably from the adult, being much darker, and, like the young of *H. Indus*, having the lower parts streaked and the upper spotted with fulvous; they have also darker-coloured eyes.

Head, neck, chest, and upper part of the abdomen snow-white; back, wings, lower part of the abdomen, thighs, upper and under tail-coverts rich chestnut-red; first six primaries chestnut at the base and black at the tip; tail-feathers chestnut-red on their upper surface, lighter beneath, the eight central feathers tipped with greyish white; irides light reddish yellow; cere pale yellowish white; orbits smoke-grey; upper mandible light ash-grey at the base, passing into sienna-yellow, and terminating at the tip in light horn-colour; under mandible smoke-grey; tarsi cream-yellow, much brighter on all the large scales of the tarsi and toes.

According to Mr. Gurney, this species has also been obtained by Mr. Wallace in Macassar, Batchian, Ternate, Timor, and Moro.

Sp. 5.   HALIASTUR? SPHENURUS.

WHISTLING EAGLE.

Milvus sphenurus, Vieill. 2nde édit. du Nouv. Dict. d'Hist. Nat., tom. xx. p. 564.
Haliaëtus canorus, Vig. and Horsf. in Linn. Trans., vol. xv. p. 187.
Milvus sphenurus, Swains. Class. of Birds, vol. ii. p. 211.
Haliaëtus (Ictinoaëtus) canorus, Kaup, Isis, 1847, p. 277.
Moru and *Wirrin*, Aborigines of New South Wales.
En-na-jook, Aborigines of the Cobourg Peninsula.
Jan-doo, Aborigines of the lowland districts of Western Australia.
Whistling Hawk, Colonists of New South Wales.
Little Swamp-Eagle, Colonists of Western Australia.

Haliastur sphenurus, Gould, Birds of Australia, fol., vol. I. pl. 5.

This species has been observed in every portion of Australia yet visited by Europeans. I did not meet with it in Tasmania; I am consequently led to believe that it rarely if ever visits that island. In New South Wales it is quite as numerous in summer as it is in winter; not that it is to be observed in the same locality at all times, the greater or less abundance of its favourite food inducing it to wander from one district to another, wherever the greatest supply is to be procured. It never attacks animals of a large size; but preys upon carrion, small and feeble quadrupeds, birds, lizards, fish, and the larvæ of insects; and while on the one hand it is the pest of the poultry-yard, on the other no species of the *Falconidæ* effects more good during the fearful visitations of the caterpillar, a scourge of no infrequent occurrence in Australia. So partial, in fact, is the Whistling Eagle to this kind of food that the appearance of one is the certain prelude to the appearance of the other. It is generally to be seen in pairs, inhabiting alike the brushes near the coast and the forests of the interior of the country. It is incessantly hovering over the harbours and sides of rivers and lagoons, for any floating animal substance that may present itself on the surface of the water or be cast on the banks; and when I visited the colony in 1839, it was nowhere more common or more generally to be seen than over the harbour of Port Jackson. Its flight is buoyant and easy, and it frequently soars to a great altitude, uttering at the same time a shrill whistling cry, from which circumstance it has obtained from the colonists the name of the Whistling Hawk.

The nest, which is constructed of sticks and fibrous roots, is frequently built on the topmost branches of the lofty *Casuarinæ* growing by the sides of creeks and rivers. The eggs, which are laid during the months of November and December, are usually two in number, but sometimes only one; they are two inches and three lines long by one inch and nine lines broad, and are of a bluish white slightly tinged with green, the few brown markings with which they are varied being very obscure and appearing as if beneath the surface of the shell. I once found a nest of this species in the side of which had been constructed that of the beautiful little Finch called *Amadina Lathami*, and both birds sitting on their respective eggs close beside each other; and both would doubtless have reared their progenies had I not robbed the nests of their contents to enrich my collection.

The Whistling Eagle, which is allied to the Kites, presents the usual difference in the size of the sexes, but in respect to colour no variation is observable; the plumage of the young, on the contrary, presents a striking contrast to that of the adult, being striated, and rendering the bird far handsomer during the first autumn of its existence.

Head, neck, and all the under surface light sandy brown, each feather margined with a darker colour; feathers of the back and wings brown, margined with greyish white; primaries blackish brown; tail greyish brown, rather long, and rounded at the end; cere and bill brownish white, gradually becoming darker towards the tip; legs bluish white; irides hazel.

Common also in New Caledonia (Gurney).

Genus PANDION.

Sp. 6.   PANDION LEUCOCEPHALUS, *Gould.*

WHITE-HEADED OSPREY.

Pandion leucocephalus, Gould, in Proc. of Zool. Soc., part v. p. 138.
—— *Gouldii*, Kaup, Isis, 1847, p. 270.—List of Birds in Brit. Mus. Coll., part i. 2nd edit. p. 22.
Yoon-door-doo, Aborigines of the lowland districts of Western Australia.
Joor-jool, Aborigines of Port Essington.
Little Fish-Hawk, Colonists of New South Wales.
Fish-Hawk, Colonists of Swan River.

Pandion leucocephalus, Gould, Birds of Australia, fol., vol. I. pl. 6.

The White-headed Osprey, though not an abundant species, is generally diffused over every portion of Australia suited to its habits; I myself shot it in Recherche Bay, at the extreme south of Tasmania; and Gilbert found it breeding both at Swan River on the western, and at Port Essington on the northern shore of Australia. Like its near allies of Europe and America (*Pandion haliæetus* and *P. carolinensis*), of which it is a beautiful representative in the southern hemisphere, it takes up its abode on the borders of rivers, lakes, inlets of the sea, and the small islands lying off the coast. Its food consists entirely of living fish, which it procures precisely after the manner of the other members of the genus, by plunging down upon its victim from a considerable height in the air with so true an aim as rarely to miss its object, although an immersion to a great depth is sometimes necessary to effect its accomplishment. Its prey when secured is borne off to its usual resting-place and devoured at leisure. Wilson's elegant description of the habits and manners of the American bird is in fact equally descriptive of those of the present species. Independently of its white head, this species differs from its near allies in the much lighter colouring of the tarsi, which are yellowish white slightly tinged with grey.

The nest being of great size is a very conspicuous object; it is composed of sticks varying from the thickness of a finger to that of the wrist, and lined with the softer kinds of sea-weed. It is usually placed on the summit of a rock, but is sometimes constructed on the top of a large *Eucalyptus*, always in the vicinity of water. A nest observed by Gilbert in Rottnest Island measured fifteen feet in circumference. The eggs are two in number, of a yellowish white, boldly spotted and blotched with deep rich reddish brown, which colour in some specimens is so dark as to be nearly black; other specimens, again, are clouded with large blotches of purple, which appear as if beneath the surface of the shell. The medium length of the eggs is two inches and five lines, and the breadth one inch and nine lines.

When near the water, its flight is heavy and flapping; but when soaring aloft at a great altitude, its actions are the most easy and graceful imaginable; at one moment it appears motionless, and at another performs a series of beautiful curves and circles, apparently for mere enjoyment; for from the great height at which they are executed it is hardly to be conceived that the bird can be watching the motions of its finny prey in the waters beneath. The velocity of the stoop made by these aquatic Eagles when in the act of capturing a fish is indeed truly wonderful; and equally surprising is the unerring aim and rapidity with which they clutch their victims.

Crown of the head, back of the neck, throat, abdomen, thighs, and under tail-coverts white; feathers of the chest mottled with brown, and with a dark brown mark down the centre; ear-coverts and sides of the neck dark brown; back, wings, and tail clove-brown, each feather of the back with a narrow circle of white at its extremity; primaries black; bill black; cere bluish lead-colour; feet pale bluish white; irides primrose-yellow in some, bright orange in others.

Mr. Gurney thinks this bird, *P. haliæetus*, and *P. carolinensis* may be one and the same species.

Genus FALCO.

The members of the genus *Falco* are perhaps more universally dispersed over the face of the globe than any other portion of the family of birds to which they belong; and I question whether the law of representation is in any case more clearly shown than by the numerous species of the present form.

Sp. 7.   FALCO HYPOLEUCUS, *Gould.*

GREY FALCON.

Falco hypoleucus, Gould in Proc. of Zool. Soc., part viii. p. 162.
Boork-ga, Aborigines of Moore's River in Western Australia.

Falco hypoleucus, Gould, Birds of Australia, fol., vol. ii. pl. 7.

"Of this rare and beautiful Falcon I have seen only four examples, three of which are in my own collection, and the fourth in that of the Earl of Derby[*]. The specimen from which my description in the 'Proceedings of the Zoological Society' was taken was presented to Mr. Gilbert by Mr. L. Burgess, who stated that he had killed it over the mountains, about sixty miles from Swan River; subsequently it was obtained by Mr. Gilbert himself in the vicinity of Moore's River in Western Australia; and my friend Captain Sturt had the good fortune to secure a male and a female during his late adventurous journey into the interior of South Australia. 'They were shot at the Dépôt in May 1845; they had been soaring very high, but at length descended to the trees on the creek, and coming within range were shot."

"The acquisition of the *Falco hypoleucus* is highly interesting, as adding another species to the true or typical Falcons,

[*] The last-mentioned specimen is now in the Derby Museum at Liverpool, to which town his Lordship bequeathed his fine collection.

and as affording another proof of the beautiful analogies which exist between certain groups of the southern and northern hemispheres,—this bird being as clearly a representative of the Jerfalcon of Europe as the *Falco melanogenys* is of the Peregrine, and the *Falco frontatus* of the Hobby.

"The adult has the whole of the upper and under surface and wings grey, with a narrow line of black down the centre of each feather; a narrow ring of black nearly surrounding the eyes; primaries brownish black, which colour assumes a pectinated form on a mottled-grey ground on the inner webs of those feathers; tail-coverts grey, barred with brownish grey; tail dark brownish grey, crossed with bars of dark brown; irides dark brown; cere, orbits, gape, base of the bill, legs and feet brilliant orange-yellow, the yellow becoming paler from the base of the bill until it meets the black of the tips.

"Total length of female, 17 inches; bill, $1\frac{1}{4}$; wing, $12\frac{1}{2}$; tail, $7\frac{1}{2}$; tarsi, $1\frac{3}{4}$.

"The young birds have the upper surface mottled brown and grey, and the under surface nearly white, and more strongly marked with black than in the adult."—*Birds of Australia*, fol., vol. i. p. at pl. 7.

Although a quarter of a century has elapsed since I first had the pleasure of characterizing this species by giving the above name to a young female then in my possession, little or no additional information has been obtained respecting the extent of the range of the bird, and still less about its habits and economy. In the folio edition of the 'Birds of Australia' I stated that four specimens were all that were then known; in the lengthened interval which has since elapsed, about the same number, and not more, have come under my notice; it must therefore still be considered as one of the rarities of the avifauna of Australia. When comparing this species with the Falcons of the northern hemisphere, *F. candicans*, *F. islandicus*, and *F. gyrfalco*, I have omitted to mention that, however identically typical in form it may be, and however similar in structure, to those noble birds, it is far more feeble and less able to cope with large-sized prey than its representatives of the snowy and glacial regions. In point of courage it will probably prove to be entitled to the first rank among the Australian *Falconidæ*, and may hereafter be found to be readily brought under command and used in falconry, should that interesting sport be taken up by the Australian offshoots from the inhabitants of Britain with the zeal with which it was followed by their ancestors. They certainly have the means for its pursuit; for no better birds than the present and the following species could be found for starting from the hand of "faire ladye."

A knowledge of the place of breeding and of the nest and eggs of this species is a matter of great interest to me, since the occurrence of a bird of this form in Australia would seem to indicate that there are yet some undiscovered high rocky lands in the interior of the country—localities which such birds principally affect.

Sp. 8.   FALCO MELANOGENYS, *Gould*.

BLACK-CHEEKED FALCON.

Falco melanogenys, Gould, in Proc. of Zool. Soc., part v. p. 139.
—— *macropus*, Swains. Anim. in Menag. p. 841.
Blue Hawk, Colonists of Western Australia.
Wolga, Aborigines of New South Wales.
Gwel-al-Bur, Aborigines of the mountain and lowland districts of Western Australia.

Falco melanogenys, Gould, Birds of Australia, fol, vol. I. pl. 6.

The bold and rapacious habits of the *F. melanogenys*, which, like the *F. hypoleucus*, may be classed among the noble Falcons, render it a favourite with the aborigines, who admire it for its courage in attacking and conquering birds much larger than itself.

What the Peregrine Falcon is to the continent of Europe

and England, the *Falco minor* is to South Africa, the *F. peregrinoïdes* to the peninsula of India, and the Black-cheeked Falcon to Australia. All these species are of the same type; but I agree with Prince Charles Lucien Bonaparte and Professor Kaup in considering them to be distinct, and representatives of each other in the respective countries they inhabit. The Duck-Hawk of America (*F. anatum*), as its trivial name implies, strikes down the *Anas obscura*; while the Peregrine (*F. peregrinus*) of Europe indulges a like taste by now and then taking a Mallard (*Anas boschas*); and Gilbert states that he has seen the Australian bird carry off a *Nyroca australis*—a species at least as heavy again as itself. To say, therefore, that this bird could not be trained and brought into use in the science of falconry would be to affirm what would probably prove to be untrue were the experiment made. Let the Australians, then, bestow some care upon this fine bird, and not, as they are doing with the Emu and the Bustard, let it be entirely eradicated from the fauna of the country. When I visited the colony in 1839, it was universally dispersed over the whole southern portion of Australia and Tasmania; and probably future research will discover that its range extends over all parts of the continent. It gives preference to steep rocky cliffs, and the sides of precipitous gullies, rather than to fertile and woodland districts, but especially seeks such rocky localities as are washed by the sea, or are in the neighbourhood of inland lakes and rivers. In such situations it dwells in pairs throughout the year, much after the manner of the Peregrine. Its breeding-season is, of course, in the spring of Australia—the autumn of Europe. Its nest is placed in those parts of the rocks that are most precipitous and inaccessible. The eggs are two in number; their ground-colour is buff; but this is scarcely perceptible, from the predominance of the blotching of deep reddish chestnut with which it is marbled all over; they are two inches and one line long, by one inch and seven and a half lines broad.

The stomach is large and membranous; and the food consists of birds, principally of the Duck tribe.

The sexes present the usual difference in size, the male being considerably smaller than the female.

The male has the head, cheeks, and back of the neck deep brownish black; the feathers of the upper surface, wings, and tail alternately crossed with equal-sized bands of deep grey and blackish brown; outer edges of the primaries uniform blackish brown, their inner webs obscurely barred with light buff; throat and chest delicate fawn-colour, passing into reddish grey on the abdomen; tail-feathers ornamented with oval-shaped spots of dark brown; abdomen, flanks, under surface of the wing, and under tail-coverts reddish grey, crossed by numerous irregular bars of blackish brown; bill light bluish lead-colour at the tip, becoming much lighter at the base; cere, legs, and feet yellow; claws black.

The female differs from the male in being larger in all her proportions, and in having the throat and chest more richly tinted with fulvous, which colour also extends over the abdomen, the feathers of which are not so strongly barred with brown as in the male.

The young of the first year have the breast longitudinally striped, instead of barred, as is the case with the young of the Peregrine.

Sp. 9.   FALCO SUBNIGER, *G. R. Gray.*
BLACK FALCON.

Falco subniger, Gray, in Ann. Nat. Hist. 1843, p. 371.
Falco (Hierofalco) subniger, Kaup, Isis, 1847, p. 70.

Falco subniger, Gould, Birds of Australia, fol., vol. I. pl. 9.

During the long interval which has elapsed since I first figured this bird in the folio edition of the 'Birds of Australia' no additional information has been obtained respecting this rare species of Falcon. Nothing is known of its habits, and as yet I have only seen four examples; all of which were killed in South Australia. It was observed by Captain Sturt during his expedition into the interior of that country, and he has favoured me with a note, in which he says, "This well-shaped and rapid bird was killed at the Dépôt, where both male and female were procured; but it was by no means common, only two others having been seen."

The occurrence of this Falcon in the interior of Australia is an additional evidence of the probability of there being mountainous districts in the unexplored portions of the north-west. Mr. Gurney states that it is also found in New Zealand.

The original specimen from which Mr. Gray took his description is in the British Museum.

The entire plumage dark sooty brown, becoming paler on the edges of the feathers of the upper surface; chin whitish; irides dark brown; cere yellow; bill lead-colour; legs and feet leaden yellow; claws black.

Total length 22 inches; bill 1 inch; wings $10\frac{1}{4}$; tarsi $2\frac{1}{4}$.

Ornithologists who may be desirous of becoming better acquainted with the *Falco subniger* will do well to consult the figure in the folio edition of the 'Birds of Australia.'

Sp. 10.   FALCO LUNULATUS, *Lath.*

WHITE-FRONTED FALCON.

Falco lunulatus, Lath. Ind. Orn. Supp., p. xiii.
Sparvius lunulatus, Vieill. Nouv. Dict. d'Hist. Nat., tom. x. p. 324.
Falco longipennis, Swains. Anim. in Menag., p. 341.
—— *frontatus*, Gould, in Proc. of Zool. Soc., part v. p. 139.
—— (*Hypotriorchis*) *frontatus*, Kaup, Isis, 1847, p. 65.
Wow-oo, Aborigines of the Murray in Western Australia.
Little Falcon, Colonists of Western Australia.

Falco frontatus, Gould, Birds of Australia, fol., vol. i. pl. 10.

This, which is the smallest of the true Falcons found in Australia, and which combines in its structure characters pertaining to the Hobby and to the Merlin of Europe, is universally spread over the southern portion of that country, Tasmania, and the islands in Bass's Straits; Gilbert also observed it on the Cobourg Peninsula, but did not succeed in obtaining a specimen. As its long pointed wings clearly indicate, it possesses great and rapid powers of flight; and I have frequently been amused by pairs of this bird following my course over the plains for days together, in order to pounce upon the Quails as they rose before me; and had I wished to witness falconry in perfection, I could not have had a better opportunity than on these occasions.

The White-fronted Falcon is a stationary species in all the colonies I visited. I succeeded in finding several of its nests, both in Tasmania and on the continent: they were all placed near the tops of the most lofty and generally inaccessible trees, and were rather large structures, being fully equal in size to that of a Crow, slightly concave in form, outwardly built of sticks, and lined with the inner bark of trees and other soft materials: the eggs were either two or three in number, of a light buff, blotched and marbled all over with dark buff, one inch and ten lines long by one inch and four lines broad.

The stomach is rather muscular and capacious, and the food consists of small birds and insects.

Forehead greyish white; crown of the head, cheeks, ear-coverts, and all the upper surface uniform dark bluish grey; internal webs of the primaries, except the tips, numerously barred with oval-shaped markings of buff; two centre tail-feathers grey, transversely barred with obscure markings of black; the remainder of the feathers on each side alternately barred with lines of dark grey and reddish chestnut; throat and chest white, tinged with buff, the feathers of the chest marked down the centre with a stripe of brown; the whole of the under surface and thighs dull reddish orange; irides blackish brown; bill bluish lead-colour, becoming black at the tip; cere, base of the upper mandible, legs, and feet yellow; claws black.

The sexes exhibit the usual difference in size, the female being much the largest. The plumage of the young differs from that of the adult in being more rusty and the markings less defined, in the feathers of the wings and tail being margined with rufous, and in the whole of the under surface being washed more deeply with rufous than the adult.

Genus HIERACIDEA.

Sp. 11.   HIERACIDEA BERIGORA.

BROWN HAWK.

Falco Berigora, Vig. and Horsf. in Linn. Trans., vol. xv. p. 184.
Ieracidea Berigora, Gould, Syn. Birds of Australia, part iii.
Berigora, Aborigines of New South Wales.
Orange-speckled Hawk of the Colonists.
Brown Hawk, Colonists of Tasmania.

Ieracidea Berigora, Gould, Birds of Australia, fol., vol. I. pl. 11.

This species is universally distributed over New South Wales and Tasmania, and is represented in Western and North-western Australia by a nearly allied species, to which I have given the name of *H. occidentalis*. In its disposition it is neither so bold nor so daring as the typical Falcons, but resembles in many of its habits and actions the Kestrels. Although it sometimes captures and preys upon birds and small quadrupeds, its principal food consists of carrion, reptiles, and insects: the crops of several that I dissected were literally crammed with the latter kind of food. It is generally met with in pairs; but at those seasons when hordes of caterpillars infest the newly sprung herbage it congregates in flocks of many hundreds—a fact I myself witnessed during the spring of 1840, when the downs near Yarrundi, on the Upper Hunter, were infested in this way to such an extent as to spread destruction throughout the entire district. By the settlers this bird is considered one of the pests of the country; but it was clear to me that whatever injury it may inflict by now and then pilfering the newly hatched chickens from the poultry-yard is amply compensated for by the havoc it commits among the countless myriads of the destructive caterpillar. To give an idea of the numbers of this bird to be met with at one time, I may state that I have frequently seen from ten to forty on a single tree, so sluggish and indisposed to fly that any number of specimens might have been secured.

So much difference occurs in the plumage of the *H. Berigora* that the changes it undergoes require to be closely studied. Professor Kaup considers it and the next species to be identical; but having had numerous opportunities of observing both birds in a state of nature, I regard them as distinct; and in confirmation of this opinion I may state that the present bird, which is from the eastern coast, is always the largest, has the cere blue-grey, and the plumage of the adult light brown, sparingly blotched with white on the breast; while the *H. occidentalis*, of the western coast, is a more delicately formed bird, has the cere yellow and the breast white, with faint lines of brown down the centre of each feather.

The sexes are nearly alike in colour, but the female is the largest in size. The *Hieracidea Berigora* breeds during the months of October and November.

The nest, which is placed on the highest branches of lofty *Eucalypti*, is similar in size to that of a Crow, is composed outwardly of sticks, and lined with strips of stringy bark, leaves, &c.; the eggs, which are very beautiful, and which are two and sometimes three in number, vary so much in colour that they are seldom found alike, even in the same nest; they are also longer or of a more oval shape than those of the generality of Falcons; the prevailing colour is, the ground buffy white, covered nearly all over with reddish brown; in some specimens an entire wash of this colour extends over nearly half the egg, while others are blotched or freckled with it in small patches over the surface generally: their medium length is two inches and two lines, and breadth one inch and six lines.

Crown of the head ferruginous brown, with a fine black line down the centre of each feather; a streak of black from the base of the lower mandible down each side of the cheek; ear-coverts brown; throat, chest, centre of the abdomen, and under tail-coverts pale buff, with a fine line of brown down each side of the shaft of every feather; flanks ferruginous, each feather crossed with spots of buffy white; thighs dark brown, crossed like the flanks, but with redder spots; centre of the back reddish brown; scapularies and wing-coverts brown, crossed with conspicuous bars and spots of ferruginous; tail brown, crossed with ferruginous bars, and tipped with light brown; primaries blackish brown, margined on their inner webs with large oval-shaped spots of buff; bill light lead-colour, passing into black at the tip; cere and orbits pale bluish lead-colour; irides dark brown; feet light lead-colour.

During the first autumn the dark markings are of a much deeper hue, and the lighter parts more tinged with yellow, than in the adult state, when the upper surface becomes of a uniform brown, and the white of the under surface tinged with yellow.

Sp. 12. HIERACIDEA OCCIDENTALIS, *Gould.*

WESTERN BROWN HAWK.

Ieracidea occidentalis, Gould, in Proc. of Zool. Soc., June 25, 1844.
Kar-gyne, Aborigines of the lowland and mountain districts of Western Australia.

Ieracidea occidentalis, Gould, Birds of Australia, fol., vol. i. pl. 12.

The *Hieracidea occidentalis*, which is very generally spread over Western and Southern Australia, loves to dwell in swampy places, where it may at all times find an abundant supply of lizards, frogs, and newts, to which are added young birds, insects, caterpillars, and carrion. As its small legs, compact body, and lengthened pointed wings would indicate, it flies with ease, making long sweeps and beautiful curves, which are often performed near the ground. The smaller size of this bird renders it a somewhat less formidable enemy to the denizens of the farm-yard than the Brown Hawk; still considerable vigilance on the part of the stock-keepers is necessary to check its depredations among the broods of poultry, ducks, &c.

The months of September and October constitute the breeding season; and the nest, which is formed of dried sticks, is usually built in thickly foliaged trees, sometimes near the ground, but more frequently on the topmost branches of the highest *Eucalypti*. The eggs, which are two, three, or four in number, differ very much in their markings, rich brown pervading the surface in some more than others: those in my collection measure two inches long by one and a half broad. Mr. White, of the Reed Beds, near Adelaide, kindly sent me some eggs of this species, accompanied by the following note:— "The nest is usually composed of sticks, and lined with leaves; the eggs, generally four in number, vary in intensity of colour, but differ in little or nothing from those of *H. Berigora*."

Crown of the head, back, and scapularies rusty brown, with a narrow stripe of black down the centre; rump deep rusty brown, crossed by broad bands of dark brown, the tip of each feather buffy white; wings very dark brown; the inner webs of the primaries with a series of large spots, assuming the form of bars of a deep rusty brown near the shaft, and fading into buffy white on the margin; wing-coverts tipped with rusty red; spurious wing with a row of rusty spots on either side of the shaft; tail dark brown, crossed by numerous broad irregular bars of rusty red, and tipped with pale buff; ear-coverts and a stripe running down from the angle of the lower mandible dark brown; chin, all the under surface, and a broad band which nearly encircles the neck pale buffy white, with a fine line of dark brown down the centre; thighs deep rust-red, each feather with a line of black down the centre and tipped with buffy white; irides reddish brown; eyelid straw-yellow; orbits bluish flesh-colour; bill bluish lead-colour, becoming black at the tip; cere pale yellow; legs and feet light ashy grey, excepting the scales in front of the tarsi, which are dull yellowish white.

Genus TINNUNCULUS.

Sp. 13. TINNUNCULUS CENCHROIDES.

NANKEEN KESTREL.

Falco Cenchroïdes, Vig. and Horsf. in Linn. Trans., vol. xv. p. 183.
Cerchneis immaculatus, Brehm, Isis.
Nankeen Hawk of the Colonists.

Tinnunculus Cenchroides, Gould, Birds of Australia, fol., vol. I. pl. 13.

Ornithologists will not fail to observe how beautifully the present bird represents in Australia the well-known Kestrel of the British Islands, which it closely resembles in many of its actions and in much of its economy; it flies over the whole of the southern parts of Australia, and that it extends far towards the northern portion of the country is proved by Gilbert having found it, as well as its eggs, during Leichardt's expedition from Moreton Bay to Port Essington.

Mr. Caley states that it is a migratory species, but I am inclined to differ from this opinion; his specimens were procured in New South Wales in May and June, while mine were obtained at the opposite season of December, when it was breeding in many of the large gum-trees on the rivers Mokai and Namoi; probably some districts are deserted for a short time and such others resorted to as may furnish it with a more abundant supply of its natural food, and this circumstance may have led Caley to consider it to be migratory.

The flight of the Nankeen Kestrel differs from that of its European ally in being more buoyant and easy, the bird frequently suspending itself in the air without the slightest apparent motion of its wings: having ascended to a great height, it flies round in a series of circles, these flights being often performed during the hottest part of the day—a circumstance which leads me to suppose that some kind of insect was the object of the search, it being well known that at midday insects ascend to a much greater altitude than at any other time.

The sexes present the usual differences in their markings, the female having all the upper surface alternately barred with buff and brown, while the male is furnished with a more uniform tint. I once took four fully-fledged young from a hole in a tree by the side of a lagoon at Brezi, in the interior of New South Wales; I also observed nests which I believe were constructed by this bird, but which may possibly have been deserted domiciles of a Crow or Crow-Shrike. Gilbert, in the journal kept by him during Dr. Leichardt's expedition, says:—"October 2. Found, for the first time, the eggs of *Tinnunculus Cenchroides*, four in number, deposited in a hollow spout of a gum-tree overhanging a creek; there was no nest, the eggs being merely deposited on a bed of decayed wood." They are freckled all over with blotches and minute dots of rich reddish chestnut on a paler ground, and are one inch and five-eighths in length by one inch and a quarter in breadth. I am indebted to Mr. S. White, of the Reed-beds, near Adelaide, in South Australia, for a fine set of eggs of this bird, which I believe were taken by himself in the interior of the country.

The male has the forehead white; head and back of the neck reddish grey, with the shaft of each feather black; back, scapularies, and wing-coverts cinnamon-red, with a small oblong patch of black near the extremity of each feather; primaries, secondaries, and greater coverts dark brown, slightly fringed with white; the base of the inner webs of these feathers white, upon which the dark colouring encroaches in a series of points resembling the teeth of a large saw; face white, with a slight moustache of dark brown from each angle of the mouth; chest and flanks buffy white, with the shaft of each feather dark brown; abdomen and under tail-coverts white; upper tail-coverts and tail-feathers for two-thirds of their length from the base grey; remaining portion of all but the two centre feathers white, crossed near the tip by a broad distinct band of deep black, the band being only on the inner web of the external feather; bill horn-colour near the base, black towards the tip; base of the under mandible yellowish; cere and orbits yellowish orange; legs orange.

The female has all the upper surface, wings, and tail cinnamon-red; each feather of the former with a dark patch of brown in the centre, assuming the shape of arrow-heads on the wing-coverts; the scapularies irregularly barred with the same, and the tail with an irregular band near the extremity; throat, vent, and under tail-coverts white; remainder of the under surface reddish buff, with a stripe of brown down the centre of each feather.

Genus LEUCOSPIZA, *Kaup.*

Sp. 14.   LEUCOSPIZA RAII.

NEW HOLLAND GOSHAWK.

Astur Raii, Vig. & Horsf. in Linn. Trans., vol. xv. p. 180.
Falco clarus, Lath. Ind. Orn. Supp., p. 13?
Fair Falcon, Lath. Gen. Syn. Supp., vol. ii. p. 54?

Astur Novæ-Hollandiæ, Gould, Birds of Australia, fol., vol. i. pl. 14.

The only part of Australia in which I met with this species was New South Wales, where it would appear to evince a preference for the dense and luxuriant brushes near the coast; but so little has at present been ascertained respecting its economy, range, and habits, that its history is nearly a blank; even whether it is migratory or not is unknown. That it breeds in the brushes of the district above-mentioned is certain; for I recollect seeing a brood of young ones in the possession of Alexander Walker Scott, Esq., of Newcastle on the Hunter, a gentleman much attached to the study of the natural productions of Australia. These young birds differed but little in colour from the fully adult specimens in my collection, except that the transverse markings of the breast were darker and of a more arrow-shaped form, which markings become fainter and more linear as the bird advances in age.

The sexes present the usual difference in size, but in colour and markings they closely assimilate.

All the upper surface grey; throat and all the under surface white, crossed with numerous irregular grey bars; cere yellowish orange; feet yellow; bill and claws black.

The irides of the young are brown.

Sp. 16. LEUCOSPIZA NOVÆ-HOLLANDIÆ.

White Goshawk.

Lacteous Eagle, Lath. Gen. Hist., vol. i. p. 216.
Astur Novæ-Hollandiæ, Vig. & Horsf. in Linn. Trans., vol. xv. p. 179.
Astur albus, Jard. & Selb. Ill. Orn., vol. i. pl. 1.
Falco Novæ-Hollandiæ, Lath. Ind. Orn., vol. i. p. 16.
Falco albus, Shaw, in White's Voy., pl. at p. 260.
Sparvius niveus, Vieill. Nouv. Dict. d'Hist. Nat., tom. x. p. 338.
Dædalion candidum, Less. Traité d'Orn., p. 60.
Astur (Leucospiza) Nov. Holl., Kaup, Class. der Säug. und Vög., p. 119.
New Holland White Eagle, Lath. Gen. Syn., vol. i. p. 40.
Goo-lou-bee, Aborigines of New South Wales (Latham).
White Hawk of the Colonists.

Astur Novæ-Hollandiæ, albino, Gould, Birds of Australia, fol., vol. i. pl. 15.

This species has perplexed ornithologists more, perhaps, than any other member of the Raptorial Order—the point at issue being whether it be distinct or merely an albino variety of the *Astur Raii*. I have seen both birds in a state of nature, and critically examined numerous examples after death with regard to size, admeasurement, &c.; and, except in colouring, I found no difference whatever between the beautiful snow-white bird and the grey-backed individuals so frequently shot in the brushes of the eastern parts of Australia. Mr. Ronald C. Gunn and the Rev. T. J. Ewing, of Tasmania, however, incline to believe them distinct, and, in support of this opinion, call attention to the fact that none but white birds have been found in that island; but while I admit this to be true, I do not fail to recollect that the most lovely individual I ever shot in Tasmania had fiery-red irides; still it is only fair to state they were not pink as in albinoes, and that most frequently the irides are bright yellow; the colouring of those organs therefore is evidently inconstant, and not to be depended upon as a characteristic. We know little or nothing of the nidification of either of the birds: could it be ascertained that the grey-backed and the white individuals mate with each other, they should be considered as identical; but until then it will be better, perhaps, to keep them distinct. Cuvier has hazarded the opinion that the white bird is an albino variety which has become permanent, and that they have the power of perpetuating their white vesture.

I think Professor Kaup is right in proposing a new generic title for this form, differing as it does both in structure and habits from the true Asturs, of which the *A. palumbarius* is the type.

The sexes differ very considerably in size, the male being scarcely more than two-thirds the size of the female.

The whole of the plumage pure white; cere and legs yellow; bill and claws black.

Genus ASTUR, *Lacépède*.

Sp. 10. ASTUR RADIATUS.

RADIATED GOSHAWK.

Falco radiatus, Lath. Ind. Orn. Supp., p. xii.
Haliaëtus Caleri, Vig. & Horsf. in Linn. Trans., vol. xv. p. 186.
Sparvius radiatus, Vieill. Nouv. Dict. d'Hist. Nat., tom. x. p. 310.
Accipiter radiatus, Gould, Ann. Nat. Hist., vol. xi. p. 393.
Astur testaceus, Kaup, Isis, 1847, p. 367.

Astur radiatus, Gould, Birds of Australia, fol., vol. i. pl. 10.

This bird, which at the period of my visit to Australia was only contained in the Linnean Society's collection, is still very rare in the museums of Europe. It inhabits the dense brushes bordering the rivers Manning and Clarence on the eastern coast of New South Wales, and that it enjoys a much greater range is more than probable. It is the largest of the Goshawks inhabiting Australia, the female nearly equalling in size that sex of the *Astur palumbarius* of Europe. In some parts of its structure the Radiated Goshawk differs considerably from the typical Asturs, particularly in the lengthened form of the middle toe, in which respect it resembles the true Accipiters; in its plumage it somewhat differs from both those forms, the markings of the feathers taking a longitudinal instead of a transverse direction. These and other slight differences may hereafter be considered of sufficient importance to warrant its separation into a distinct genus; but for the present I have retained it in that of *Astur*. Of its habits and economy nothing whatever is known.

The male, which is considerably smaller than the female, has the whole of the upper surface blackish brown, each feather broadly margined with rust-red; wings brown, crossed by narrow bands of darker brown; tail greyish brown, crossed by irregular bands of dark brown; shafts of the quills and tail buffy brown; throat buff, deepening into the rich rust-red of the under surface of the shoulder and the whole of the under surface; all the feathers of the under surface with a narrow stripe of black down the centre; thighs and under tail-coverts rust-red without stripes.

The female has the striæ of the under surface broader and more conspicuous.

Sp. 17. ASTUR APPROXIMANS, *Vigors and Horsfield*.

AUSTRALIAN GOSHAWK.

Falco radiatus, Temm. Pl. Col. 123, young.
Astur radiatus, Vig. and Horsf. in Linn. Trans., vol. xv. p. 181, young male.
Astur fasciatus, Id. ib., adult male and female.
Astur approximans, Id. ib., young female.—Gould in Syn. Birds of Australia, part iii.
Bilbil, Aborigines of New South Wales.
Nisus (Urospiza) radiatus, Kaup, Mus. Senckenb., 1845, p. 259.
—— (———) *approximans*, Kaup, Isis, 1847, p. 182.
Accipiter approximans, List of Birds in Brit. Mus. Coll., part i. 2nd edit. p. 74.

Astur approximans, Gould, Birds of Australia, fol., vol. i. pl. 17.

Among the whole of the Australian *Falconidæ* there is no species the scientific appellation of which is involved in so much confusion as that of the present bird. This has arisen from two causes,—first, from its having been erroneously considered to be identical with the *Falco radiatus* of Latham, from which it is entirely distinct; and secondly, from the difference which exists between the plumage of the adult and young being so great as to have led to a multiplication of specific names. Seven specimens of this Hawk formed part of the collection of the Linnean Society, now dispersed, and were those from which Messrs. Vigors and Horsfield took their descriptions of *Astur radiatus*, *A. fasciatus*, and *A. approximans*: from the careful examination I made of these specimens, I was satisfied that they were all referable to the present bird,—*A. radiatus* being the young male, *A. fasciatus* the adult, and *A. approximans* the young female. I have retained the term *approximans* in preference to either of the others, because *radiatus* actually belongs to another species, and the employment of *fasciatus* might hereafter lead to its being confounded with the "Fasciated Falcon," an Indian species described under that name by Dr. Latham.

This bird is one of the most abundant and generally dispersed of the Hawks inhabiting New South Wales and Tasmania. It is a species which ranges pretty far north; but on the western coast its place appears to be supplied by the *Astur cruentus*. The country between South Australia and Moreton Bay may be considered its true habitat; and there it is a stationary resident.

The Australian Goshawk is a bold, powerful, and most sanguinary species, feeding upon birds, reptiles, and small quadrupeds. It may often be seen lurking about the poultry-yard of the settler, and dealing destruction among the young stock of every kind.

Its nest is usually built on a large swamp-oak (*Casuarina*) growing on the side of a brook, but I have occasionally met with it on the gum-trees (*Eucalypti*) in the forest at a considerable distance from water; it is of a large size, and is composed of sticks and lined with gum-leaves. The eggs are generally three in number, of a bluish white, smeared over with blotches of brownish buff; they are one inch and ten lines long by one inch and five lines broad. The nesting-season commences in August, and continues till November.

The male, which is considerably less than the female in size, has the crown of the head and nape of the neck leaden grey; on the back of the neck an obscure collar of rufous brown; the remainder of the upper surface, wings, and tail deep greyish brown; the latter numerously barred with brown of a deeper tint; inner webs of the primaries and secondaries greyish white, barred with dark brown; throat greyish brown; breast and all the under surface rufous brown, crossed with numerous white fasciæ, which are bounded on each side with an obscure line of dark brown; thighs rufous, crossed by numerous irregular white lines; irides bright yellowish orange, surrounded by a yellowish lash; gape and base of the bill olive green; tip and the cere greenish yellow; legs and feet yellow; claws black.

The young differ considerably from the adult, having the feathers of the head and back of the neck dark brown, margined with rufous brown; the remainder of the upper surface deep brown, each feather with a crescent-shaped mark of rufous at the extremity; tail brown, crossed with obscure bars of a darker tint, and tipped with whitish brown; inner webs of the primaries fawn-colour, barred with dark brown; throat buffy white, with a stripe of dark brown down the centre of each feather; breast buffy white, each feather crossed by two bands of dark brown, the last of which assumes a triangular form; abdomen and flanks buffy white, crossed by irregular bands of dark brown, which are blotched with rufous brown in the centre; thighs and under tail-coverts pale rufous, crossed by similar bands; irides beautiful yellow; cere, base of the bill, and gape bluish lead-colour; point of the bill blackish brown; legs gamboge-yellow.

Sp. 18. ASTUR CRUENTUS, *Gould*.

WEST-AUSTRALIAN GOSHAWK.

Astur cruentus, Gould in Proc. of Zool. Soc., part x. 1842, p. 113.
Kil-lin-gil-lee and *Mal-wel-itch*, Aborigines of the mountain districts of Western Australia (Gilbert).
Good-jee-lum, Aborigines around Perth, Western Australia (Gilbert).

Astur cruentus, Gould, Birds of Australia, fol., vol. 1. pl. 18.

This Hawk is intermediate in size between the *Astur ap-*

proximans and *Accipiter torquatus*, has a more grey or blue coloured back, and has the transverse lines on the breast narrower and of a more rufous tint. It precisely resembles the first-mentioned bird in the rounded form of the tail, in the short powerful tarsus, and in the more abbreviated middle toe. I have been surprised by observing that the late Mr. Strickland considered this bird and the *A. approximans* to be identical; no two birds of the same genus can be more distinct.

The *Astur cruentus* is a very common species in Western Australia, particularly in the York district. Since the publication of the folio edition of the 'Birds of Australia,' I have seen a specimen of this bird from Port Essington; I believe it also occurs at Lombok, Batchian, and Timor, which proves that the species is found far beyond the limits of the colony of Western Australia. Like its congener, it is a remarkably bold and sanguinary species, often visiting the farmyard and carrying off fowls and pigeons with much apparent ease.

It breeds in October and the two following months, making a nest of dried sticks on the horizontal fork of a gum or mahogany tree.

The male has the crown of the head and occiput dark slate-colour; sides of the face grey; at the back of the neck a collar of chestnut-red; back, wings, and tail slaty brown, the brown hue predominating on the back, and the slate-colour upon the other parts; inner webs of the primaries fading into white at the base, and crossed by bars of slate-colour, the interspaces freckled with buff; the inner webs of the tail-feathers are marked in a precisely similar manner; chin buffy white; the whole of the under surface rust-red, crossed by numerous narrow semicircular bands of white; irides bright yellow; cere dull yellow; bill black at the tip, blue at the base; legs and feet pale yellow; claws black.

The female differs in having all the upper surface brown; the chestnut band at the back of the neck wider, but not so rich in colour; in all other respects she resembles her mate.

Genus ACCIPITER, *Brisson.*

Sp. 19. ACCIPITER TORQUATUS, *Vig. and Horsf.*

COLLARED SPARROW-HAWK.

Falco torquatus, Cuv.—Temm. Pl. Col., 43 (adult), 93 (young).
Accipiter torquatus, Vig. and Horsf. in Linn. Trans., vol. xv. p. 182.
Nisus australis, Less. Traité d'Orn., p. 61.
Sparvius cirrhocephalus, Vieill. Nouv. Dict. d'Hist. Nat., tom. x. p. 326.
—— *tricolor,* Vieill., ibid. p. 329.
Falco melanops, Lath. Ind. Orn. Supp., p. 12?
Sparvius melanops, Vieill. Nouv. Dict. d'Hist. Nat., tom. x. p. 230?
Astur (Micronisus) torquatus, Kaup, Mus. Senckenb., 1845, p. 259.
Nisus (Uruspiza) torquatus, Kaup, Isis, 1847, p. 181.
Accipiter cirrhocephalus, List of Birds in Brit. Mus. Coll., part 1, 2nd edit., p. 73.
Bilbil, Aborigines of New South Wales.
Jil-lee-jil-lee, Aborigines of the lowland, and
Min-min of the Aborigines of the mountain districts of Western Australia.
Little Hawk, Colonists of Swan River.

Accipiter torquatus, Gould, Birds of Australia, fol., vol. i. pl. 19.

This species is especially abundant in Tasmania and New South Wales, and would appear to enjoy a wide extent of range, since I have either seen or received specimens of it from every part of Australia, with the single exception of the north coast. Gilbert's notes inform me that he saw it there, but he did not obtain a specimen.

In its habits and disposition it has all the characteristics of its European ally the *Accipiter nisus,* whose boldness and daring spirit while in pursuit of its quarry have been so often described that they are familiar to every one; the sexes also exhibit the same disparity of size, the female being nearly as large and powerful again as her mate; hence the Quails and the numerous species of Honey-eaters find in her a most powerful enemy. For rapidity of flight and unerring aim, however, she is even surpassed by her more feeble mate, who may frequently be observed at one moment skimming quietly over the surface of the ground, and the next impetuously dashing through the branches of the trees in fearless pursuit of his prey, which, from the quickness of his abrupt turns, rarely eludes the attack. Mr. Caley mentions as an instance of its boldness, that he once witnessed it in the act of darting at a Blue Mountain-Parrot, which was suspended in a cage from the bough of a mulberry-tree, within a couple of yards of his door.

The breeding-season lasts from August to November, and the nest, which is rather a large structure, composed of sticks, and lined with fibrous roots and a few leaves of the gum-tree, is usually placed in the fork of a swamp-oak (*Casuarina*) or other trees growing on the banks of creeks and rivers, but is occasionally to be met with in the depths of the forests. The eggs are generally three in number, of a bluish white, in some instances stained and smeared over with blotches of buff; in others I have observed square-formed spots, and a few hair-like streaks of deep brown: their medium length is one inch and six lines by one inch and two lines in breadth.

Head, all the upper surface, wings, and tail deep brownish grey, the tail indistinctly barred with deep brown; on the back of the neck an obscure collar of reddish brown; throat, the under surface, and thighs rufous, crossed by numerous narrow bars of white, the red predominating on the thighs; under surface of the wings and tail grey, distinctly barred with dark brown, which is deepest on the former; irides and eyelashes yellow; cere and gape yellowish green; base of the bill lead-colour, tip black; legs yellow, slightly tinged with green.

The young have the usually striated plumage of the immature European Sparrow-Hawk, and, as is the case with the young of all the other members of the genus, have the irides darker than those of the adults.

Genus BUTEO, *Cuvier*.

Sp. 20. GYPOICTINIA MELANOSTERNON, *Gould*.

BLACK-BREASTED BUZZARD.

Buteo melanosternon, Gould in Proc. of Zool. Soc., part viii. p. 162.
Gypoictinia melanosternon, Kaup, Bonap. Consp. Gen. Av. tom. i. p. 19.
Goo-dap, Aborigines of the mountain districts of Western Australia.

Buteo melanosternon, Gould, Birds of Australia, fol., vol. i. pl. 20.

If we examine the Australian members of the family *Falconidæ*, we cannot fail to observe that they comprise representatives of most of the forms inhabiting similar latitudes in the northern hemisphere; and the bird now under consideration, if not a true *Buteo*, is more nearly allied to the members of that form than to those of any other genus; still it does differ somewhat from the typical Buzzards, and I have therefore considered it advisable to adopt Professor Kaup's generic title of *Gypoictinia*. This fine species does not appear to be common in any of the Australian colonies. I have, however, received it from Swan River, and procured it myself during my journey into the interior of New South Wales, about two hundred miles northwards of Sydney; I have also a specimen which was killed on the Liverpool Plains by one of the natives of New South Wales.

The Black-breasted Buzzard generally flies high in the air, through which it soars in large circles, much after the manner of the Wedge-tailed Eagle, its black breast and the large white mark at the base of the primaries being very conspicuous when seen from beneath. In these soaring actions it differs slightly from the typical species of the genus *Buteo*—an additional reason for its separation from those birds, and for the adoption of the distinctive generic appellation assigned to it by Dr. Kaup.

A most singular story respecting this bird has been transmitted to me, and is here given as I received it; without vouching for its truth, I may remark that the testimony of the natives from whom it was derived may generally be relied upon.

"The natives, Mr. Drummond, and his son Mr. Johnson Drummond tell me," says Mr. Gilbert, "that this bird is so bold, that upon discovering an Emu sitting on her eggs it will attack her with great ferocity until it succeeds in driving her from the nest, when, the eggs being the attraction, it takes up a stone with its feet, and while hovering over the nest lets it fall upon and crush them, and then descends and devours their contents. I have had numerous opportunities of observing the bird myself, and can bear testimony to its great powers of scent or vision; for upon several occasions, when the natives had placed a small kangaroo or kangaroo rat in the fork of a tree or on the top of a *Xanthorrhœa* with the intention of taking it again on our return, we have found that the Black-breasted Buzzard had discovered, and during our short absence had devoured every part of it except the skin, which was left so perfect, that at first I could not believe it had not been done by the hand of man."

The sexes are alike in colouring, but present the usual difference in size, the male being the smallest.

Crown of the head, face, chin, chest, and centre of the abdomen deep black, passing into chestnut-red on the flanks, thighs, and under tail-coverts; back of the head chestnut-red, becoming black in the centre of each feather; shoulders whitish buff; all the upper surface deep brownish black, margined with chestnut-red; primaries white at the base, deep black for the remainder of their length; cere and base of the bill purplish flesh-colour, passing into black at the tip; irides wood-brown; feet white, tinged with lilac.

I may remark that specimens of this bird are much required by the museums of Europe; it is to be wished also that persons favourably situated would ascertain if the story of the birds breaking the eggs of the Emu be correct, or if it be one of the numerous myths of the Aborigines.

Genus MILVUS, *Cuv.*

Asia, Europe, and North Africa are the great strongholds of the Kites or the members of the restricted genus *Milvus*; but at least two are natives of Australia. One of these, the *Milvus affinis*, is so like the *M. ater* of Europe, that some ornithologists consider them identical; but they are really quite distinct. I do not affirm this without having first consulted my friend, J. H. Gurney, Esq., than whom there is no more competent authority with regard to Raptorial birds.

Sp. 21.   MILVUS AFFINIS, *Gould.*

ALLIED KITE.

Milvus affinis, Gould, in Proc. of Zool. Soc., part v. p. 140.
—— (*Hydroictinia*) *affinis*, Kaup, Isis, 1847, p. 118.
B-le-nid-jul, Aborigines of Port Essington.

Milvus affinis, Gould, Birds of Australia, fol., vol. i. pl. 21.

The Allied Kite appears to enjoy a very wide distribution, since it not only inhabits Australia, but appears to extend its range through the Indian Islands to the peninsula of India. Mr. Gurney informs me that it occurs in Macassar, and certainly in India as far north as Nepaul, though it is generally confounded in the latter country with its larger relative *M. Govinda*. With the single exception of Tasmania, it is universally dispersed over all the Australian colonies; it is quite as common on the Cobourg Peninsula as it is in the southern portions of the country; and that it is as abundant in the centre of Australia as it is near the coasts is shown by Captain Sturt having observed it flying in great numbers over the far interior; but Mr. W. Allan informs me, in a letter dated August 8, 1859, "that there is an uncertainty, or rather an irregularity, in its appearance in different parts of the country. During a residence of nine years on the River Manning I never saw a single example of the bird until a few months ago, when it appeared on the flats bordering the river in flocks of forty or fifty in number. As far as I can learn, they previously appeared on the banks of the Hunter; they have now left, but are plentiful at Port Macquarie; they seem, therefore, to be journeying northward. I am told that they appeared in a similar manner about twenty years ago."

The confident and intrepid disposition of this bird renders it familiar to every one, and not unfrequently leads to its destruction, as it fearlessly enters the farm-yard of the settler, and, if unopposed, impudently plays havoc among the young poultry, pigeons, &c. It is also a constant attendant at the camps of the aborigines and the hunting-parties of the settlers, perching on the small trees immediately surrounding them, and patiently waiting for the refuse or offal. The temerity of one individual was such, that it even disputed my right to a Bronze-winged Pigeon that had fallen before my gun, for which act it paid the penalty of its life.

The flight of this bird, which is closely allied in character to that of the *Milvus ater* of Europe, is much less protracted and soaring than that of the typical Kites; the bird is also much more arboreal in its habits, skulking about the forest after the manner of the true Buzzards. Great numbers have been observed hovering over the smoke of the extensive bush-fires so common in Australia, closely watching for lizards and any of the smaller mammalia that may have fallen victims to the flames, or have been driven by the heat from their lurking-places.

The sexes are nearly alike in size and colouring.

Feathers of the head and the back and sides of the neck reddish fawn-colour, with a central stripe of dark blackish brown; all the upper surface glossy brown inclining to chocolate, and passing into reddish brown on the wing-coverts, the shaft of each feather being black, and the extreme tip pale brown; primaries black; secondaries blackish brown; tail,

which is slightly forked, brown, crossed by several indistinct
bars of a darker tint, and each feather tipped with greyish
white; throat brownish fawn-colour, with the stem of each
feather black; the remainder of the under surface rufous
brown, with a central line of dark brown on each feather,
which is broadest and most conspicuous on the chest; cere,
gape, and base of the lower mandible yellow; upper mandible
and point of the lower black; tarsi and toes yellow; claws
black; irides very dark brown.

Sp. 22.   MILVUS ISURUS, *Gould.*

SQUARE-TAILED KITE.

Milvus isurus, Gould, in Proc. of Zool. Soc., part v. (1837) p. 140.
Ge-dora-mal-uk and *Mar-ari,* Aborigines of the mountain districts of
    Western Australia (Gilbert).
Kite of the Colonists.

Milvus isurus, Gould, Birds of Australia, fol., vol. i. pl. 22.

This species, although possessing the short feet, long wings,
and other characters of the true Kites, particularly of the
Milvus regalis of the British Islands, may at once be distinguished from that bird by the square form of its tail. I met
with it in various parts of New South Wales, both in the
wooded districts near the coast and on the plains bordering
the interior; still it is by no means abundant, and persons who
had been long resident in the colony knew but little about it.
I had, however, the good fortune, in one instance, to find its
nest, from which I shot the female. I have received two
specimens from Swan River, and Mr. Gurney states that it
also inhabits New Zealand. It is a true Kite in all its
manners, at one time soaring high above the trees of the
forest, and at others hunting over the open wastes in search
of caterpillars, reptiles, and young birds.

The nest, which I found near Scone on the Upper Hunter,

in the month of November, was of a large size, built exteriorly of sticks, and lined with leaves and the inner bark of the gum-trees: it contained two eggs, the ground-colour of which was buffy white; one was faintly freckled with rufous, becoming much deeper at the smaller end, while the other was very largely blotched with reddish brown; they were somewhat round in form, one inch and eleven lines long by one inch and seven lines broad.

In his notes from Western Australia, Gilbert remarks that it is there "always found in thickly wooded places. Its flight at times is rapid, and it soars high for a great length of time. I found a nest on the 10th of November, 1839; it contained two young ones scarcely feathered, and was formed of sticks on a lofty horizontal branch of a white gum-tree, in a dense forest about four miles to the eastward of the Avon. I have not observed it in the lowlands, but it appears to be tolerably abundant in the interior. The stomach is membranous and very capacious: the food mostly birds."

Forehead and space over the eye buffy white, each feather tipped and marked down the shaft with black; crown of the head, back and sides of the neck, throat, shoulders, both above and beneath, and the under surface generally reddish orange; the feathers on the crown and the back of the head, like those of the forehead, marked longitudinally and tipped with black; but in no part are these markings so widely spread as on the chest, whence they suddenly diminish, and are altogether lost on the abdomen, the uniformity of which, particularly on the flanks, is broken by obscure transverse bands of a lighter colour; upper part of the back and scapularies deep blackish brown; tips of the primaries on the upper surface dark brown, obscurely banded with black; internal web of the basal portion of the primaries, together with the stem and under surface generally, greyish white; secondaries dark brown banded with black, the remainder of the wing light brown, the edges of the feathers being still lighter; rump and upper tail-coverts white, with transverse bands of brown and buff; tail brownish grey, and nearly square in form, all the feathers, except the two outer on each side, marked with about four obscure narrow bands of black, the whole tipped with black; irides very pale yellow, freckled with light rufous; cere, base of the bill and feet greyish white; culmen and tip of the bill, and claws black.

The female has the same character of markings as the male, but is readily distinguished by her greater size.

Genus ELANUS, *Savigny*.

The avifaunas of Europe, Asia, Africa, and the northern portion of America are enriched by one, two, or more species of this interesting form; another and a truly elegant species inhabits the Celebes and Java; and Australia is tenanted by two others (*E. axillaris* and *E. scriptus*), which appear to perform very important offices in the parts of the country they frequent: both are denizens of the warmer parts of Australia, and consequently do not proceed so far south as Tasmania. They hawk for insects in the air, and are truly beautiful when seen from beneath, their silvery-white under surface offering a pleasing contrast to the conspicuous markings of jet-black.

Sp. 23. ELANUS AXILLARIS.
BLACK-SHOULDERED KITE.

Falco axillaris, Lath. Ind. Orn. Supp., vol. ii. p. 42.
Circus axillaris, Vieill. Encyc. Méth. Orn., part iii. p. 1212.
Elanus melanopterus, Vig. & Horsf. in Linn. Trans., vol. xv. p. 185.
—— *notatus*, Gould, in Proc. of Zool. Soc., part v. p. 141.

Elanus axillaris, Gould, Birds of Australia, fol., vol. i. pl. 23.

The *Elanus axillaris* is a summer visitant to the southern portions of the Australian continent, over which it is very widely but thinly dispersed, being found at Swan River on the west coast, at Moreton Bay on the east, and over all the intervening country.

In its disposition it is much less courageous than the other members of the Australian *Falconidæ*, and, as its feeble bill and legs would indicate, lives more on insects and reptiles than on birds or quadrupeds.

I very often observed it flying above the tops of the highest trees, and where it appeared to be hawking about for insects; it was also seen perched upon the dead and leafless branches of the *Eucalypti*, particularly such as were isolated from the other trees of the forest, whence it could survey all around.

While under the Liverpool range I shot a young bird of this species which had not long left the nest; it is probable, therefore, that it had been bred within the colony of New South Wales; but I could never obtain any information respecting the nest and eggs.

The sexes closely assimilate to each other in colouring. The young differ in having the feathers of the upper surface tipped with buffy brown.

The adults have the eye encircled by a narrow ring of black; forehead, sides of the face, and under surface of the body pure white; back of the neck, back, scapularies, and upper tail-coverts delicate grey; a jet-black mark commences at the shoulders and extends over the greater portion of the wing; under surface of the shoulders pure white, below which is an oval spot of jet-black; primaries dark grey above, brownish black beneath; tail greyish white; bill black; cere and legs pale yellow; irides reddish orange.

On reference to the synonymy given above, it will be seen that neither Mr. Vigors nor myself had sufficiently studied the Australian Raptorial Birds described by the venerable Latham to be aware that he had assigned the specific designation of *axillaris* to this Kite; the terms *melanopterus* of Vigors, and *notatus* of myself, must therefore be reduced to synonyms.

Sp. 24.   ELANUS SCRIPTUS, *Gould.*

LETTER-WINGED KITE.

Elanus scriptus, Gould, in Proc. of Zool. Soc., June 28, 1842.

Elanus scriptus, Gould, Birds of Australia, fol., vol. I. pl. 24.

The principal character by which the *Elanus scriptus* is distinguished from the *E. axillaris* is the great extent of the black mark on the under surface of the wing, which, following the line of the bones from the body to the pinion, assumes when the wing is spread the form of the letter V, or, if both wings are seen from beneath at the same time, that of a W, divided in the centre by the body,—which circumstance has suggested the specific name I have applied to it.

It will be admitted by every one that this new species is an interesting addition to the Australian *Falconidæ*. Little or nothing was known respecting it when I published my figure in the folio edition of the 'Birds of Australia'; but we now know that it is a denizen of the interior of the country, Captain Sturt having obtained it at the Dépôt, and Mr. White, of the Reed-beds, South Australia, informing me that he found this species "in great numbers on Cooper's Creek, between lat. 27° and 28°, always in companies of from ten to twenty or thirty in number. It flies when near the ground with a heavy flapping motion, but occasionally soars very high, when its movements are very graceful. It is rather inquisitive, but not so bold as *Milvus affinis*. It nests in companies, as near each other as possible. The nest is composed of sticks, lined with the pellets ejected from their stomachs, which are principally composed of the fur of the rats upon which they chiefly subsist. The eggs, which are four or five in number, have a white ground, blotched and marked with reddish brown, darkest at the smaller end; they are one inch and three-quarters long, by one inch and three-eighths broad. The markings are easily removed by wetting."

Forehead and line over the eye white; head and all the upper surface dark grey, washed with reddish brown; wing-coverts deep glossy black; primaries greyish brown, becoming nearly white on their webs, all but the first two or three margined with white at the tip; secondaries brownish grey on the outer web, white on the inner and at the extremity; tertiaries brownish grey; two centre tail-feathers grey; the remaining tail-feathers pale brown on their outer webs, and white on the inner; lores black; all the under surface and edge of the shoulder white; on the under surface of the wing, following the line of the bones, a broad mark of black, assuming the form of the letter V; bill black; cere and legs yellow; claws black; irides reddish orange, and not yellow as represented in my figure.

Genus BAZA, *Hodgson*.

Of this genus four species are known; three of which inhabit India and the Indian islands, and the fourth Australia.

Sp. 25.   BAZA SUBCRISTATA, *Gould*.
CRESTED HAWK.

Lepidogenys subcristatus, Gould, in Proc. of Zool. Soc., part v. p. 140.
Avicida subcristata, Lafresn. Rev. Zool. 1846, p. 127.
Baza subcristata, G. R. Gray, List of Birds in Brit. Mus. Coll., part i. p. 19, 2nd edit. p. 41.
Pernis (Hyptiopus) subcristatus, Kaup, Isis, 1847, p. 343.

Lepidogenys subcristatus, Gould, Birds of Australia, fol, vol. i. pl. 25.

I am not sufficiently acquainted with this singular species to give any account of its habits and economy; but, judging from the feebleness of its bill and talons and the shortness of its tarsi, I conceive that it principally preys upon insects and their larvæ; and it is not improbable that honey and the larvæ of bees and ants, which abound in Australia, may form a portion of its food. Any information on this head that may have been ascertained by residents in Australia would, if made known, be of the highest interest to ornithologists, as an addition to the history of this singular form among the *Falconidæ*. Its extreme rarity, however, will, I fear, tend much to prevent the acquirement of this desirable information.

I saw it soaring high in the air over the plains in the neighbourhood of the Namoi, but never sufficiently near to admit of a successful shot. All the specimens I have seen were collected either at Moreton Bay or on the banks of the Clarence.

As little or no difference exists in the plumage of the specimens I have examined, I presume the sexes are very similar.

The only remark I have to make in addition to the above meagre account is, that I have lately received an egg procured in the brushes of the Clarence, and kindly sent to me by Mr. Allan, which is said to be of this bird. It is of a pure white, about an inch and five-eighths in length and an inch and a quarter in breadth. Without doubting Mr. Allan's intention to send me the egg of this species, I think it only right to say that I give the size and colouring on his authority; unfortunately the letter which accompanied it contained no remark on the subject.

Crown of the head, sides of the face, ear-coverts, and upper part of the back brownish grey; occiput and lengthened occipital plumes blackish brown; back and scapulars brown; wings uniform dark brownish grey above, beneath silvery grey; primaries and secondaries crossed by several bands, and largely terminated with black; rump and upper tail-coverts chocolate-brown; tail brownish grey above, lighter beneath, crossed by three narrow bands of black near the base, and deeply terminated with the same colour; throat, chest, part of the shoulder, and under tail-coverts greyish white tinged with rufous; abdomen, flanks, and thighs buffy white, crossed with conspicuous narrow bands of reddish chestnut; bill bluish horn-colour; tarsi yellowish.

Genus CIRCUS, *Lacépède.*

Two, if not three, Harriers inhabit Australia; consequently the number of species is nearly equal in Europe, Asia, Africa, America, and Australia. Those inhabiting the latter country are precisely of the same form, and perform the same offices as their near allies do in the other parts of the world.

Sp. 26. CIRCUS ASSIMILIS, *Jardine and Selby.*

ALLIED HARRIER.

Circus assimilis, Jard. & Selb. Ill. Orn., vol. ii. pl. 51.
—— *Gouldi,* Bonap. Consp. Gen. Av., tom. i. p. 34 (young ?).
Swamp Hawk of the Colonists.

Circus assimilis, Gould, Birds of Australia, fol., vol. i. pl. 26.

The *Circus assimilis* may be regarded as the commonest of the Harriers inhabiting New South Wales and South Australia; it also occurs, but in smaller numbers, in Tasmania. A Harrier is also rather abundantly dispersed over all the localities suitable to its existence in Western Australia, and it is just possible that it may prove to be the same species; if such should be the case, the whole of the southern portion of that vast country, from east to west, must be included within the range of its habitat. In size the *Circus assimilis* is but little inferior to the Marsh Harrier (*C. æruginosus*) of Europe, to which it offers a great resemblance in its habits and economy—being generally seen flying slowly and somewhat heavily near the surface of the ground, evincing a partiality to lagoons and marshy places, situations which offer it an abundance of food consisting of reptiles, small mammalia, and birds. I believe this bird also inhabits New Zealand, and that it is the *C. Gouldi* of Bonaparte.

That the Allied Harrier breeds in the localities in which I observed it I have little doubt, from the circumstance of the adults paying regular and hourly visits to the marshes in search of food, which was doubtless borne away to their young. When in a state of quiescence, this species, like the other Harriers, perches on some elevation in the open plain rather than among the trees of the forest—the trunk of a fallen tree, a large stone, or small hillock being among its favourite resting-places.

The sexes offer the usual differences in the larger size of the female; her markings are also rather less well-defined, and have not so much of the grey colouring as the male. The young resemble the young of the Marsh Harrier of Europe.

Head and all the upper surface rich dark brown; the feathers at the back of the neck margined with reddish buff; face light reddish brown; facial disk buffy white, with a dark stripe down the centre of each feather; all the under surface buffy white, which is deepest on the lower part of the abdomen and thighs, each feather with a streak of brown down the centre; upper tail-coverts and base of the tail-feathers white; remaining length of the tail-feathers brownish grey; irides yellow; eyelash and cere pale greenish yellow; bill dark brown, becoming light blue at the base; tarsi greenish white; feet yellowish buff; claws dark brown.

Mr. White, of Adelaide, informs me that "this bird is very numerous in South Australia during the summer months, and is generally found in swampy situations. I have seen it on the Murray, and in many other places. It feeds on eggs, birds, reptiles, and indeed on almost everything. I have often observed it flying close over the tops of the reeds, when quite dark. Its cry is a kind of loud shrill whistle of one note. At times it will fly very high. It varies much in colour; the two sexes are much alike, but the female is the larger bird of the two." I possess eggs which I have no doubt belong to this species; they are of a pure white, about one inch and seven-eighths long by one inch and a half in breadth.

Sp. 27.   CIRCUS JARDINII, *Gould*.

JARDINE'S HARRIER.

Circus Jardinii, Gould, in Proc. of Zool. Soc., part v. p. 141; and in Syn. Birds of Australia, part iii.
—— (*Spilocircus*) *Jardinii*, Kaup, Isis, 1847, p. 102.

Circus Jardinii, Gould, Birds of Australia, fol., vol. I. pl. 27.

This very beautiful Harrier, which is distinguished from every other species of the genus at present known by the spotted character of its plumage, is plentifully dispersed over every portion of New South Wales, wherever localities favourable to the existence of the Harrier tribe occur, such as extensive plains, wastes, and luxuriant grassy flats between the hills in mountainous districts. The extent of its range over the Australian continent has not yet been ascertained, and I have never observed it from any other portion of the country than that mentioned above; it is probable, however, that it extends all along the east coast. Mr. Wallace has obtained examples in Macassar.

To describe the economy of the Jardine's Harrier would be merely to repeat what has been said respecting that of the former species. Like the other members of the genus, it flies lazily over the surface of the plains, intently seeking for lizards, snakes, small quadrupeds, and birds; and when not pressed by hunger, reposes on some dried stick, elevated knoll, or stone, from which it can survey all around. Although I observed this species in all parts of the Hunter in summer, when others of the *Falconidæ* were breeding, I did not succeed in procuring its eggs, or obtain any satisfactory information respecting its nidification; in all probability its nest is constructed on or near the ground, on the scrubby crowns of the low, open, sterile hills that border the plains. An egg sent to me by Mr. White of Adelaide, and taken by him at Lake

Hope in the interior of South Australia, is white, one inch and seven-eighths long by one inch and a half broad.

The sexes present considerable difference in size, but are very similar in their markings; both are spotted; but the female is by far the larger and finer bird in every respect.

Crown of the head, cheeks, and ear-coverts dark chestnut, each feather having a mark of brown down the centre; facial disk, back of the neck, upper part of the back, and chest uniform dark grey; lower part of the back and scapulars dark grey, most of the feathers being blotched and marked at the tips with two faint spots of white, one on each side of the stem; shoulders, under surface of the wing, abdomen, thighs, and under tail-coverts rich chestnut, the whole of the feathers beautifully spotted with white, the spots, which are regularly disposed down each web, being largest and most distinct on the abdomen; greater and lesser wing-coverts brownish grey, irregularly barred and tipped with a lighter colour; secondaries dark grey, crossed with three narrow lines of dark brown, and tipped with a broad band of the same colour, the extreme tips being paler; primaries black for two-thirds of their length, their bases brownish buff; upper tail-coverts brown, barred and tipped with greyish white; tail alternately barred with conspicuous bands of dark brown and grey, the brown band nearest the extremity being the broadest, the extreme tips greyish white; irides bright orange-yellow; cere olive-yellow; bill blue at the base, black at the culmen and tips; legs yellow.

Those ornithologists who are in favour of a more minute division of the *Falconidæ* than myself may be inclined to adopt Professor Kaup's generic term of *Spilocircus* for this bird; but the propriety of separating it from the other Harriers appears to me very questionable, since it does not differ from them in structure in any respect.

Family STRIGIDÆ.

Genus STRIX, *Linn.*

In my remarks on the Raptores generally, I have mentioned that the birds of that order are but feebly represented in Australia as compared with their numbers in other parts of the globe; and I may now state, with regard to the Owls, that they are even less numerous than the *Falconidæ*; for, according to the present state of our knowledge, there appear to be but two, or at the most three forms in the country—*Strix, Hieracoglaux,* and *Spiloglaux.* The first of these genera comprises the true nocturnal Owls; the second the huge birds I have characterized under the specific appellations of *strenua, rufa,* and *connivens*; and the third the smaller species, *maculata, marmorata,* and *boobook.*

While as a general rule other great countries are only inhabited by a single species of the restricted genus *Strix,* the fauna of Australia comprises no less than four, all of which appear to be necessary to prevent an inordinate increase of the smaller quadrupeds which there abound.

Sp. 28.   STRIX CASTANOPS, *Gould.*

CHESTNUT-FACED OWL.

Strix castanops, Gould, in Proc. of Zool. Soc., part iv. p. 140.
Dactylostrix castanops, Kaup, Monog. Strig. in Jard. Cont. to Orn. 1852, p. 119.

Strix castanops, Gould, Birds of Australia, fol., vol. I. pl. 28.

Tasmania and probably the brushes of the opposite coasts of Victoria and New South Wales are the native countries of this Owl, a species distinguished from all the other members of its genus by its great size and powerful form. Probably few of the Raptorial birds, with the exception of the Eagles, are more formidable or more sanguinary in disposition.

Forests of large but thinly scattered trees, skirting plains and open districts, constitute its natural habitat. Strictly nocturnal in its habits, as night approaches it sallies forth from the hollows of the large gum-trees, and flaps slowly and noiselessly over the plains and swamps in search of its prey, which consists of rats and small quadrupeds generally.

I regret that the brevity of my stay in Tasmania did not admit of sufficient opportunities for observing this bird in its native haunts, and of my making myself acquainted with the various changes which take place in the colouring of its plumage. Considerable variety in this respect occurred among the specimens I collected—not so much in the form of the markings, as in the hue which pervades the face, neck, under surface, and thighs. In some these were deep rusty yellow; in others the same parts were slightly washed with buff, while others, again, had the face of a dark reddish buff approaching to chestnut, and the under surface much lighter; I have also seen others with the facial feathers lighter than those of the body, and, lastly, some with the face and all the under surface pure white, with the exception of the black spots which are to be found in all. Whether the white or the tawny plumage is the characteristic of the adult, or whether these changes are influenced by season, are points that might be easily cleared up by persons resident in Tasmania; and I would invite those who may be favourably situated for observation to fully investigate the subject, and make known the results.

The sexes differ very considerably in size, the female being by far the larger, and in every way more powerful than the male.

Facial disk deep chestnut, becoming deeper at the margin, and encircled with black; upper surface, wings, and tail fine rufous brown, each feather irregularly and broadly barred with dark brown, with a few minute white spots on the head and shoulders; under surface uniform deep sandy brown; sides of the neck and flanks sparingly marked with round blackish spots; thighs and legs the same, but destitute of spots; bill yellowish brown; feet light yellow.

Total length of the female 18 inches; bill 2¼; wing 15; tail 7; tarsi 3¼.

Sp. 29. STRIX NOVÆ-HOLLANDIÆ, *Steph.*

Masked Owl.

Strix ? Novæ Hollandiæ, Steph., Cont. of Shaw's Gen. Zool., vol. xiii. pt. ii. p. 61.

—— *personata,* Vig. in Proc. of Com. of Sci. and Corr. of Zool. Soc., part i. p. 60.

—— *Cyclops,* Gould, in Proc. of Zool. Soc., part iv. p. 140.

Strix personata, Gould, Birds of Australia, fol., vol. i. pl. 29.

This bird, although nearly allied to the preceding, differs in so many essential characters as to leave little doubt in my mind of its being specifically distinct. It is confined to the continent of Australia, over which it enjoys a wide range. With the exception of the north coast, I have received specimens from every part of the country. During my visit to the interior of South Australia, numerous individuals fell to my gun, which upon comparison presented no material variation from others killed in New South Wales and Western Australia.

If I were puzzled with respect to the changes to which the *Strix castanops* is apparently subject, I am not less so with those of the present bird; for although I find the tawny and buff colouring of the face and under surface is generally lighter, I also find a diversity in the colouring of the different parts of the under surface: in some specimens the face, all the under surface, and the ground-colour of the upper are pure white. Prior to my visit to Australia I characterized specimens thus coloured as a distinct species under the name of *Strix Cyclops,* but I now believe them to be one of the states of plumage of the present bird, which ornithologists are inclined to consider was first described by Stephens under the name of *S. Novæ-Hollandiæ.* I may remark that, out of the numerous examples I killed in South Australia in the month of June, I did not meet with one in the white plumage. Those who are desirous of making themselves acquainted with the differences in these nearly allied species of Owls will do well to consult the plates of the different species in the folio edition, which will render them more readily perceptible than the most lucid description.

The *Strix Novæ Hollandiæ* is almost a third smaller than the *S. castanops,* and as the sexes of both species bear a relative proportion in size, the male of the one is about equal to the female of the other. The white spottings of the upper surface of the former are larger than those of the latter, and the surrounding patches of dark brown and buff are not so deep, giving the whole of that part of the bird a more marbled or speckled appearance.

General colour pale buff; the upper part of the head, the back, and the wings variegated with dark brown, and sparingly dotted with white; under surface paler, with a few brown spots; tail buff, undulated with a brown facia; facial disk purplish, but margined with deep brown spots; bill pale horn-colour; toes yellow.

Sp. 30.   STRIX TENEBRICOSUS, *Gould.*

SOOTY OWL.

Strix tenebricosus, Gould, in Proc. of Zool. Soc., part xiii. p. 80.
Megastrix tenebricosa, Kaup, Monog. Strig. in Jard. Cont. to Orn., 1852, p. 120.

Strix tenebricosus, Gould, Birds of Australia, fol. vol. i. pl. 30.

Although I cannot possibly affirm that such is the case, I

believe this fine Owl to be an inhabitant of the great brushes of New South Wales, those of the Clarence, Richmond, &c.; for since the publication of my figure in the folio edition of the 'Birds of Australia,' I have received an example said to have been procured in one of those districts.

A fine specimen is comprised in the collection of the British Museum, and a second example in that of the Academy of Sciences at Philadelphia. It is a very powerful bird, and the rarest in our collections of the Australian members of the genus to which it belongs, from all of which it is conspicuously distinguished by the dark sooty hue of its plumage, and by the primaries being of a uniform colour, or destitute of the bars common to all the other species.

Facial disk sooty grey, becoming much deeper round the eyes; upper surface brownish black, with purplish reflexions, and with a spot of white near the tip of each feather; wings and tail of the same hue but paler, the primaries of a uniform tint, without bars, those of the tail faintly freckled with narrow irregular lines of white; under surface brownish black, washed with buff, and with the white marks much less decided; legs mottled brown and white; irides dark brown; bill horn-colour; feet yellowish.

Total length, 16 inches; bill, 1¾; wing, 12; tail, 5½; tarsi, 3.

Sp. 31.   STRIX DELICATULUS, *Gould*.

DELICATE OWL.

Strix delicatulus, Gould, in Proc. of Zool. Soc., part iv., 1836, p. 140.
Yon-ja, Aborigines of the Lowlands of Western Australia.

Strix delicatulus, Gould, Birds of Australia, fol. vol. i. pl. 31.

This is the least of the Australian Owls belonging to that section of the group to which the generic term of *Strix* has been retained; it is also the one most generally distributed.

I observed it in almost every part of New South Wales that
I visited; it is a common bird in South Australia, and I have
also seen specimens from Port Essington. It has not yet been
found in the colony of Swan River, nor can it be included in
the fauna of Tasmania. Although good specific differences
are found to exist, it is very nearly allied to the Barn Owl
(*Strix flammea*) of our own island, and the *S. javanica* of
India, and, as might be naturally expected, the habits, actions,
and general economy of the three species are as similar as is
their outward appearance: mice and other small mammals,
which are very numerous in Australia, are preyed upon as its
natural food. To attempt a description of its noiseless flight, its
mode of capturing its prey, or of its general habits, would be
merely to repeat what has been so often and so ably written
relative to the Barn Owl of Europe.

Although the plumage of youth and that of maturity do
not differ so widely in this species as in the other Australian members of the genus, the fully adult bird may always be
distinguished by the spotless and snowy whiteness of the
breast, and by the lighter colouring of the upper surface.

Facial disk white, margined with buff; upper surface light
greyish brown tinged with yellow, very thickly and delicately
pencilled with spots of brownish black and white; wings pale
buff lightly barred with pale brown, marked along the outer
edge and extremities with zigzag pencillings of the same, each
primary having a terminal spot of white; tail resembles the
primaries, except that the terminal white spot is indistinct,
and the outer feathers are almost white; under surface white,
sparingly marked about the chest and flanks with small
brownish dots; legs and thighs white; bill horn-colour; feet
yellowish.

Total length, 14 inches; bill, 1¾; wing, 11; tail, 4; tarsi, 2¼.

Mr. Gurney informs me that this species is also found in
New Caledonia, and in Aniteum, one of the New Hebrides.

Genus HIERACOGLAUX, *Kaup.*

All the species of this and the following genus are partially diurnal. They all have very large eyes; which in some are pale yellow, while in others they are light brown. I shall commence with the largest member of the present form, *H. strenuus*, thus reversing the order of the species as arranged in the folio edition of the Birds of Australia.

Sp. 32.   HIERACOGLAUX STRENUUS.

GREAT OWL OF THE BRUSHES.

Athene? strenua, Gould, in Proc. of Zool. Soc., part v. p. 142.
Ieraglaux strenua, Kaup, Monog. Strig. in Jard. Cont. to Orn., 1852, p. 100.

Athene strenua, Gould, Birds of Australia, fol. vol. I. pl. 35.

With the exception of the Eagles, *Aquila audax* and *Polioaëtus leucogaster*, this is the most powerful Raptorial bird yet discovered in Australia. Its strength is prodigious, and woe to him who ventures to approach it when wounded. So far as I have been able to ascertain, it is an inhabitant of the brushes, particularly those of Victoria and New South Wales which extend along the coast from Port Philip to Moreton Bay. I did, however, obtain it on the precipitous sides of the cedar brushes of the Liverpool range; in all such situations the silence of night is frequently broken by its hoarse loud mournful note, which more resembles the bleating of an ox than any other sound I can compare it to. During the day it reposes under the canopy of the thickest trees, from which however it is readily roused, when it glides down the gulleys with remarkable swiftness; the manner in which so large a bird threads the trees while flying with such velocity is indeed truly astonishing.

Its food consists of birds and quadrupeds, of which the

brushes furnish a plentiful supply. In the stomach of one I dissected in the Liverpool range were the remains of a bird and numerous green seed-like berries, resembling small peas; whether the latter had formed the contents of the stomach of a bird or quadruped which the Owl had devoured, or had been eaten by the Owl itself, I could not satisfactorily ascertain.

The bill of this species stands out from the face very prominently; it has also a smaller head and more diminutive eyes than the *Hieracoglaux consivens*, although it is a much larger bird.

The sexes differ but little in size or in the colouring of the plumage, which may be thus described:—

Crown of the head, all the upper surface, wings, and tail dark clove-brown, crossed by numerous bars of broccoli-brown, which become much larger, lighter, and more conspicuous on the lower part of the back, the inner edges of the secondaries and of the tail; face, throat, and upper part of the chest buff, with a large patch of dark brown down the centre of each feather; the remainder of the under surface white, slightly tinged with buff, and crossed with irregular bars of brown; bill light blue at the base, passing into black at the tip; feet pale gamboge-yellow; irides yellow; cere greenish olive.

Total length, 24 inches; bill, 2; wing, 15; tail, 10½; tarsi, 2¼.

Sp. 33. HIERACOGLAUX RUFUS.

RUFOUS OWL.

Athene rufa, Gould, in Proc. of Zool. Soc., part xiv. p. 18.
Ieraglaux rufa, Kaup, Monog. Strig. in Jard. Cont. to Orn., 1852, p. 109.
Ngor-gork, Aborigines of Port Essington.

Athene rufa, Gould, Birds of Australia, fol. vol. i. pl. 36.

What the *Hieracoglaux strenuus* is to the brushes of New South Wales, the *H. rufus* is to the primitive forests of the

Cobourg Peninsula. That this powerful Owl has a very extensive range over that part of the country is probable, as it also is that the numerous Vampires (*Pteropus funereus*) which suspend themselves from the trees along the north coast are not free from its attacks.

A single specimen was obtained at Port Essington by Gilbert, who shot it in a thicket amidst the swamps in the neighbourhood of the settlement. It is a large species, nearly equalling in size the *Hieracoglaux strenuus*, from which however it is at once distinguished by the more rufous tint of its plumage and by the more numerous and narrower barring of the breast. No other specimen was procured during Gilbert's residence in the colony, neither have the collections transmitted from that locality since his untimely death furnished us with additional examples.

Facial disk dark brown; all the upper surface dark brown, crossed by numerous narrow bars of reddish brown; the tints becoming paler and the barrings larger and more distinct on the lower part of the body, wings, and tail; all the under surface sandy red, crossed by numerous bars of reddish brown; the feathers of the throat with a line of brown down the centre; vent, legs, and thighs of a paler tint, with the bars more numerous but not so decided; bill horn-colour; cere, eyelash, and feet yellow, the latter slightly clothed with feathers; irides light yellow.

Total length, 20 inches; bill, $1\frac{3}{4}$; wing, $13\frac{1}{4}$; tail, $8\frac{1}{4}$; tarsi, $2\frac{1}{4}$.

It is not to be expected that Gilbert, almost unaided by any one, either settler or native, could make himself acquainted with all the birds of a primitive country like the Cobourg Peninsula; when that portion of Australia becomes better known, much additional information respecting species already characterized as well as many novelties will doubtless be acquired.

Sp. 34.   HIERACOGLAUX CONNIVENS.
WINKING OWL.

Falco connivens, Lath. Ind. Orn. Supp., p. 12.
Buteo connivens, Vieill. Nouv. Dict. d'Hist. Nat., tom. iv. p. 481.
Noctua frontata, Less. Traité d'Orn. p. 106.
Athene frontata, Gray and Mitch. Gen. of Birds, vol. i. p. 35, *Athene*, sp. 34.
Ieraglaux connivens, Kaup, Mon. Strig. in Jard. Cont. to Orn., 1852, p. 109.
Athene? fortis, Gould, in Proc. of Zool. Soc., part v. p. 141.
Goora-a-gang, Aborigines of New South Wales.
Wool-boo-gle, Aborigines of the mountain district of Western Australia.

Athene connivens, Gould, Birds of Australia, fol. vol. i. pl. 34.

This is a far more common species than either of the two last described; it is also much less in size and very different in colour; its range appears to extend over the whole of the southern coast of Australia. I have received examples from Western Australia, Victoria, and nearly every part of New South Wales; specimens from these distant localities differ a little in their plumage, those obtained in the West being rather lighter in colour, and having the markings less clear and defined, than those from the eastern portion of the country. There is no difference in the plumage of the sexes; but the female is somewhat the larger in size.

Brushes, wooded gullies, and the sides of creeks are its favourite places of resort; it is consequently not so restricted in the localities it chooses as the *Hieracoglaux strenuus*, which I have never known to leave the brushes. It sallies forth early in the evening, and even flies with perfect use of vision during the mid-day sun, when roused and driven from the trees upon which it has been sleeping. I have frequently observed it in the daytime among the thick branches of the *Casuarinæ* which border the creeks. Gilbert procured an egg of this species in Western Australia; it was pure white, somewhat round in form, and large for the size of the bird; measuring two inches in length by one and five-eighths in breadth.

It will be seen, on reference to the synonyms, that I described this bird, in the " Proceedings of the Zoological Society," under the specific name of *fortis*; but I have since ascertained, through the kindness of the late Earl of Derby in affording me the use and inspection of the three volumes of drawings of Australian Birds, formerly in the possession of the late A. B. Lambert, Esq., that it is identical with the Winking Falcon of Latham; any seeming inattention on my part in describing an apparently new Owl without consulting that author will I hope be readily excused, as few ornithologists would think of looking for the description of this bird under the genus *Falco*.

It is due to the acumen of the late Mr. Strickland that, by means of the drawings above alluded to, the present and other species described by Latham have been identified, a circumstance which has caused *Aquila fucosa* to become *A. audax*; *Falco frontatus*, *F. lunulatus*; *Strix personata*, *S. Novæ-Hollandiæ*, &c.; unfortunately I did not obtain the loan of these drawings until my work was far advanced, otherwise the errors I now correct would not have occurred.

Face and throat greyish white; crown of the head and all the upper surface dark brown, tinged with purple; scapularies, secondaries, and greater wing-coverts spotted with white; primaries alternately barred with dark and greyish brown, the light marks on the outer edges approaching to white; tail dark brown, transversely barred with six or seven lines of greyish white, the extreme tips of all the feathers terminating with the same; the whole of the under surface mottled brown and white, the latter occupying the outer edges of the feathers; tarsi clothed to the toes, and mottled brown and fawn-colour; irides bright yellow; cere yellowish olive; bill light yellowish horn-colour; toes long, yellow, and covered with fine hairs.

Genus SPILOGLAUX, *Kaup*.

The members of this form are very diminutive when compared with those forming the genus *Hieracoglaux*; they are all clothed in a thick fluffy kind of plumage, in which respect they differ from their allies, the true *Athenes*. They are both diurnal and nocturnal in their habits, but fly less by day than they do by night. I commence with the largest species of the genus, *S. marmoratus*, which has not yet been figured either in the folio edition of the 'Birds of Australia' or in the 'Supplement.'

Sp. 35. SPILOGLAUX MARMORATUS.

Athene marmorata, Gould, in Proc. of Zool. Soc., part xiv. p. 18.
Spiloglaux marmoratus, Kaup, Monog. Strig. in Jard. Cont. to Orn., 1852, p. 108.

All the upper surface, wings, and tail dark brown, obscurely spotted with white round the back of the neck, on the wing-coverts and scapularies; inner webs of the primaries at their base, and the inner webs of the lateral tail-feathers, crossed by bands, which are buff next the shaft, and white towards the extremity of the webs; face and chin whitish; under surface dark brown, blotched with white and sandy brown; legs and thighs fawn-colour; bill horn-colour; feet yellow.

Total length, 14 inches; bill, 1½; wing, 9¼; tail, 6; tarsi, 2.

This bird so far exceeds in size the *S. maculatus* that, notwithstanding the resemblance in its markings, I have no doubt of its being a distinct species. Besides those in my own, there are specimens of the *S. marmoratus* in the national collection; all of which have been sent from South Australia.

Sp. 36.   SPILOGLAUX BOOBOOK.

Boobook Owl.

Strix Boobook, Lath. Ind. Orn. Suppl., p. xv. no. 9.
Noctua Boobook, Vig. and Horsf. in Linn. Trans., vol. xv. p. 188.
Spiloglaux bubuk, Kaup, Monog. Strig. in Jard. Cont. to Orn., 1852, p. 108.
Buck-buck, Aborigines of New South Wales.
Goor-goor-da, Aborigines of Western Australia.
Mel-in-de-ye, Aborigines of Port Essington.
Koor-koo, Aborigines of South Australia.
Brown or *Cuckoo-Owl* of the Colonists.

Athene Boobook, Gould, Birds of Australia, fol. vol. 1. pl. 32.

I have seen individuals of this Owl from every one of the Australian colonies, all presenting similar characters, with the exception of those from Port Essington, which differ from the others in being a trifle smaller in size and paler in colour.

In Tasmania this species is seldom seen, while it is very common throughout the whole of the southern portion of the continent. It appears to inhabit alike the brushes and those plains which are studded with belts of trees. It is no unusual occurrence to observe it on the wing in the daytime in search of insects and small birds, upon which it mainly subsists. It may be readily distinguished from *Spiloglaux maculatus* by its smaller size, and by the spotted markings of its plumage.

The flight of this bird is tolerably rapid; and as it passed through the shrubby trees that cover the vast area of the belts of the Murray, it strongly reminded me of a woodcock. In such places travellers frequently flush it from off the ground, to which, after a flight of one or two hundred yards, it either descends again or takes shelter in any thickly foliaged trees that may be at hand.

It breeds in the holes of the large gum-trees, during the

months of November and December, and lays its eggs on the rotten surface of the wood, without any kind of nest. Three eggs procured on the 6th of November, by my useful native companion Natty, were in a forward state of incubation; their contour was unusually round, the medium length of the three being one inch and seven lines, and the breadth one inch and four lines. They were perfectly white, as is ever the case with the eggs of owls.

"The native name of this bird," says Mr. Caley, "is *Buck-buck*, and it may be heard nearly every night during winter uttering a cry corresponding with the sound of that word. Although this cry is known to every one, yet the bird itself is known but to few; and it cost me considerable time and trouble before I could satisfy myself of its identity. The note of the bird is somewhat similar to that of the European *Cuckoo*, and the colonists have hence given it that name. The settlers in New South Wales are led away by the idea that everything is the reverse in that country to what it is in England; and the *Cuckoo*, as they call this bird, singing by night is one of the instances they point out." I believe that its note is never uttered during the daytime.

The sexes offer but little difference in the colouring of their plumage, but the female is the largest in size. A great diversity is found to exist in the colouring of the irides, some being yellowish white, others greenish yellow, and others brown.

Its food is very varied, but consists principally of small birds and insects of various orders, particularly locusts and other *Neuroptera*.

Fore part of the facial disk greyish white, each feather tipped with black; hinder part dark brown; head, all the upper surface, wings, and tail reddish brown; the wing-coverts, scapularies, and inner webs of the secondaries spotted with white; primaries and tail-feathers irregularly barred with light reddish brown, the spaces between the bars becoming buffy white on the under surface; breast and all the under surface rufous, irregularly blotched with white, which predominates on the abdomen; thighs deep tawny buff; irides light brown in some, greenish brown inclining to yellow in others; cere bluish grey; feet lead-colour.

Sp. 37.   SPILOGLAUX MACULATUS.

Spotted Owl.

Noctua maculata, Vig. and Horsf. in Linn. Trans., vol. xv. p. 189.
Spiloglaux maculatus, Kaup, Monog. Strig. in Jard. Cont. to Orn., 1852, p. 108.
Athene maculata, Gould, Birds of Australia, fol. vol. i. pl. 33.

This species is very generally distributed over Tasmania; it also inhabits South Australia and New South Wales, but in far less numbers. It generally takes up its abode in the thickly-foliaged trees of the woods and gullies, usually selecting those that are most shielded from the heat and light of the sun, spending, like the diurnal species, the entire day in a state of drowsiness, from which, however, it can be easily aroused. Its visual powers are sufficiently strong to enable it to face the light, and even to hunt for its food in the daytime. Like other members of the genus, it preys chiefly upon small birds and insects, which, from the more than ordinary rapidity of its movements, are captured with great facility.

The sexes are precisely alike in colour, and differ but little in size; the female is, however, the largest.

The drawing in the folio edition was made from a pair of living examples which I kept for some time during my stay at Hobart Town, and which bore confinement so contentedly, that had an opportunity presented itself I might easily have sent them alive to England.

Facial disk white, each of the feathers immediately above

the bill with the shafts and tips black; head and all the upper surface brown, the scapularies and secondaries numerously spotted with white; tail brown, crossed by irregular bands of a lighter tint, which become nearly white on the outer feathers; chest and all the under surface brown, blotched and spotted with tawny and white; primaries brown, crossed with bands of a lighter tint; thighs tawny buff; bill dark horn-colour; irides yellow; feet yellowish.

I have now enumerated all the Raptorial Birds of Australia at present known; but I have no doubt that when the northern portions of that great country have been duly explored their number will be greatly increased; indeed such a result may be looked for with a degree of certainty; especially with regard to the family we have just left—the Strigidæ—for there is no knowing what Owls exist in the brushes of the Cape York district, or those of the north coast lying immediately opposite that *terra incognita*, New Guinea and its numerous islands. Where insect life is abundant, small quadrupeds and birds are sure to occur in sufficient numbers to keep them in check or within the necessary bounds. The next Order—the Insessores—which commences with the *Caprimulgidæ*, will afford ample evidence of this being always the case, for in no other country is there a greater proportion of insectivorous birds, and certainly none in which nocturnal species, such as the *Podargi*, are more numerous.

Order INSESSORES.

If the Raptores inhabiting Australia are few in number, such is not the case with those next in succession—the Insessores; for the birds of this Order are not only numerous in species, but comprise many forms peculiar to that country. These will all be ranked, in the following pages, as near to each other as an arrangement of the birds of one portion of the globe will admit. I commence with the *Caprimulgidæ*, to which succeed the *Cypselidæ* or Swifts, the *Hirundinidæ* or Swallows, the *Meropidæ* or Bee-eaters, the *Halcyonidæ* or Kingfishers, the members of that singular genus *Artamus* or Wood Swallows, the *Pardaloti*, the *Gymnorhinæ*, *Graucali*, *Pachycephalæ*, *Colluricinclæ*, *Rhipiduræ*, *Gerygones*, *Petroicæ*, *Menuræ*, *Psophodes*, *Maluri*, *Acanthizæ*, *Cinclorhamphi*, *Estreldæ*, *Cinclosomæ*, *Ptilinorhynchi*, *Sericulus*, *Orioli*, *Corcorax*, *Pomatorhini*, *Struthidea*, the great family of *Meliphagidæ* or Honey-eaters, the *Cuculi*, *Climacteres*, *Ptilores*, and *Sittellæ*; followed by the cream of the Australian avifauna, the *Psittacidæ* or Parrots,—the whole comprising many genera which it would be out of place to particularize here, but which will be commented upon as they may require in due succession.

Family CAPRIMULGIDÆ.

The members of this group of birds inhabit nearly every portion of the known world; but none occur in New Zealand nor, I believe, in the Polynesian Islands.

Genus ÆGOTHELES, *Vigors and Horsfield.*

Two species of this singular form inhabit Australia; one its southern, the other its northern portions. They are both very Owl-like in their habits, actions, and dispositions;

remaining by day within the hollow branches of trees, in which situations, without any nest, their four or five round white eggs are deposited. The sexes are alike in colouring.

38. ÆGOTHELES NOVÆ-HOLLANDIÆ, *Vig. and Horsf.*

OWLET NIGHTJAR.

Crested Goat-sucker, Phill. Bot. Bay, pl. in p. 270.
Caprimulgus Novæ-Hollandiæ, Lath. Ind. Orn., vol. ii. p. 588.
—— *cristatus*, Shaw in White's Voy., pl. in p. 241.
New-Holland Goat-sucker, Lath. Gen. Syn. Supp., vol. ii. p. 261.
Bristled Goat-sucker, Lath. Gen. Hist., vol. vii. p. 342.
Caprimulgus vittatus, Ib. Ind. Orn. Supp., p. lviii.
Banded Goat-sucker, Ib. Gen. Syn. Supp., vol. ii. p. 262, pl. 136.
Ægotheles Novæ-Hollandiæ, Vig. and Horsf. in Linn. Trans., vol. xv. p. 197.
—— *lunulatus*, Jard. and Selby, Ill. Orn., vol. iii. pl. 149.
—— *Australis*, Swains. Class. of Birds, vol. ii. p. 338.
—— *cristatus*, G. R. Gray, List of Gen. of Birds, p. 7.
Little More-pork, Colonists of Tasmania. *Teringing*, Aborigines of the coast of New South Wales.

Ægotheles Novæ-Hollandiæ, Gould, Birds of Australia, fol., vol. ii. pl. 1.

This very interesting little Nightjar possesses a great range of habitat, being found in every part of Tasmania, and throughout the southern portion of Australia, from Swan River on the western coast to Queensland on the eastern; time, and the continued exploration of that vast country, can alone determine how far it may be found to the northward: it is a stationary species, inhabiting alike the densest brushes near the coast, and the more thinly-wooded districts of the interior.

While rambling in the Australian forests I had the good fortune to meet with more than an ordinary number of specimens of this curious bird. I also procured its eggs, and considerable information respecting its habits and actions, which differ most remarkably from those of the other members of the family, and, on the other hand, assimilate so closely to those of the smaller Owls, that the English name of Owlet Nightjar has been assigned to it.

During the day the bird resorts to the hollow branches or spouts as they are called, and the boles of the gum-trees, sallying forth as night approaches in quest of insects, particularly small *Coleoptera*. Its flight is straight, and not characterized by the sudden turns and descents of the *Caprimulgi*. On driving it from its haunts I have sometimes observed it to fly direct to a similar hole in another tree, but more frequently to alight on a neighbouring branch, perching across and never parallel to it. When assailed in its retreat it emits a loud hissing noise, and has the same stooping motion of the head observable in the Owls; it also resembles that tribe of birds in its erect carriage, the manner in which it sets out the feathers round the ears and neck, and in the power it possesses of turning the head in every direction, even over the back, a habit it is constantly practising. A pair I had for some time in captivity frequently leapt towards the top of the cage, and had a singular mode of running or shuffling backwards to one corner of it.

While traversing the woods, the usual mode of ascertaining its presence is by tapping with a stone or a tomahawk at the base of the hollow trees, when the little inmate will almost invariably ascend to the outlet and peep over to ascertain the cause of disturbance. If the tree be lofty or its hole inaccessible, it will frequently retire again to its hiding-place, and there remain until the annoyance be repeated, when it flies off to a place of greater security. In these holes, without forming any nest, it deposits its eggs, which are four or five in number, perfectly white, nearly round, and about one inch and a line in length and eleven lines in breadth. At least two broods are reared by each pair of birds during the year. I have known the young to be taken in Tasmania in October, and in New South Wales I have procured eggs in January.

Specimens from Tasmania, Swan River, South Australia, and New South Wales present considerable difference in the colour and markings of the plumage, but none, so far as I have yet seen, of sufficient importance to justify their separation into distinct species: in some the nuchal band and the circular mark on the head are very conspicuous, while in others scarcely a trace of these markings is observable; these variations do not appear to occur in certain localities only, but are generally found in all.

Little or no difference is apparent in the size or plumage of the sexes. In all the irides are blackish brown.

Sp. 39. ÆGOTHELES LEUCOGASTER, *Gould*.

WHITE-BELLIED OWLET-NIGHTJAR.

Ægotheles leucogaster, Gould in Proc. Zool. Soc., part xii. p. 106.

Ægotheles leucogaster, Gould, Birds of Australia, fol., vol. ii. pl. 2.

This is altogether a larger and more powerful bird than the *Ægotheles Novæ-Hollandiæ*; besides which, the white colouring of the lower part of the belly will at all times serve to distinguish it from that species.

Gilbert states that it is abundant in most parts of the settlement at Port Essington, "where it is frequently seen flying about at twilight, and occasionally during the day. On the approach of an intruder it flies very heavily from tree to tree, and on alighting invariably turns round on the branch to watch his approach, moving the head all the time after the manner of the Hawk tribe."

The White-bellied Owlet-Nightjar feeds on insects; and as the bird is strictly a nocturne, they are, as a matter of course, procured at night.

The sexes when fully adult will not, I expect, be found to differ in plumage; but whether the red or the grey varieties are the most mature birds, or if the difference in colour be sexual, I have not had sufficient opportunities of ascertaining.

Head black; the crown, a lunar-shaped mark at the back of the head, and a collar surrounding the back of the neck freckled with grey; back freckled black and white; wings brown, crossed by numerous bands of lighter brown freckled with dark brown; primaries margined externally with buff, interrupted with blotchings of dark brown; tail dark brown, crossed by numerous broad irregular bands of reddish buff freckled with dark brown; ear-coverts straw-white; chin, abdomen, and under tail-coverts white; breast and sides of the neck white, crossed by numerous freckled bars of black; irides dark brown; upper mandible dark olive-brown, lower mandible white with a black tip; legs very pale yellow; claws black.

Total length, $9\frac{1}{4}$ inches; bill, 1; wing, $5\frac{3}{4}$; tail, 5; tarsi, 1.

Genus PODARGUS, *Cuvier*.

With no one group of the Australian birds have I had so much difficulty in discriminating the species as with those of the genus *Podargus*. It is almost impossible to determine with certainty those described by Latham; could this have been done satisfactorily, it would have greatly facilitated their investigation.

The species are much more numerous than those of the genus *Ægotheles*, and unlike them are not so exclusively confined to Australia; for although that country constitutes their head-quarters, some are found in New Guinea and the adjacent islands, where they unite with the *Batrachostomi*.

Six species of this form were described in the folio edition; during the twenty years which have elapsed since its completion, two others have been discovered; and thus we now know that Australia is inhabited by eight species of these large nocturnal birds to keep in check the great families of

Cicadæ and *Phasmidæ*, upon which they mainly subsist: but they do not refuse other insects, and even berries have been found in their stomachs. They are an inanimate and sluggish group of birds, and depend for their supplies less upon their power of flight than upon the habit they are said to have of traversing the branches of the various trees upon which their favourite insects reside; at intervals during the night they sit about in open places, on rails, stumps of trees, on the roofs of houses.

In their nidification the *Podargi* differ in a most remarkable manner from all the other *Caprimulgidæ*, inasmuch as while the eggs of the *Ægothelæ* are deposited in the holes of trees, and those of the members of the other genera of this family on the ground, these birds construct a flat nest of small sticks on the horizontal branches of trees for the reception of theirs, which are moreover of the purest white.

Although I have no satisfactory evidence that the *Podargi* resort to a kind of hybernation for short periods during some portions of the year, I must not omit to mention that I have been assured that they do occasionally retire to and remain secluded in the hollow parts of the trees; and if such should prove to be the case, it may account for the extreme obesity of many of the individuals I procured, which was often so great as to prevent me from preserving their skins. I trust that these remarks will cause the subject to be investigated by those who are favourably situated for so doing; for my own part, I see no reason why a bird should not pass a portion of its existence in a state of hybernation; at the same time the notion of its so doing is very like a repetition of the old assertion respecting the Swallows, for which there is no foundation.

I would also ask the Australians to ascertain if the difference in colour which occurs in these birds be distinctive of their sex, and if so, to which the respective tints of red and grey pertain.

Sp. 40. PODARGUS STRIGOIDES.

TAWNY-SHOULDERED PODARGUS.

Caprimulgus strigoides, Lath. Ind. Orn. Supp., p. 58.
—— *gracilis* ? Ib., p. 58.
—— *podargus* ? Dumont, Dict. Sci. Nat., tom. xiv. p. 504.
Gracile Goatsucker ? Ib. Gen. Syn. Supp., vol. ii. p. 203.
Podargus ? *gracilis* ? Steph. Cont. of Shaw's Zool., vol. xiii. p. 93.
—— *Australis* ? Ib., vol. xiii. p. 92.
—— *cinereus* ? Vieill. Nouv. Dict. d'Hist. Nat., tom. xxvii. p. 151, pl. G. 37. fig. 3.
Cold-River Goatsucker, Lath. Gen. Hist., vol. vii. p. 369.
Podargus humeralis, Vig. and Horsf. in Linn. Trans., vol. xv. p. 198.

Podargus humeralis, Gould, Birds of Australia, fol., vol. ii. pl. 3.

The Tawny-shouldered Podargus is plentifully dispersed over New South Wales, where it is not restricted to any peculiar character of country, but inhabits alike the thick brushes near the coast, the hilly districts, and the thinly wooded plains of the interior. I found it breeding on the low swampy islands studding the mouth of the Hunter, and on the Apple-tree (*Angophora*) flats of Yarrundi, near the Liverpool Range.

Like the rest of the genus, this species is strictly nocturnal, sleeping throughout the day on the dead branch of a tree, in an upright position across, and never parallel to, the branch, and which it so nearly resembles as scarcely to be distinguishable from it. I have occasionally seen it beneath the thick foliage of the *Casuarinæ*, and I have been informed that it sometimes shelters itself in the hollow trunks of the *Eucalypti*, but I could never detect one in such a situation; I mostly found them in pairs, perched near each other on the branches of the gums, in situations not at all sheltered from the beams of the midday sun. So lethargic are its slumbers, that it is almost impossible to arouse it, and I have frequently shot one without disturbing its mate sitting close by; it may also be knocked off with sticks or stones, and sometimes is even taken with the hand: when aroused, it flies lazily off with heavy flapping wings to a neighbouring tree, and again resumes its slumbers until the approach of evening, when it becomes as animated and active as it had been previously dull and stupid. The stomach of one I dissected induced me to believe that it does not usually capture its prey while on the wing, or subsist upon nocturnal insects alone, but that it is in the habit of creeping among the branches in search of such as are in a state of repose. The power it possesses of shifting the position of the outer toe backwards, as circumstances may require, is a very singular feature, and may also tend to assist them in their progress among the branches. A bird I shot at Yarrundi, in the middle of the night, had the stomach filled with fresh-captured mantis and locusts (*Phasmida* and *Cicadæ*), which seldom move at night, and the latter of which are generally resting against the upright boles of the trees. In other specimens I found the remains of small coleoptera, intermingled with the fibres of the roots of what appeared to be a parasitic plant, such as would be found in decayed and hollow trees. The whole contour of the bird shows that it is not formed for extensive flight or for performing those rapid evolutions that are necessary for the capture of its prey in the air, the wing being short and concave in comparison with those of the true aërial Nightjars, and particularly with the Australian form to which I have given the name of *Eurostopodus*.

Of its mode of nidification I can speak with confidence, having seen many pairs breeding during my rambles in the woods. It makes a slightly-constructed flat nest of sticks carelessly interwoven together, and placed at the fork of a horizontal branch of sufficient size to ensure its safety; the trees most frequently chosen are the *Eucalypti*, but I have occasionally seen the nest on an Apple-tree (*Angophora*) or a Swamp-Oak (*Casuarina*). In every instance one of the birds was sitting on the eggs and the other perched on a neighbouring bough, both invariably asleep: that the male participates in the duty of incubation I ascertained by having shot a bird on the nest, which on dissection proved to be a male. The eggs are generally two in number, of a beautiful immaculate white, and of a long oval form, one inch and ten lines in length by one inch and three lines in diameter.

Like the other species of the genus, it is subject to considerable variation in its colouring; the young, which assume the adult livery at an early age, being somewhat darker in all their markings. In some a rich tawny colour predominates, while others are more grey.

The night-call of this species is a loud hoarse noise, consisting of two distinct sounds, which cannot be correctly described.

The stomach is thick and muscular, and is lined with a thick hair-like substance like that of the Common Cuckoo.

All the upper surface brown, speckled with greyish white and darker brown, the feathers of the crown having a blackish-brown stripe on the centre terminating in a minute spot of white; wings similar to the upper surface, but lighter and with bolder black and buff spots, the coverts having an irregular spot of white and tawny on the outer web near the tip, which, as they lie over each other, form indistinct bands across the wing; primaries brownish black, with light-coloured shafts, and with a series of whitish spots on the outer webs, between which they are margined with tawny; their inner webs irregularly barred with the same; tail tawny brown, sprinkled with lighter brown, and crossed with a series of irregular bands of blackish brown, sprinkled with dusky white, each feather having a spot of brownish black near the extremity, and tipped with white; face and all the under surface greyish white, crossed by numerous narrow and irregular bars of tawny, and with a stripe of brown down the centre of each feather, the latter colour being most conspicuous and forming a kind of semilunar mark down each side of the chest; bill light brown, tinged with purple; inside of the mouth pale yellow; tongue long, transparent, and of the same colour with the inside of the mouth; irides brownish orange; feet light brownish olive.

Sp. 41. PODARGUS CUVIERI, *Vig. and Horsf.*
Cuvier's Podargus.

Podargus Cuvieri, Vig. and Horsf. in Linn. Trans., vol. xv. p. 200.
More-pork of the residents in Tasmania.

Podargus Cuvieri, Gould, Birds of Australia, fol., vol. ii. pl. 4.

This species is readily distinguished from the *Podargus humeralis* by the bill being much less robust and of a more adpressed form, while the culmen is sharp and elevated; the bird itself is also of a smaller size and altogether more slender than its near ally. Tasmania, if not its exclusive habitat, is certainly its great stronghold, it being there very numerous, as evidenced by the frequency with which I encountered it during my rambles over the country. I observed it both among the thick branches of the *Casuarinæ* and on the dead limbs of the *Eucalypti*; it appeared however to evince a greater partiality for the latter, which it closely resembles in colour, and, from the position in which it rests, looks so like a part of the branch itself as frequently to elude detection; it is generally seen in pairs sitting near each other, and frequently on the same branch. Like the other members of the genus, this bird is strictly nocturnal, and feeds almost exclusively on insects, of which coleoptera form a great part. It is frequently captured and kept in captivity in Tasmania, where it excites attention more from the sluggishness of its nature and the singular position it assumes than from any other cause. It will pass the entire day in sleep on the back of a chair or any other piece of furniture on which it can perch. Like the owl, it is considered by superstitious people a bird of ill omen, principally from the extraordinary sound of its hoarse, unearthly cry, which resembles the words *morepork*; it approaches the immediate vicinity of the houses, and frequently emits this sound while perched in their verandahs.

The *Podargus Cuvieri* builds a neatly formed flat nest, about seven inches in diameter, in the fork of an horizontal branch; the exterior formed of small sticks, and the interior of the fibrous portions of various plants; the eggs are white, and nearly of a true oval in form, being one inch and nine lines long by one inch and three lines broad.

Considerable variation occurs in the colouring of individuals, the prevailing tint being a dull ashy grey, while others are of a rich chestnut hue; but whether this be indicative of immaturity, or characteristic of the fully adult plumage of the two sexes, I have not been able to satisfy myself.

Lores brown, each feather tipped with mealy white, forming a line before and above the eye; feathers of the forehead mealy white, blending into the dull ashy grey of the head and back, all the feathers of which have a stripe of blackish brown down the centre, terminating in a small spot of white, and are moreover minutely freckled with greyish white and dark brown; wing-coverts chestnut, each tipped with an oval spot of white bounded posteriorly with black, forming a line across the wing; remainder of the wing brown, mottled with greyish white, arranged, particularly on the primaries, in the form of irregular bars; scapularies washed with buff and with a broad stripe of blackish brown down the centre; under surface brownish grey, minutely freckled with white, and with a narrow line of blackish brown down the centre; sides of the neck washed with chestnut; tail grey, minutely freckled with greyish white and black, assuming the form of broad irregular bands, each feather with a small spot of white at the tip; irides varying from yellow to reddish yellow and hazel; feet olive-brown.

Other examples have the general tint rich chestnut-brown, with all the markings larger and more decided.

Sp. 42.   PODARGUS MEGACEPHALUS.

Caprimulgus megacephalus, Lath. Ind. Orn. Supp., p. lviii.
Great-headed Goatsucker, Lath. Gen. Syn. Supp., vol. ii. p. 265.
Wedge-tailed Goatsucker, Lath. Gen. Hist., vol. vii. p. 368 ?
Podargus Stanleyanus, Lath. MSS., Vig. & Horsf. in Linn. Trans., vol. xv. p. 197 ?

Podargus megacephalus, Gould, Birds of Australia, fol., vol. i. Introd., p. xxviii.

I believe I have good grounds for regarding the *Podargus megacephalus* as distinct from *P. humeralis*. For many years two birds of this form have lived in the Gardens of the Zoological Society, one of which is doubtless the *P. humeralis*; the other, which is much larger and possesses greatly developed mandibles, is to all appearance distinct, and is so considered by the keeper who has charge of these Nocturnes—an opinion in which the learned Secretary also, I believe, coincides but I must admit that the question is an open one, and one to which I would call the attention of those persons resident in Australia who pay attention to ornithology, that they may aid us in its solution.

The habitat of this species is the brushes of the eastern coast, whence I have received specimens.

Sp. 43.   PODARGUS BRACHYPTERUS, *Gould.*

SHORT-WINGED PODARGUS.

Podargus brachypterus, Gould in Proc. of Zool. Soc., part viii. p. 168.

Podargus brachypterus, Gould, Birds of Australia, fol., Introd., p. xxviii.

In its general appearance this bird closely resembles the *P. humeralis*, but is even smaller in size than *P. Cuvieri*; at the same time the bill is larger than that of the former species, and projects much further from the face than in any other of its congeners; it also differs in the shortness of its wings, which circumstance suggested the specific appellation I have assigned to it.

It is a native of Western Australia.

Sp. 44. PODARGUS PHALÆNOÏDES, *Gould.*

MOTH-PLUMAGED PODARGUS.

Podargus Phalænoïdes, Gould in Proc. of Zool. Soc., part vii. p. 142.

Ny̆-ane? and *In-ner-jin-ert,* Aborigines of the neighbourhood of Port Essington.

Podargus phalænoides, Gould, Birds of Australia, fol., vol. ii. pl. 5.

The present bird, which is from Port Essington, may be readily distinguished from every other Australian species of *Podargus* by its small size, by the beautiful, delicate, and moth-like painting of its plumage, and by the colouring of the thighs, which are light brown instead of black; its tail also is rather more lengthened than that of the common species, *P. humeralis* and *P. Cuvieri.* Like the other members of the genus, it exhibits considerable variation in size and colouring; in some a rusty-red tint pervades the whole plumage, while in others no trace of this hue occurs. The difference in the colouring of the *Podargi* may be sexual, as we find to be the case in many of the Owls.

I have several specimens of the Moth-plumaged Podargus from the north-west coast of Australia, and Gilbert states that it is abundant in every part of the Coburg Peninsula.

Like the rest of the genus, it is strictly nocturnal; its whole economy in fact, as far as known, so closely resembles that of the *Podargus humeralis* that one description would serve for both.

Forehead, sides of the face, and all the under surface brownish grey, minutely freckled with black; the feathers of the under surface with a stripe of blackish brown down the centre, these stripes being broadest and most conspicuous on the sides of the chest; all the upper surface brown, minutely freckled with grey, each feather with a broad stripe of black down the centre; shoulders dark brown; coverts freckled with greyish white and with a spot of white, the centre of which is fawn-colour at the tip; primaries dark brown, crossed on their outer webs with an irregular bar of white, the interspaces on the outer primaries rufous; inner webs of the primaries crossed by irregular bands of freckled brown and fawn-colour; tail brown, crossed by numerous broad bands of freckled grey, bounded on either side by irregular blotchings of black; irides orange or reddish hazel; bill horn-colour.

In the other state, to which I have alluded, the whole of the upper surface is of a dark rust-red, freckled on the forehead, wing-coverts, and scapularies with white; the bands on the tail less apparent; a rufous tint pervades the grey of the under surface, and the striæ are much narrower than in the specimen above described.

Sp. 45. PODARGUS PAPUENSIS, *Quoy et Gaim.*

PAPUAN PODARGUS.

Podargus papuensis, Quoy et Gaim. Voy. de l'Astrol., Ois. t. 13.

Podargus papuensis, Gould, Birds of Australia, Supplement, pl.

Of this fine bird several specimens were procured during the voyage of Her Majesty's Ship Rattlesnake, under the command of Captain Owen Stanley, with Mr. Macgillivray as Naturalist, whose names will ever hold a prominent place in the annals of Australian zoology. All the specimens were obtained at Cape York, the contiguity of which to New Guinea induced me to believe the bird to be identical with the one described and figured by MM. Quoy and Gaimard in the Voyage of the Astrolabe under the name of *Podargus Papuensis*; and this belief proved to be correct on a comparison of Australian examples with the New Guinea birds in the Museums of Paris and Leyden.

The *P. Papuensis* is the largest species of the genus yet discovered; the beauty of its markings and the extreme length of its cuneate tail render it also one of the most graceful. The only specimen that came into my possession from Mr. Macgillivray, for the purpose of figuring, before being deposited in the National Collection, was a male. This is of a light brown colour, beautifully marbled on the under surface with large blotches of white. I have another specimen from Cape York, which is said to be the female; and such, judging from its redder colouring and smaller size, I believe to be the case, for a similar difference exists between the sexes of *P. marmoratus*.

The male has the whole of the upper surface mottled with greyish white, brown, and black, presenting a very close resemblance to some of the larger kinds of moths, the lighter tints prevailing in some parts and the darker in others; on the primaries the marks assume the form of bars, and are of a redder hue; tips of the coverts white, forming irregular bars across the wing; tail very similar, but here also the markings assume the form of alternate darker and lighter bands with a rufous tint on the edges of the feathers; the under surface is much lighter than the upper; the greyish white assumes a larger and more blotch-like form, and the darker marks that of an irregular gorget across the breast; bill and feet olive.

The female is altogether of a more sandy hue; the dark marks proceed down the centre of the feathers, and terminate in a round spot of buff; the wing-coverts are tipped with white, and the lighter blotches on the wing are very conspicuous; the under surface, like the upper, is also of a redder hue than in the male, and the markings are of a smaller and more freckled character.

Sp. 46.   PODARGUS PLUMIFERUS, *Gould.*

PLUMED PODARGUS.

Podargus plumiferus, Gould in Proc. of Zool. Soc., part xiii. p. 104.

Podargus plumiferus, Gould, Birds of Australia, fol., vol. ii. pl. 6.

The only information I have to communicate respecting this beautiful Podargus is, that it is a native of the brushes of the Clarence and neighbouring rivers in New South Wales, and that several examples have come under my notice, of which one is deposited in the Museum at Dublin, another in the Museum at Manchester, and a third was sent to me by the late Mr. Strange of Sydney. The *Podargus plumiferus* is readily distinguished from all the other Australian members of the genus by the more lengthened form of tail, and by the remarkable and conspicuous tufts of feathers which spring from immediately above the nostrils: considerable variation is found to exist in the colouring of the various specimens, some being much redder than the others, and having the markings on the under surface much less distinct and of a more chestnut tint.

Tuft of feathers covering the nostrils alternately banded with blackish brown and white; all the upper surface mottled brown, black, and brownish white, the latter predominating over each eye, where it forms a conspicuous patch; the markings are of a larger but similar kind on the wings, and on the primaries and secondaries assume the form of bars; tail similar, but paler, and with the barred form of the markings still more distinct; centre of the throat and chest brownish white, minutely freckled with brown; sides of the neck and breast and all the under surface similar, but with a dark line of brown down the centre, and two large nearly square-shaped spots of brownish white near the tip of each feather; bill and feet horn-colour.

Sp. 47. PODARGUS MARMORATUS, *Gould*.

MARBLED PODARGUS.

Podargus marmoratus, Gould in App. to Macgillivray's Voy. of Rattlesnake, vol. ii. p. 856.

Podargus marmoratus, Gould, Birds of Australia, Supplement, pl. .

On carefully comparing examples of this species with the original example of MM. Quoy and Gaimard's *Podargus ocellatus* in the Museum of the Jardin des Plantes, I found them to differ so greatly that I could come to no other conclusion than they were distinct. The *P. ocellatus* is a smaller bird, has a redder tail, and very conspicuous large round white spots on the wing, arranged in the form of three distinct semicircular bars—characters which do not exist in the Australian bird; I had, therefore, no alternative but to give the latter a distinctive appellation.

The present species is particularly elegant in form, and is, in fact, a miniature representative of the *P. Papuensis*, and, like that bird, has a lengthened cuneate tail—a feature which adds much to its gracefulness.

Much difference exists in the colouring of the sexes, the female being of a deep rusty hue, while the male is beautifully marbled with pearl-white, interspersed with freckles of brown and black, particularly on the under surface.

Both the specimens from which my descriptions were taken were shot by Mr. Macgillivray on the Cape York Peninsula, one on the 14th, the other on the 19th of November 1849. These examples now grace the National Collection, where they will be available for comparison should any nearly allied species be discovered.

The male has the whole of the upper surface and wings minutely mottled with brown, grey, and buff, the buffy tint prevailing over the eyes, on the scapularies, and on the tips of the wing-coverts; on the outer webs of the primaries the markings assume the form of bars of mingled buffy, buffy white, and rufous; tail light brown, crossed with numerous defined bands of grey, freckled with black, and with a rufous hue on the lateral feathers; under surface pearly white, minutely freckled with brown, and with a line of brown down the stem; a series of these darker marks, forming an irregular line, down each side of the neck; bill and feet brownish olive.

The markings of the female are similar, but her general tint is very much darker, and of a more rufous hue; the under surface, too, is dark brown, with here and there large blotches of buffy white; a series of nearly quadrangular blotches, bordered with dark brown, descends down each side of the neck.

Genus EUROSTOPODUS.

This genus, so far as is yet known, comprises but two species, both of which are probably confined to Australia. They differ considerably in their habits from the other *Caprimulgi*. Their wing-powers being enormous, they pass through the air with great rapidity, and while hawking for insects during the twilight of the early dawn and evening, they make the most abrupt and sudden turns in order to secure their prey. Like the typical Nightjars, they rest on the ground during the day. In every instance in which the site employed for incubation by the *Eurostopodus guttatus* has been discovered, a single egg, deposited on the bare ground, has alone been found.

The members of this genus are very nearly allied to the *Lyncorni*, a genus of Nightjars inhabiting the Indian Islands, of which *L. cerviniceps* and *L. macrotis* are typical examples.

Sp. 48.   EUROSTOPODUS ALBOGULARIS.

WHITE-THROATED NIGHTJAR.

Caprimulgus albogularis, Vig. and Horsf. in Linn. Trans., vol. xv. p. 194, note.
——— *mystacalis*, Temm. Pl. Col. 410.

Eurostopodus albogularis, Gould, Birds of Australia, fol, vol. ii. pl. 7.

During my visit to Australia I had frequent opportunities of observing this species. How far it may range over the Australian continent is not known: the south-eastern are the only portions in which it has yet been discovered. I have seen specimens in collections formed at Moreton Bay, and I have killed three or four individuals of an evening on the cleared lands in the neighbourhood of the Upper Hunter, which shows that it is far from being a scarce bird in that part of New South Wales. In all probability it is only a summer visitant in the colony, for it was at this season only that I observed it. In the daytime it sleeps on the ground on some dry knoll or open part of the forest, and as twilight approaches sallies forth to the open glades and small plains or cleared lands in search of insects; its flight, which is much more powerful than that of any other species of the family I have seen, enabling it to pass through the air with great rapidity, and to mount up and dart down almost at right angles whenever an insect comes within the range of its eye, which is so large and full that its powers of vision must be very great. Most of those I shot were gorged with insects, principally coleoptera and locusts, some of which were entire, and so large as to excite surprise how they could be swallowed; in several instances they were so perfect that I preserved them as specimens for the cabinet.

Of its nidification I have no reliable information to furnish; but that it deposits a single egg on the bare ground is very probable.

Contrary to what might have been expected, I found that although the sexes are nearly alike in colour, the females always exceed the males in size and in the brilliance of the tints; the males, on the other hand, have the two white spots on the third and fourth primaries more conspicuous than in the female.

This species has very large and lustrous black eyes, which clearly indicate that it is a night-flier; its wings are very long; its tarsi short, and partially feathered; and the stiff rictal bristles of the typical *Caprimulgi* are absent.

All the upper surface very minutely freckled grey and brown; the feathers on the crown of the head and at the occiput with a large patch of black down the centre; behind the ear-coverts a patch of dark brown sprinkled with brownish buff; from the angle of the mouth passing round the back of the neck an indistinct collar of intermingled buff, chestnut, and black; scapularies variegated with dark brown on their outer webs and margined with bright fulvous; wing dark brown, variegated with fulvous and grey; secondaries dark brown, with a regular series of bright fulvous spots along each web; primaries blackish brown, the two first without any spot, the remainder spotted like the secondaries, the third having a spot of white on its inner and outer web about the centre of the feather, the fourth with a large white spot on its outer web; two centre and outer webs of the remaining tail-feathers dark brown, marbled with irregular bars of grey; the inner webs of the lateral feathers dark brown, crossed with irregular bands of light buff; throat blackish brown, spotted with bright buff; on each side of the throat a large oval spot of white; breast dark brown, spotted above with dull buff, and broadly freckled with dull buff and grey; abdomen and under tail-coverts bright fulvous, crossed with bars of dark brown; irides dark brown; feet mealy reddish brown.

Sp. 49.   EUROSTOPODUS GUTTATUS.

SPOTTED NIGHTJAR.

Caprimulgus guttatus, Vig. and Horsf. in Linn. Trans., vol. xv. p. 192.
Ficktel's Goatsucker, Lath. Gen. Hist., vol. vii. p. 345.
Kal-ga, Aborigines of the lowland districts of Western Australia.
Goatsucker of the Colonists.

Eurostopodus guttatus, Gould, Birds of Australia, fol., vol. ii. pl. 8.

As the similitude of its form would lead us to suspect, this species closely resembles the preceding, both in its habits and in the whole of its economy; unlike that species, however, whose range of habitat would appear to be very limited, the present bird is universally, but thinly, distributed over the whole of the southern portion of Australia. I killed it in South Australia and in New South Wales; the collection formed by Gilbert at Swan River contained specimens which presented no difference whatever, either in size or markings, and I have since seen examples from the north-west coast.

During my rambles in New South Wales I more than once flushed this bird in open day, when, after mounting rapidly in the air, it performed a few zigzag evolutions and pitched again to the earth at a distant spot. That it breeds on the ground there can be no doubt, as I found a newly hatched young one on the precise spot from which I had flushed the adult; the little helpless creature, which much resembled a small mass of down or wool, was of a reddish-brown colour, not very dissimilar from the surface of the ground where it had been hatched: my utmost endeavours to find the broken shell were entirely unavailing; but I have since obtained undoubted eggs of this species from two or three sources. They differ both in form and colour from those of any of the typical *Caprimulgi*, and also from those of the *Podargi* and *Ægotheles*. They may be described in a few words. In size they are about an inch and three-eighths in length by nearly an inch in breadth; in colour nearly uniform olive stone-colour, with here and there a roundish purple blotch or spot. In confirmation of the opinion I have expressed that the birds of this form lay but one egg, I may cite the following note received from Mr. White, of the Reed-Beds, near Adelaide:—

"I have several times found the female sitting on the ground or rock with only a single egg under her; the one sent to you was placed on a bare piece of stony ground, and the bird was sitting so close that she allowed me to approach within a few feet of her without moving. The egg is dusky green, spotted with black, and is of equal size at both ends."

The sexes are so nearly alike in colour and size that they are not to be distinguished except by dissection; the young, on the contrary, is clothed in a more buffy-brown dress until it has attained the size of the adult.

Forehead and centre of the head brownish black, each feather spotted and margined with bright buff; over each eye the feathers are pearly white, very finely pencilled with brownish black; lores and sides of the face brown, spotted with buff; collar at the back of the head reddish chestnut; back grey, freckled with black; scapularies light grey freckled with brownish black, largely tipped with bright buff, with an irregular diagonal patch of black; wing-coverts grey, spotted and freckled with brown, each feather largely tipped with bright buff; primaries and secondaries brownish black, marked on both webs with buff, the buff on the outer webs being in the form of round spots, on the inner webs irregular bars; on the inner web of the first primary is a large spot of pure white, on the second primaries a similar but larger spot, and a small one on the outer web; the third and fourth crossed by a large irregular patch of white; middle tail-feathers light grey, marbled and finely freckled with dark brown; lateral feathers light grey, barred with blackish brown and bright buff, and freckled with dark brown, the buff on the outer web of the outside feather forming a regular row of spots; on each side of the throat an oblique line of white; chest dark brown, each feather broadly barred and spotted with light buff; abdomen bright buff, finely and irregularly barred with black; under tail-coverts sandy; bill black; irides very dark brown; feet mealy reddish brown.

Genus CAPRIMULGUS, *Linnæus*.

Europe, Asia, and Africa are the great strongholds of the members of this genus as at present restricted. A single species only has yet been discovered in Australia, where it frequents the northern or intertropical parts of the country.

Sp. 50. CAPRIMULGUS MACRURUS, *Horsfield*.

LARGE-TAILED NIGHTJAR.

Caprimulgus macrurus, Horsf. in Linn. Trans., vol. xiii. p. 142.

Caprimulgus macrurus, Gould, Birds of Australia, fol., vol. ii. pl. 9

This, the only true *Caprimulgus* known to inhabit Australia, is I believe identical with the *C. macrurus* of Horsfield, whose specimens were procured in Java, while those I possess were obtained at Port Essington, where the bird is moderately plentiful; it is also found in Southern India, hence it has an unusually wide range of habitat. It frequents the open parts of the forest, and is strictly nocturnal; it mostly roosts on the ground on the shady side of a large tree close to the roots, and if disturbed several times in succession takes to the branch of one of the largest trees. I have never seen the eggs of this species, but I possess a young bird apparently only a few days old, which Gilbert found lying under a shrubby tree, without any nest or even a blade of grass near it; the little creature was so similar in colour to that of the ground upon which it was lying, that it was with difficulty detected, and he was only induced to search for it from the very peculiar manner in which the old bird rose, the reluctance it evinced to leave the spot, and its hovering over the place it had risen from, instead of flying off to the distance of nearly a hundred yards, as it usually does.

The male is distinguished by the greater extent of the white mark on the primaries and outer tail-feathers; in the other parts of their plumage and in size the sexes do not differ.

Head brownish grey, very minutely freckled with black; the feathers down the middle of the head and occiput with a large broad stripe of black down the centre; lores, space surrounding the eyes and ear-coverts reddish brown; on each side of the neck a broad stripe of rich buff barred with black; a narrow line of white passes below the angle of the mouth; chin brown; across the throat a band of white bounded below by black, the extremities of the white feathers being of that hue; centre of the back dark brown, freckled with black and buff; shoulders blackish brown; wing-coverts freckled grey, buff, and black, each with a large spot of buff at the tip; primaries and secondaries blackish brown, the former crossed at their base, and the latter throughout their entire length, with reddish buff; the second and third primaries crossed near their base with a broad band of white, stained with buff on the outer margin; the first primary with a spot of white only on the margin of the inner web; the first three primaries freckled at their tips, and the remainder for the entire length of their inner webs, with brownish grey; scapularies freckled grey and brown, with a large patch of deep dull black on their outer webs, margined externally with buff; rump freckled with dark brown and grey, and with an interrupted line of darker brown down the centre of each feather; two centre tail-feathers minutely and coarsely freckled with very dark brown; the next on each side very dark brown, crossed by irregular bands of freckled brownish grey and black; the next on each side similar, but the bands narrower and less conspicuous; the two outer ones on each side very dark brown for three parts of the length, the apical portion being white, stained with freckled buff and black on the outer webs; the basal or dark portion crossed by narrow indistinct and irregular bars of deep buff; breast freckled buff, grey, and brown, some of the feathers in the centre of the breast largely tipped with buff; abdomen and under tail-coverts deep buff, crossed by narrow regular bands of dark brown; irides blackish brown; bill black; feet and claws reddish brown.

Specimens of this species were brought from the Aru Islands by Mr. Wallace.

Family CYPSELIDÆ.

Whether the Swifts and the Swallows are naturally separated by the interposition of numerous other genera of birds is a point respecting which it is not necessary for me to enter into in a handbook on the 'Birds of Australia.' I place them next each other here, because they were so in the folio edition.

Of the Cypselines two very distinct forms or genera are found in Australia — *Chætura* and *Cypselus*; both are migrants, and at present it is uncertain whether either of them breed in that part of the world. The power of flight enjoyed by both is enormous, and it is probable that their migratory movements extend from India and China to the extreme southern limits of the mainland of Australia; one of them, the Spine-tailed Swift, even crosses Bass's Straits to Tasmania, and occasionally appears there in great numbers. Other Spine-tailed Swifts are found in America; but these differ somewhat in form; it was, however, to a species inhabiting that country that the generic term *Chætura* was first applied.

Genus CHÆTURA, *Stephens*.

The type of this genus is an American bird, the *Hirundo spinicauda* of authors. Mr. Hodgson considered the Indian *H. caudacuta* to differ sufficiently to warrant its separation, and proposed for it the generic appellation of *Hirundapus*; but such a division does not appear to me advisable, and I have not therefore adopted it. The Spine-tailed Swifts are inhabitants both of the Old and New Worlds.

Sp. 51.   CHÆTURA CAUDACUTA.

Spine-tailed Swift.

Hirundo caudacuta, Lath. Ind. Orn. Supp., p. 57.
—— *fusca*, Steph. Cont. of Shaw's Gen. Zool., vol. xiii. p. 76.
—— *pacifica*, Lath. Ind. Orn. Supp., p. 58 ?
Needle-tailed Swallow, Lath. Gen. Syn. Supp., vol. ii. p. 307.
Pin-tailed Swallow, Lath. Gen. Hist., vol. vii. p. 308.
Chætura Australis, Steph. Cont. Shaw's Gen. Zool., vol. xiii. p. 76.
—— *macroptera*, Swains. Zool. Ill., 2nd ser. pl. 42.
—— *nudipes*, Hodgs. Journ. Asiat. Soc. Beng. 1836, p. 770.
Acanthylis caudacuta, G. R. Gray, Ann. Nat. Hist. 1843, p. 194.
—— *nudipes*, Gray and Mitch. Gen. of Birds, vol. i. p. 55, *Acanthylis*, sp. 4.
Pallene macroptera, caudacuta, et *leuconota*, Boie, Isis, 1844, p. 168.
Cypselus leuconotus, Delcss. Mag. de Zool. 1840, Ois. t. 20.
Hirundapus nudipes, Hodgson.

Acanthylis caudacuta, Gould, Birds of Australia, fol., vol. ii. pl. 10.

This noble species, one of the largest of the Cypselines yet discovered, is a summer visitant of the eastern portions of Australia, proceeding as far south as Tasmania; but its visits to this island are not so regular as to New South Wales. During the months of January and February it appears in large flocks, which, after spending a few days, disappear as suddenly as they arrived. I am not aware of its having been observed in Western Australia, neither has it occurred in any of the collections formed at Port Essington, although it doubtless pays that colony passing visits during its migrations. I believe it will be found that Indian and Australian examples do not differ. It is supposed to have been known to Pallas, and if so, it is the bird described by that author as *Hirundo ciris*. Von Schrenck found it in Amoorland, and it is also said to have once occurred in England. Mr. Jerdon states that it breeds among the huge wall-like crags of the Himalayas, and under the snow-level.

The keel or breast-bone of this species is more than ordinarily deep, and the pectoral muscles more developed than in any bird of its weight with which I am acquainted. Its whole form is especially and beautifully adapted for extended flights; hence it readily passes from one part of the world to another, and, if so disposed, may be engaged in hawking for flies on the continent of Australia at one hour, and in the next be similarly employed in Tasmania.

So exclusively is this bird a tenant of the air, that I never, in any instance, saw it perch, and but rarely sufficiently near the earth to admit of a successful shot; it is only late in the evening and during lowery weather that such an object can be accomplished. With the exception of the Crane, it is certainly the most lofty as well as the most vigorous flier of the Australian birds. I have frequently observed in the middle of the hottest days, while lying prostrate on the ground with my eyes directed upwards, the cloudless blue sky peopled at an immense elevation by hundreds of these birds, performing extensive curves and sweeping flights, doubtless attracted thither by the insects that soar aloft during serene weather; on the contrary, the flocks that visit the more humid climate of Tasmania, necessarily seek their food near the earth.

The sexes offer no perceptible difference in their outward appearance; but the female, as is the case with the other members of the family, is a trifle smaller than her mate.

Crown of the head, back of the neck, and ear-coverts deep shining green, strongly tinged with brown; a small space immediately before the eye deep velvety black; band across the forehead, throat, inner webs of the secondaries nearest the back, a patch on the lower part of the flanks and the under tail-coverts white; wings and tail deep shining green, with purple reflexions; centre of the back greyish brown, becoming darker towards the rump; chest and abdomen dark clove-brown; bill black; feet brown.

Genus CYPSELUS, *Illiger*.

Of this genus, as now restricted, many species inhabit Europe, Asia, Africa, and the Indian Islands; and one is found in Australia.

Sp. 52. CYPSELUS PACIFICUS.

AUSTRALIAN SWIFT.

Hirundo pacifica, Lath. Ind. Orn. Suppl., p. 58.
Cypselus pacificus, Steph. Cont. of Shaw's Gen. Zool., vol. x. p. 132.
—— *australis*, Gould in Proc. of Zool. Soc., part vii. 1839, p. 141.
—— *vittatus*, Jard. Ill. Orn., ser. 2, pl. 39.
Micropus australis et *vittata*, Boie, Isis, 1844, p. 165.
Hirundo apus, var. β, Pall. Zoog. Ross.-Asiat., tom. i. p. 540.

Cypselus Australis, Gould, Birds of Australia, fol. vol. ii. pl. 11.

As I had never seen or heard of a true Swift in Australia, I was no less surprised than gratified when I discovered this species to be tolerably numerous on the Upper Hunter, during my first visit to that district in 1839. Those I then observed were flying high in the air, and performing immense sweeps and circles, while engaged in the capture of insects. I succeeded in killing six or eight individuals, among which were adult examples of both sexes; but I was unable to obtain any particulars as to their habits and economy. It would be highly interesting to know whether this bird, like the Swallow, returns annually to spend the months of summer in Australia. I think it likely that this may be the case, and that it may have been frequently confounded with the *Acanthylis caudacuta*, as I have more than once seen the two species united in flocks, hawking together in the cloudless skies, like the Martins and Swallows of England.

Throat and rump white; upper and under surface of the body brown; the back tinged with a bronzy metallic lustre; each feather of the under surface margined with white; wings and tail dark brown; irides, bill, and feet black.

It is considered by some ornithologists that this bird and the Swift with crescentic markings of white on the breast, which inhabits China and Amoorland, are the same. If this supposition be correct, this species ranges very widely over the surface of the globe.

Family HIRUNDINIDÆ.

I wish it to be understood that, although I unite the Swifts and the Swallows, I am not unaware of the difference which exists in the structure of these two groups; but, as I have stated in the 'Birds of Great Britain,' I consider it desirable that they should follow each other in an arrangement of the birds of a single country. The Swifts being disposed of in their two genera, *Chætura* and *Cypselus*, I proceed with the true Hirundines, three or four forms of which, with many others not Australian, compose the extensive family of the *Hirundinidæ*. It may not be out of place if I say a few words on the almost general distribution of these aërial birds over the face of the globe. In America, Africa, China, India, the islands of the Eastern Archipelago, and Australia, Swallows and Martins of varied forms are numerous, and species abound; yet, strange to say, neither Swifts nor Swallows are found in New Zealand, or in any of the islands adjacent to that country.

INSESSORES. 107

At least two members of the genus *Hirundo*, or true Swallows, are found in Australia; of these one is very common there; the other, in all probability, is merely a transient visitor to its northern portions. Independently of these, Australia has two other species, one allied to the Swallows, the other to the Martins, to each of which I have been constrained to give new generic names: one of these lays its eggs on the bare wood in the holes of trees; while the other constructs a singular nest under the eaves of the house and verandahs of the settlers. This last beautifully represents the *Chelidon urbica* of Britain, from which it mainly differs in being destitute of feathers on the tarsi. There is no true *Cotyle*, or Sand-Martin, in Australia; but there is a bird whose habits and economy are very similar, for it occasionally drills a hole in a bank-side in which to nidify, like to our *C. riparia*. This is the only species known of M. Cabanis's genus *Cheramœca*.

Genus HIRUNDO, *Linnæus.*

The members of the genus *Hirundo*, or true Swallows, inhabit Europe, Asia, Africa, North America, the Indian Islands, and Australia.

Sp. 53. HIRUNDO FRONTALIS, *Quoy et Gaimard.*

WELCOME SWALLOW.

Hirundo frontalis, Quoy & Gaim. Voy. de l'Astrol., Ois. tab. 12. fig. 1.
—— *neoxena*, Gould in Proc. of Zool. Soc., part x. 1842, p. 131.
—— (*Herse*) *frontalis*, Less. Compl. Buff., tom. viii. p. 497.
Cecropis frontalis, Boie, Isis, 1844, p. 174.
Kun-na-meet, Aborigines of the lowland districts of Western Australia.
Ber-rin-nin, Aborigines of New South Wales.

Hirundo neoxena, Gould, Birds of Australia, fol. vol. ii. pl. 13.

The arrival of this bird in the southern portions of Australia is hailed as a welcome indication of the approach of spring, and is associated with precisely the same ideas as those popularly entertained respecting our own pretty Swallow in England. The two species are in fact beautiful representatives of each other, and assimilate not only in their migratory movements, but also most closely in their whole habits, actions, and economy. It arrives in Tasmania about the middle or end of September, and, after rearing at least two broods, departs again northwards in March; but it is evident that the migratory movement of the Swallow, and doubtless that of all other birds, is regulated entirely by the temperature, and the more or less abundant supply of food necessary for its existence; for I found that in New South Wales, and every country in Australia within the same latitude, it arrived much earlier and departed considerably later than in Tasmania; and Mr. Caley, who resided in New South Wales for several years, and whose valuable notes on the birds of that part of the country have been so often quoted, states that "the earliest period of the year that I noticed the appearance of *Swallows* was on the 12th of July 1803, when I saw two; but I remarked several towards the end of the same month in the following year (1804). The latest period I observed them was on the 30th of May 1806, when a number of them were twittering and flying high in the air. When I missed them at Paramatta, I have sometimes met with them among the north rocks, a romantic spot about two miles to the northward of the former place." A few stragglers remain in New South Wales during the winter, but their numbers cannot for a moment be compared with those observed in the summer, which have passed the colder months in a warmer climate. This Swallow having been found by the naturalists of the 'Astrolabe' in the Eastern Islands, and more recently by Mr. Wallace in New Guinea, it is evident that its range extends beyond the northern limits of Australia.

The natural breeding-places of this bird are the deep clefts of rocks and dark caverns, but since the colonization of Australia it has in a remarkable degree imitated its European prototype, by selecting for the site of its nest the smoky chimneys, the chambers of mills and out-houses, or the corner of a shady verandah; the nest is also similarly constructed, being open at the top, formed of mud or clay, intermingled with grass or straw to bind it firmly together, and lined first with a layer of fine grasses, and then with feathers. The shape of the nest depends upon the situation in which it is built, but it generally assumes a rounded contour in front. The eggs are usually four in number, of a lengthened form; their ground colour pinky white, with numerous fine spots of purplish brown, the interspaces with specks of light greyish brown, assuming in some instances the form of a zone at the larger end; they are from eight to nine lines long by six lines broad. At Swan River the breeding-season is in September and October. In the spring of 1862 two nests of this species were sent to me by George French Angas, Esq. These very closely resembled those of our own bird, both in form and materials; they were, however, somewhat more square and more stoutly built. The interior was composed of the usual plastered mud strengthened with a little hair, and thickly lined with the downy feathers of various domestic birds. These nests are now in the British Museum.. The following note by Mr. Angas was attached to one of them:—" Built on a rafter of my stable at Collingrove, South Australia: taken Oct. 3, 1861."

Forehead, chin, throat, and chest rust-red; head, back of the neck, back, scapularies, wing-coverts, rump, and upper tail-coverts deep steel-blue; wings and tail blackish brown, all but the two centre feathers of the latter with an oblique mark of white on the inner web; under surface very pale brown; under tail-coverts pale brown, passing into an irregular crescent-shaped mark near the extremity, and tipped with white; irides dark brown; bill and legs black.

Sp. 64.   HIRUNDO FRETENSIS, *Gould.*

TORRES STRAIT'S SWALLOW.

The only specimen I possess of the bird now to be noticed was shot by Mr. Rayner, Surgeon of H.M.S. Herald, on the northern shore of Australia. As it is somewhat immature, I am unable to institute a rigid comparison between it and other known species, as I could wish. In size and general appearance it is very like an English Swallow at the end of its first autumn; but its bill is larger and longer than that of any adult specimen, either of our own island or from India, that I have seen. I have a fully adult Swallow from the Island of Java, which I believe to be a mature individual of the present species. It is very like our *H. rustica*, but is somewhat smaller in the body, has a very large bill, and but a faint indication of the black pectoral band.

Throat rusty red, bounded below by an indistinct band of dull bluish black; under surface white; tail forked, but the outer feathers, which I consider to be imperfectly developed, do not exceed the central one by more than three-quarters of an inch; all the tail-feathers, except the two middle ones, with an oval spot of white on the inner web, about half an inch from the tip; crown of the head brownish black, with steel reflexions; back and upper tail-coverts glossy steel-bluish black; wings black, glossed with green; bill and feet black.

Total length, from tip of bill to end of tail, 5 inches; bill, gape to lip, $\frac{9}{16}$; breadth at base $\frac{1}{8}$; wing $4\frac{1}{2}$; outer tail-feathers $2\frac{1}{4}$; middle tail-feathers $1\frac{3}{4}$; tarsi $\frac{1}{2}$.

Genus HYLOCHELIDON, *Gould.*

I have not instituted a new generic appellation for the following bird without maturely considering the propriety of so doing, after carefully comparing it with the various forms already characterized of this extensive family; which, whenever it may be monographed by a scientific ornithologist, will be found to comprise ample materials for the formation of more genera than has yet been proposed, as well as numerous species with which we are at present unacquainted; and I have no doubt that Mr. Blyth's notion of dividing them into sections in accordance with the forms of their nests will be found a very happy suggestion—saucer-builders, retort-builders, bank-burrowers, builders in the holes of trees, &c.

The species of this form are part of a small section of the Swallows which nidify in the holes of trees, without any nest for the deposition of their delicate eggs. Their bare tarsi at once separate them from the Chelidons, and they also differ from the American Petrochelidons. Of these birds, which appear to be an offshoot from the typical or true Hirundines, my collection contains at least two species, one from Australia, the other from Timor; I say at least, because it is a question whether the birds from Australia do not constitute two in themselves,—specimens from Tasmania being very much larger than those from the main land.

Sp. 55. HYLOCHELIDON NIGRICANS.

Tree Swallow.

Chelidon arborea, Gould, Birds of Australia, vol. I. Introd. p. xxix.
Cecropis pyrrhonota, Boie, Isis, 1844, p. 175.
Hirundo (Herse) nigricans et *pyrrhonota*, Less. Compl. Buff., tom. viii. p. 497.
Dun-rumped Swallow, Lath. Gen. Hist., vol. vii. p. 309.
Hirundo pyrrhonota, Vig. and Horsf. in Linn. Trans., vol. xv. p. 190.
Hirundo nigricans, Vieill. Ency. Méth., part ii. p. 525.
Cecropis nigricans, Boie, Isis, 1844, p. 175.
Petrochelidon nigricans, Cab. Mus. Hein. Theil i. p. 47.
Gab-by-kal-lan-goo-rong, Aborigines of the lowlands of Western Australia.
Martin of the Colonists.

Collocalia arborea, Gould, Birds of Australia, fol., vol. ii. pl. 14.

The Tree Swallow is a very common summer visitant to

the southern portions of Australia and Tasmania, arriving in August and retiring northwards as autumn approaches. It is a very familiar species, and frequents the towns in company with the Swallow. I observed it to be particularly numerous in the streets of Hobart Town, where it arrives early in September; the more southern and colder situation of the island rendering all migratory birds later in their arrival there.

It breeds during the month of October in the holes of trees, making no nest, but laying its eggs on the soft dust generally found in such places: the eggs are from three to five in number, of a pinky white faintly freckled at the larger end with fine spots of light reddish brown; they are eight lines long by six lines broad.

Considerable difference exists both in size and in the depth of colouring of specimens killed in New South Wales, Swan River, and Tasmania; but as there exists no distinctive character of marking, I regard them as local varieties rather than as distinct species. Tasmanian specimens are larger in all their admeasurements, and have the fulvous tint of the under surface and the band across the forehead much deeper than in those killed in New South Wales; individuals from the latter locality again exceed in size those from Western Australia.

Specimens of this bird, identical with others from New South Wales, were brought from the Aru Islands by Mr. Wallace.

Genus LAGENOPLASTES, *Gould.*

The little Fairy Martin of Australia, the constructor of a singular retort-shaped nest, is the type of the present genus; in which I think must also be placed another species in my collection, which I received from India, and which precisely resembles it in form and greatly in colour. What the members of the genus *Hylochelidon* are to the Swallows, those of the present are to the Martins, from which they differ in their diminutive and bare tarsi, and from the American Hylochelidons in their more feeble structure and colouring.

Sp. 50.   LAGENOPLASTES ARIEL, *Gould.*

FAIRY MARTIN.

Collocalia Ariel, Gould in Proc. of Zool. Soc., part x. 1842, p. 132.
Chelidon Ariel, Gould, Birds of Australia, vol. i. Introd. p. xxix.
Hirundo Ariel, Gray and Mitch. Gen. of Birds, vol. i. p. 66, *Hirundo*, sp. 17.

Collocalia ariel, Gould, Birds of Australia, fol., vol. ii. pl. 15.

The Fairy Martin is dispersed over all the southern portions of Australia, and, like every other member of the genus, it is strictly migratory. It usually arrives in the month of August, and departs again in February or March; during this interval it rears two or three broods. The Fairy Martin, unlike the favourite Swallow of the Australians, although enjoying a most extensive range, appears to have an antipathy to the country near the sea, for neither in New South Wales nor at Swan River have I ever heard of its approaching the coast-line nearer than twenty miles; hence, while I never observed it at Sydney, the town of Maitland on the Hunter is annually visited by it in great numbers. In Western Australia it is common between Northam and York, while the towns of Perth and Fremantle on the coast are, like Sydney, unfavoured with its presence. I observed it throughout the district of the Upper Hunter, as well as in every part of the interior, breeding in various localities, wherever suitable situations presented themselves; sometimes their nests are constructed in the cavities of decayed trees; while not unfrequently clusters of them are attached to the perpendicular banks of rivers, the sides of rocks, &c., generally in the vicinity of water. The long bottle-shaped nest is composed of mud or clay, and, like that of our Common Martin, is only worked at in the morning and evening, unless the day be wet or lowery. In the construction of the nests these birds appear to work in small companies, six or seven assisting in the formation of each nest, one remaining within and receiving the mud brought by the others in their mouths: in shape these nests are nearly round, but vary in size from four to six or seven inches in diameter; the spouts of some being eight or nine inches in length. When built on the sides of rocks or in the hollows of trees, they are placed without any regular order, in clusters of thirty or forty together, some with their spouts inclining downwards, others at right angles, &c.; they are lined with feathers and fine grasses. The eggs, which are four or five in number, are sometimes white, at others spotted and blotched with red; eleven-sixteenths of an inch long by half an inch broad.

The sexes are alike in colour.

Crown of the head rust-red; back, scapularies, and wing-coverts deep steel-blue; wings and tail dark brown; rump buffy white; upper tail-coverts brown; under surface white, tinged with rust-red, particularly on the sides of the neck and flanks; the feathers of the throat with a fine line of dark brown down the centre; irides blackish brown; bill blackish grey; legs and feet olive-grey.

Genus CHERAMŒCA, *Cabanis*.

In the "Introduction" to the folio edition, I remarked that I was not fully satisfied of the propriety of placing the White-breasted Swallow in the genus *Atticora*; and that I erred in so doing has since been shown by M. Cabanis having deemed it necessary to make it the type of a new one, which I here adopt.

Sp. 57. CHERAMŒCA LEUCOSTERNA, *Gould.*
WHITE-BREASTED SWALLOW.

Hirundo leucosternus, Gould in Proc. of Zool. Soc., part viii. p. 172.
Cheramœca leucosterna, Cab. Mus. Hein., Theil i. p. 49.
Boo-de-boo-de of the Aborigines of the mountain districts of Western Australia.
Black and White Swallow of the Colonists.

Atticora leucosternon, Gould, Birds of Australia, fol., vol. ii. pl. 12.

The White-breasted Swallow is a very wandering species, never very numerous, and is generally seen in small flocks of from ten to twenty in number, sometimes in company with the other Swallows. It usually flies very high, a circumstance which renders it difficult to procure specimens.

In Western Australia this bird chooses for its nest the deserted hole of either the Dalgyte (*Peragalea lagotis*) or the Doodee (a species of *Bettongia*), but more generally drills holes in the sides of banks, like the Sand-Martin of Europe.

These holes are perfectly round, about two inches in diameter, run horizontally for three feet from the entrance, and then expanding into a chamber or receptacle for the nest, which is constructed of the broad portions of dried grasses and the dry dead leaves of trees. Mr. Johnson Drummond informed Gilbert that he had frequently found seven, eight, or nine eggs in a single nest, from which he inferred that more than one female lays in the same nest: the eggs are white, somewhat longthoned, and pointed in form. It would seem that the holes are not constructed exclusively for the purpose of nidification, for upon Gilbert's inserting a long grass stalk into one of them, five birds made their way out, all of which he succeeded in catching; upon his digging to the extremity, in the hope of procuring their eggs, no nest was found, and hence he concludes that their holes are also used as places of resort for the night.

Since this information was transmitted, I have received notices of this bird from many other sources, which enable me to state with tolerable certainty that it is spread during summer at least over the whole of the southern portion of the interior, from Queensland to Swan River. Strange to say, however, I have never seen examples of this species in any collection formed out of Australia; yet the occurrence of a bird whose wing-powers are so great might naturally be expected in New Guinea or some of the adjacent islands.

Crown of the head light brown, surrounded by a ring of white; lores black; a broad band commencing at the eye, and passing round the back of the neck, brown; centre of the back, throat, chest, and under surface of the shoulder white; wings and tail brownish black; rump, upper tail-coverts, abdomen, and under tail-coverts black; irides dark reddish brown; bill blackish brown; legs and feet greenish grey.

Family MEROPIDÆ.

Like all other extensive families of birds, the varied members of the Meropidæ or Bee-eaters are divisible into many genera. In India, we find the beautiful *Nyctiornis amictus* and its two allies; and in Africa, several other genera, comprising birds of considerable size and gaiety of colouring.

These aërial birds live almost exclusively on insects, and it is while engaged in the capture of these that the very beautiful colours with which they are adorned are shown to the greatest advantage. In their mode of nidification and in the colouring of their eggs they are allied to the Kingfishers.

Generally speaking, the sexes are alike in plumage, and differ but little in size.

Genus MEROPS, *Linnæus.*

India and Africa may be said to be the great nursery of this lovely group of birds; of which one, common in the southern parts of Europe, is beautifully represented in Australia by the *Merops ornatus*, the only species inhabiting that country.

Sp. 58.   MEROPS ORNATUS, *Latham.*

AUSTRALIAN BEE-EATER.

Merops ornatus, Lath. Ind. Orn., Supp. p. xxxv.
Mountain Bee-eater, Lewin, Birds of New Holl., pl. 18.
Variegated Bee-eater, Lath. Gen. Syn., Supp. vol. ii. p. 155, pl. 128.
Merops melanurus, Vig. and Horsf. in Linn. Trans., vol. xv. p. 203.
Philemon ornatus, Vieill. Nouv. Dict. d'Hist. Nat., tom. xvii. p. 423.
Merops Thouini, sp. tenuipennis, Dumont, Id., et Levrault, p. 52.
Melittophagus ornatus, Reich. Handb. tom. i, p. 82.
Cosmaërops ornatus, Cab. et Hein. Mus. Hein., Theil ii. p. 138.
Dee-weed-gang, Aborigines of New South Wales.
Bée-roo-bée-roo-long, Aborigines of the lowland, and
Beʺr-rin-beʺr-rin, Aborigines of the mountain districts of Western Australia.
Bee-eater of the Colonists.

Merops ornatus, Gould, Birds of Australia, fol., vol. ii. pl. 18.

This bird has so many attractions that it will doubtless be always regarded as a general favourite with the Australians; the extreme beauty of its plumage, the elegance of its form, and the graceful manner of its flight all combining to render it especially worthy of their notice; besides which, many pleasing associations are connected with it, for, like the Swallow and the Cuckoo of Europe, it arrives in New South Wales and in all the colonies lying within the same degree of latitude in August, and departs in March, the intervening period being employed in the duties of incubation and of rearing its progeny. During the summer months it is universally spread over the whole southern portion of the continent from east to west; and in winter the northern.

In South Australia and at Swan River it is equally numerous as in New South Wales, generally giving preference to the inland districts rather than to those near the coast; hence it is rarely to be met with in the neighbourhood of Perth, while in the York district it is very common. In New South Wales I found it especially abundant on the

Upper Hunter, and all other parts towards the interior, as far as I had an opportunity of exploring. Its favourite resorts during the day are the open, arid, and thinly-timbered forests; and in the evening the banks and sides of rivers, where numbers may frequently be seen in company. It almost invariably selects a dead or leafless branch whereon to perch, and from which it darts forth to capture the passing insects. Its flight somewhat resembles that of the *Artami*, and although it is capable of being sustained for some time, the bird more frequently performs short excursions, and returns to the branch it had left.

The eggs are deposited and the young reared in holes made in the sandy banks of rivers or any similar situation in the forest favourable for the purpose. The entrance is scarcely larger than a mouse-hole, and is continued for a yard in depth, at the end of which is an excavation of sufficient size for the reception of the four or five beautiful pinky-white eggs, which are ten lines long by eight or nine lines broad.

The stomach is tolerably muscular, and the food consists of various insects, principally Coleoptera and Neuroptera.

The sexes are alike in plumage, and may be thus described:—

Forehead, line over the eye, back, and wing-coverts brownish green; crown of the head and nape orange-brown; wings orange-brown, passing into green on the extremities of the primaries, and broadly tipped with black; two or three of the scapularies, lower part of the back, rump, and upper tail-coverts cærulean blue; tail black, most of the feathers, particularly the two centre ones, slightly margined with blue; lores, line beneath and behind the eye and ear-coverts velvety-black; beneath which is a stripe of cærulean blue; throat rich yellow, passing into orange on the sides of the neck; beneath this a broad band of deep black; under surface like the back, becoming green on the lower part of the abdomen; under tail-coverts light blue; irides light brownish red; bill black; legs and feet mealy greenish grey.

The young, until after their first autumn, are destitute of the black on the throat, and of the blue line beneath the eye, and their two central tail-feathers very short.

The range of this species appears to extend to some of the islands in the Eastern Archipelago, since specimens brought by Mr. Wallace from the Island of Lombock are identical with the birds found in Australia.

Family CORACIDÆ.

Genus EURYSTOMUS, *Vieillot*.

One species of this genus is found in Australia, and others inhabit India, the Indian islands, and Africa. They are closely allied to the Rollers, and not very distantly related to the Kingfishers.

Sp. 60.   EURYSTOMUS PACIFICUS.

AUSTRALIAN ROLLER.

Coracias pacifica, Lath. Ind. Orn. Supp., p. xxvii.
―― (*Galgulus*) *pacifica*, Vieill. Nouv. Dict. d'Hist. Nat., tom. xxix. p.
Eurystomus orientalis, Vig. and Horsf. in Linn. Trans., vol. xv. p. 202.
―― *Australis*, Swains. Anim. in Menag., p. 320.
―― *pacificus*, G. R. Gray, Ann. Nat. Hist. 1843, p. 190.
Galgulus pacificus, Vieill. Ency. Méth., part ii. p. 870.
Colaris pacificus, Bonap. Consp. Vol. Anis., p. 7.
Pacific Roller, Lath. Gen. Syn. Supp., vol. ii. p. 371 ?
Naty-kin, Aborigines of New South Wales.
Dollar Bird of the Colonists.

Eurystomus Australis, Gould, Birds of Australia, fol., vol. ii. pl. 17.

In Australia the Roller would appear to be a very local species, for I have never seen it from any other part of the country than New South Wales; but the late Mr. Elsey informed me that he found it very common in the Victoria basin, and that it became very numerous about the head of the Lynd. It arrives early in spring, and, after having brought forth its progeny, retires northwards on the approach of winter. It appeared to be most active about sunrise and sunset; in sultry weather it was generally perched upon some dead branch in a state of quietude. It is a very bold bird at all times, but particularly so during the breeding season, when it attacks with the utmost fury any intruder that may venture to approach the hole in the tree in which its eggs are deposited.

When intent upon the capture of insects it usually perches upon the dead upright branch of a tree growing beside and overhanging water, where it sits very erect, until a passing insect attracts its notice, when it suddenly darts off, secures its victim, and returns to the same branch; at other times it may constantly be seen on the wing, mostly in pairs, flying just above the tops of the trees, diving and rising again with many rapid turns. During flight the silvery-white spot in the centre of each wing shows very distinctly, and hence the name of Dollar Bird bestowed upon it by the colonists.

It is a very noisy bird, particularly in dull weather, when it often emits its peculiar chattering note during flight.

It is said to take the young Parrots from their holes and kill them, but this I never witnessed; the stomachs of the many I dissected contained the remains of Coleoptera only.

The breeding-season lasts from September to December; and the eggs, which are three and sometimes four in number, are deposited in the hole of a tree without any nest; they are of a beautiful pearly white, considerably pointed at the smaller end; their medium length is one inch and five lines, and breadth one inch and two lines.

The sexes are alike in plumage.

Head and neck dark brown, passing into the sea-green of the upper surface, and deepening into black on the lores; spurious wing, outer webs of the basal half of the quills, outer webs of the secondaries, and the basal half of the outer webs of the tail-feathers vivid blue; six of the primaries with a greenish-white basal band; extremities of the primaries black; tail green at the base, black at the tip; throat vivid blue, with a stripe of lighter blue down the centre of each feather; under surface of the shoulder and abdomen light green; under surface of the inner webs of the primaries and of all but the two centre tail-feathers deep blue, the former interrupted by the greenish-white band; irides dark brown; eyelash, bill, and feet red; inside of the mouth yellow.

Mr. Wallace found this species in the Aru Islands.

Family ALCEDINIDÆ.

But few of the various families, into which birds have been divided, are more clearly or more distinctly defined than those composing the *Alcedinidæ*. The bony structure of the whole is very much alike; and they are all clothed in a similar kind of plumage, differing only in colour." In some genera, as in *Dacelo*, it is of a sombre character; while in others, as in *Alcedo* and *Alcyone*, the plumage is very beautiful. Some, as the members of the first-mentioned genus, are of large size; while others are equally diminutive. The various members of the family are dispersed over all parts of the globe, but are most numerous in its tropical and temperate regions. Those inhabiting Australia pertain to four or five very distinct genera, namely, *Dacelo, Todirhamphus, Syma, Tanysiptera,* and *Alcyone.*

Genus DACELO, *Leach.*

The members of the genus *Dacelo* are among the largest species of the great family *Alcedinidæ*, and form a conspicuous feature in the ornithology of Australia, but, remarkably enough, are confined to the south-eastern and northern portions of the country, the south-western parts being uninhabited by any species of this group. I believe that water is not essential to their existence, and that they seldom or ever drink. They feed almost exclusively upon animal substances, small quadrupeds, birds, snakes, lizards, and insects being equally acceptable.

At least three species inhabit Australia.

Sp. 00.   DACELO GIGAS.

GREAT BROWN KINGFISHER.

Alcedo gigas, Bodd. Tabl. des Pl. Enl. d'Aubent., p. 40, pl. 663.
——— *gigantea*, Lath. Ind. Orn., vol. i. p. 245.
——— *fusca*, Gmel. edit. of Linn. Syst. Nat., vol. i. p. 454.
Grand Martin-pêcheur de la Nouvelle Guinée, Son. Voy., p. 171, pl. 106.
Martin Chasseur, Temm. Man. d'Orn., 2nd edit. p. lxxxviii.
Giant Kingfisher, Shaw, Gen. Zool., vol. viii. p. 53.
Great Brown Kingfisher, Lath. Gen. Syn., vol. ii. p. 609.
Dacelo gigantea, Leach, Zool. Misc., vol. ii. p. 126, pl. cvi.
Choucalcyon australe, Less. Traité d'Orn., p. 248.
Paralcyon gigas, Gloger.
Dacelo gigas, G. R. Gray, List of Gen. of Birds, 2nd edit. p. 14.
Gogo-bera, Aborigines of New South Wales.
Laughing Jackass of the Colonists.

Dacelo gigantea, Gould, Birds of Australia, fol., vol. ii. pl. 18.

The *Dacelo gigas* is a bird with which every resident and traveller in New South Wales is more or less familiar, for, independently of its large size, its voice is so extraordinary as to be unlike that of any other bird. In its disposition it is by no means shy, and when any new objects are presented to its notice, such as a party traversing the bush or pitching their tent in the vicinity of its retreat, it becomes very prying and inquisitive, often perching on the dead branch of some neighbouring tree, and watching with curiosity the kindling of the fire and the preparation of the meal; its presence, however, is seldom detected until it emits its extraordinary gurgling, laughing note, which generally calls forth some exclamation according with the temper of the hearer, such as "There is our old friend the Laughing Jackass," or an epithet of a less friendly character. So remarkable are the sounds emitted by the bird that they have been noted by nearly every writer on New South Wales and its productions. Mr. Calcy states that its "loud noise, somewhat like laughing, may be heard at a considerable distance, from which circumstance, and its uncouth appearance, it probably received the extraordinary appellation given to it by the settlers on their first arrival in the colony." Captain Sturt says, "Its cry, which resembles a chorus of wild spirits, is apt to startle the traveller who may be in jeopardy, as if laughing and mocking at his misfortune"; and Mr. Bennett, in his 'Wanderings,' says, "Its peculiar gurgling laugh, commencing in a low, and gradually rising to a high and loud tone, is often heard in all parts of the colony, the deafening noise being poured forth while the bird remains perched upon a neighbouring tree; it rises with the dawn, when the woods re-echo with its gurgling laugh; at sunset it is again heard; and as that glorious orb sinks in the west, a last 'good night' is given in its peculiar tones to all within hearing."

It frequents every variety of situation; the luxuriant brushes stretching along the coast, the more thinly-timbered forest, the belts of trees studding the parched plains, and the brushes of the higher ranges being alike favoured with its presence; over all these localities it is rather thinly dispersed, being nowhere very numerous.

Its food, which is of a mixed character, consists exclusively of animal substances; reptiles, insects, and crabs, however, appear to be its favourite diet: it devours lizards with avidity, and it is not an unfrequent sight to see it bearing off a snake in its bill to be eaten at leisure; it also preys on small mammalia. I recollect shooting a Great Brown Kingfisher in South Australia in order to secure a fine rat I saw hanging from its bill, and which proved to be a rare species. The *Dacelo gigas* breeds during the months of August and September, generally selects a hole in a large gum-tree for the purpose, and deposits its beautiful pearl-white eggs, which are one inch and nine lines long by one inch and five lines broad, on the decomposed wood at the bottom. When the young are hatched, it defends its breeding-place with great courage and daring, darting down upon any intruder who may attempt to ascend the tree.

The sexes present so little difference in the colouring of their plumage, that they are scarcely distinguishable from each other; neither do the young at a month old exhibit any great variation from the adult, the only difference being that the markings are somewhat darker and the brown more generally diffused.

It bears confinement remarkably well, and is one of the most amusing birds for the aviary with which I am acquainted; many examples have been brought alive to England; and several are now living in the Gardens of the Zoological Society of London.

Sp. 61. DACELO LEACHII, *Vigors and Horsfield.*

LEACH'S KINGFISHER.

Dacelo Leachii, Lath. MSS. Vig. and Horsf. in Linn. Trans., vol. xv. p. 205.

Dacelo Leachii, Gould, Birds of Australia, fol., vol. ii. pl. 19.

Since the completion of the folio edition, in which I was only enabled to mention the existence of a few specimens of this Kingfisher, so many examples have been sent to England that it has now become common, and may be found in every collection. Its habitat may be stated to be the Cape York Peninsula and the northern part of Queensland.

The habits, actions, food, and indeed the whole of the economy of the *Dacelo Leachii* are so like those of the *D. gigas* that a separate description of them is unnecessary.

The male has the head and back of the neck striated with brown and white; sides of the neck and under surface white, crossed with very narrow irregular markings of brown, these markings becoming much broader and conspicuous on the under surface of the shoulder; back brownish black; wing-coverts and rump shining azure-blue; wings deep blue; primaries white at the base, black on their inner webs, and blue on the outer; tail rich deep blue, all but the two centre feathers irregularly barred near the extremity and largely tipped with white; upper mandible brownish black, under mandible pale buff; irides dark brown; feet olive.

The female differs but little from the male in the colouring of the plumage, except that the tail-feathers, instead of being of a rich blue barred and tipped with white, are of a light chestnut-brown conspicuously barred with bluish black.

Sp. 62.   DACELO CERVINA, *Gould.*

Fawn-breasted Kingfisher.

Dacelo cervina, Gould, Birds of Australia, part ii. cancelled.
—— *cervicalis*, Kaup, Fam. Eisv., p. 8.
—— *Saturni*, Homb. et Jacq. Voy. au Pôle Sud, pl. 28. fig. 1.
Paralcyon cervina, Cab. et Hein. Mus. Hein., Theil ii. p. 164.
Lä-rool, Aborigines of Port Essington.

Dacelo cervina, Gould, Birds of Australia, fol, vol. ii. pl. 20.

The north-western portions of Australia constitute the true habitat of this species; it was observed in tolerable abundance by Sir George Grey during his expedition to that part of the country, and specimens of it have also formed a part of every collection of any extent made at Port Essington. In disposition it appears to be more shy and wary than the *Dacelo gigas* of New South Wales, of which it is a representative. Gilbert, who observed it on the Coburg Peninsula, states that it "inhabits well-wooded forests, generally in pairs, is extremely shy and very difficult to procure; is very fond of perching on the topmost dead branch of a tree, whence it can have an uninterrupted view of everything passing around, and where it pours out its loud discordant tones. Sometimes three or four pairs may be heard at one time, when the noise is so great that no other sound can be heard. The natives assert that it breeds in the honey-season, which is during the months of May, June, and July.

In his 'Journal of an Overland Expedition from Moreton Bay to Port Essington,' Dr. Leichardt states that "The Laughing Jackass (*Dacelo cervina*, Gould) observed near the Gulf of Carpentaria is of a different species from that of the eastern coast, is of a smaller size, and speaks a different language; but the noise is by no means so ridiculous as that of *Dacelo gigas*; he is heard before sunrise and immediately after sunset, like his representative of the eastern coast; the latter was observed as far as the upper Lynd, where the new one made his appearance."

The food of this Kingfisher is doubtless similar to that of the *Dacelo gigas*. The stomachs of those examined by Gilbert were tolerably muscular, and contained the remains of coleopterous and other kinds of insects.

As in the case with the preceding species, the male, when fully adult, differs from his mate in having the tail-feathers of a deep and splendid blue instead of brown.

The male has the feathers of the head buffy white, with a central stripe of dark brown, the latter colour becoming most conspicuous on the occiput; throat white; cheeks, ear-coverts, back of the neck, chest, and all the under surface sienna-yellow, crossed on the flanks with very minute irregular zig-zag bands of brown; primaries black at the tip, white at the base; the base of their external webs, the secondaries, and spurious wing rich china blue; greater and lesser wing-coverts, lower part of the back, and upper tail-coverts shining light blue; tail and the longest of the upper tail-coverts rich deep blue, the former broadly tipped with white; irides greenish white; upper mandible blackish brown, the cutting edges greenish white; lower mandible greenish white, the base dark brown on the sides, and blue on the under surface; tarsi and feet emerald green; claws black.

The female has the feathers of the head, cheeks, and ear-coverts buffy white, with a central stripe of dark brown; throat white; back of the neck, chest, and all the under surface sienna-yellow; the chest, flanks, and abdomen crossed by fine zigzag lines of brown; upper part of the back and scapularies umber-brown; primaries blackish brown at the tip and white at the base; the basal portion of their external webs, the secondaries, and the spurious wing rich china blue; greater and lesser wing-coverts and upper tail-coverts light shining blue; tail and the longest of the upper coverts rich chestnut-brown, which passes into buff at the tip, the whole transversely marked with eight or nine bands of rich blue-black.

Genus TODIRHAMPHUS, *Lesson*.

The members of this genus are more numerous and more widely dispersed than the *Dacelo*. The range of the various species extends from Asia, through the Indian Islands, to Australia; but I believe no one of them has yet been found in Africa. In making this statement, I wish it to be understood I do not intend to say that there are no Kingfishers in that country; on the contrary, they are very numerous there, but not of this particular form. Some of them bear a very general resemblance to it, and one of these is the type of the genus *Halcyon*, in which the Australian *Todirhamphi* have hitherto been placed. In their habits and mode of life the *Todirhamphi* resemble the *Dacelæ*, and must have the power, like those birds, of sustaining themselves for a long time without water, since they are frequently found in the driest parts of the country.

Sp. 03.   TODIRHAMPHUS SANCTUS.

SACRED KINGFISHER.

Sacred Kingfisher, Phill. Hot. Bay, pl. in p. 156.
Halcyon Sanctus, Vig. and Horsf. in Linn. Trans., vol. xv. p. 206.
—— *sacra*, Steph. Cont. Shaw's Gen. Zool. vol. xiii. p. 98.
—— *sancta*, G. R. Gray, List of Spec. of Birds in Coll. Brit. Mus., part ii. sec. i. p. 56.
Dacelo chlorocephala, var. β, Less. Traité d'Orn., p. 240.
Todirhamphus sanctus, Bonap. Consp. Gen. Av., tom. i. p. 156, *Todirhamphus*, sp. 3.
—— *Australasia*, Cass. Cat. Halc. in Coll. Acad. Nat. Sci. Mus. Philad., p. 13.
Sauropatis sancta, Cab. et Hein. Mus. Hein., Theil ii. p. 158.
Kingfisher of the Colonists.
Kŭn-yee-muk of the Aborigines, Western Australia.

Halcyon sanctus, Gould, Birds of Australia, fol., vol. ii. pl. 21.

The Sacred Kingfisher is very generally dispersed over the Australian continent. I have specimens from nearly every locality: those from Port Essington on the north are precisely identical with those of the south coast; on the other hand, those inhabiting Western Australia are a trifle larger in all their measurements, but otherwise present no differences of sufficient importance to warrant their being considered as distinct. It does not inhabit Tasmania.

It is a summer resident in New South Wales and throughout the southern portion of the continent, retiring northwards after the breeding-season. It begins to disappear in December, and by the end of January few are to be seen: solitary individuals may, however, be met with even in the depth of winter. They return again in spring, commencing in August, and by the middle of September are plentifully dispersed over all parts of the country, inhabiting alike the most thickly wooded brushes, the mangrove-forests which border, in many parts, the armlets of the sea, and the more open and thinly timbered plains of the interior, often in the most dry and arid situations far distant from water; and it would appear that, as is the case with many of the insectivorous birds of Australia, a supply of that element is not essential to its existence, since, from the localities it is often found breeding in, it must necessarily pass long periods without being able to obtain it.

The gaiety of its plumage renders it a conspicuous object in the bush: its loud piercing call, also, often betrays its presence, particularly during the season of incubation, when the bird becomes more and more clamorous as the tree in which its eggs are deposited is approached by the intruder. The note most frequently uttered is a loud *pee-pee*, continued at times to a great length, resembling a cry of distress. It sits very upright, generally perching on a small dead branch for hours together, merely flying down to capture its prey, and in most instances returning again to the site it has just left. Its food is of a very mixed character, and varies with the nature of the localities it inhabits. It greedily devours mantes, grasshoppers, caterpillars, lizards, and very small snakes, all of which are swallowed whole, the latter being killed by beating their heads against a stone or other hard substance, after the manner of the Common Kingfisher. Specimens killed in the neighbourhood of salt-marshes had their stomachs literally crammed with crabs and other crustaceous animals; while intent on the capture of which it may be observed sitting silently on the low mangrove-bushes skirting the pools which every receding tide leaves either dry or with a surface of wet mud, upon which crabs are to be found in abundance. I have never seen it plunge into the water after fish like the true Kingfishers, and I believe it never resorts to that mode of obtaining its prey. On the banks of the Hunter its most favourite food is the larvæ of a species of ant, which it procures by excavating holes in the nests of this insect which are constructed around the boles and dead branches of the *Eucalypti*, and which resemble excrescences of the tree itself.

The season of nidification commences in October and lasts till December, the hollow spouts of the gum- and boles of the apple-trees (*Angophoræ*) being generally selected as a receptacle for the eggs, which are four or five in number, of a pinky-white, one inch and a line in length, and ten lines in diameter.

The sexes differ but little either in their size or colouring, and the young are only distinguished by being of a less brilliant hue, and by the wing-coverts and feathers of the breast being edged with brown.

Crown of the head, back, and scapularies dull green; wings and tail green, slightly tinged with blue; ear-coverts, and an obscure circle bounding the green of the head, greenish black; rump verditer green; throat white; line from the nostrils over the eye, nuchal band, and all the under surface buff, becoming deeper on the flanks; bill black, the basal portion of the under mandible flesh-white; feet flesh-red, tinged with brown; irides dark brown.

Sp. 64. TODIRHAMPHUS PYRRHOPYGIUS, *Gould*.

RED-BACKED KINGFISHER.

Halcyon pyrrhopygia, Gould in Proc. of Zool. Soc., part viii. 1840, p. 113.
Todirhamphus pyrrhopygius, Bonap. Consp. Gen. Av., tom. i. p. 157,
  Todirhamphus, sp. 1.
Cyanalcyon pyrrhopygia, Bonap. Consp. Vol. Anis., p. 9, gen. 119.
Sauropatis pyrrhopygia, Cab. et Hein. Mus. Hein., Theil ii. p. 161.

Halcyon pyrrhopygia, Gould, Birds of Australia, fol., vol. II. pl. 22.

This Kingfisher is an inhabitant of the interior, but over what extent of country it may range is not yet known. The only parts where I myself observed it were the myall-brushes (*Acacia pendula*) of the Lower Namoi, particularly those growing on the edge of the large plain skirting the Nundawar range. It was usually seen sitting very upright on the dead branches of the myall- and gum-trees, sometimes on those growing out on the hot plains, at others on those close by the river-side. I succeeded in obtaining both old and young birds, which, judging from the plumage of the latter, I should suppose had left their breeding-place about a month before I arrived in the neighbourhood of the Namoi, in December. I also saw in this district the common or Sacred Kingfisher, but in far less abundance than between the ranges and the coast. This latter species may be hereafter found to be an inhabitant of the country bordering the sea, while the Red-backed Kingfisher may be exclusively a denizen of the interior. The unusual colouring of the back at once distinguishes it from all the other members of the genus inhabiting Australia, but in its general economy and mode of living it presents no observable difference.

Gilbert procured examples of this species during Dr. Leichardt's overland expedition; Captain Sturt found it at the depôt in South Australia, and I have received specimens from the interior of Swan River; consequently it has a very wide range.

Crown of the head dull green, intermingled with white, giving it a striated appearance; a broad black stripe commences at the base of the bill, passes through the eye, and encircles the back of the head; upper part of the back and scapularies green; remainder of the wings bluish green; lower part of the back, rump, and upper tail-coverts red; tail green, tinged with blue; throat, a broad collar encircling the back of the neck, and all the under surface white; bill black, the base of the lower mandible flesh-white; irides blackish brown; feet dark olive-brown.

Since the above account of this species was published in the folio edition, I have been informed by the late Mr. Elsey that he saw the Red-backed Kingfisher on the Macarthur River, about one hundred miles from the coast, in lat. 26° 15′ S. Two eggs in my collection are very round in form, and of the usual white colour; but they were doubtless pinky white before they were blown. They are one inch long, by seven-eighths broad.

Sp. 65. TODIRHAMPHUS SORDIDUS, *Gould*.

SORDID KINGFISHER.

Halcyon sordidus, Gould in Proc. of Zool. Soc., part x. p. 72.
Todirhamphus sordidus, Bonap. Consp. Gen. Av., tom. i. p. 157, *Todirhamphus*, sp. 8.
Sauropatis sordida, Cab. et Hein. Mus. Hein., Theil ii. p. 159 (note).

Halcyon sordidus, Gould, Birds of Australia, fol., vol. II. pl. 23.

This fine Kingfisher, the largest species of the genus *Todirhamphus* inhabiting Australia, is rather plentifully dispersed over the north-eastern portion of Queensland, or from Moreton Bay to Cape York. Its discovery is due to the late Mr. Bynoe, R.N., who obtained two examples on the north coast, but the exact locality is unknown to me.

Head, back, scapularies, and wing-coverts brownish oil-green; wings greenish blue, gradually changing into green on the tips of the tertiaries; collar round the back of the neck and all the under surface buffy white; tail greenish blue; upper mandible and tip of the lower one black; base of the latter flesh-white.

Genus CYANALCYON, *Verreaux*.

The species of this form at present known are *C. Macleayi, C. diops, C. lazuli*, and perhaps *C. lazulinus*. They are all highly coloured, and differ but little from the *Todirhamphi*, with which they have been hitherto associated Australia, New Guinea, and the adjacent islands are the countries frequented by the members of this section of the *Alcedinidæ*.

Sp. 66.   CYANALCYON MACLEAYI.

MacLeay's Kingfisher.

Halcyon MacLeayi, Jard. and Selb. Ill. Orn., vol. iii. pl. 101.
—— *incinctus*, Gould in Proc. of Zool. Soc., part v. p. 142, female.
Todirhamphus Macleayi, Bonap. Consp. Gen. Av., tom. i. p. 157, *Todiramphus*, sp. 13.
Cyanalcyon Macleayi, Cab. et Hein. Mus. Hein., Theil ii. p. 158.
Bush Kingfisher, Residents at Port Essington.

Halcyon Macleayi, Gould, Birds of Australia, fol., vol. ii. pl. 24.

There has not yet been discovered in Australia a more beautiful Kingfisher than the one dedicated to the late Alexander MacLeay by the authors of the 'Illustrations of Ornithology' as a tribute of respect for his scientific labours, in the propriety of which I entirely concur.

The extreme brilliancy of the plumage of this bird would indicate that it is a native of the hotter portions of the country, and the correctness of this inference is borne out by the fact that it inhabits all the eastern provinces from Moreton Bay to the extreme northern portions of the continent; it is tolerably abundant at Port Essington, and it is also spread over nearly every part of the Cobourg Peninsula suited to its habits; like the other members of the genus to which it belongs, it is rarely, if ever, seen near water, and evinces so decided a preference for the forests of the interior that it has obtained the name of "Bush Kingfisher" from the residents at Port Essington; it is generally dispersed about in pairs, and feeds on small reptiles, insects, and their larvæ; its general note is a loud *pee-pee* uttered with considerable rapidity. It incubates in November and December, sometimes forming its nest in the hollow trunks of trees, and at others excavating a hole for itself in the nest of the tree-ants, which presents so prominent and singular a feature in the scenery of the country: the nest of the *C. Macleayi* is easily discovered, for on the approach of an intruder the birds immediately commence flying about in a very wild manner, uttering at the same time a loud piercing cry of alarm; the eggs are three or four in number, of a pearly white, and nearly round in form, being eleven lines long by ten broad.

So much difference exists in the plumage of the sexes that Gilbert states he was for some time induced to regard them as specifically distinct—an error into which I had myself previously fallen when describing the female as a new species in the 'Proceedings of the Zoological Society' quoted above; "but upon closer observation," adds Gilbert, "I soon satisfied myself that the difference of plumage was merely sexual, the dissection of a large number of specimens fully proving that those with a ring round the neck are males and those without it females."

The male has a line under the eye and ear-coverts deep glossy black; head, occiput, wings, and tail rich deep prussian blue; primaries and secondaries white at the base, forming a conspicuous spot when the wings are spread; for the remainder of their length these feathers are black, margined externally with light prussian blue; immediately before the eye an oval spot of white; collar surrounding the back of the neck and all the under surface white, tinged with buff on the lower part of the flanks; back and upper tail-coverts verditer blue; scapularies verditer green, both these colours bounded near the white collar with prussian blue; under surface of the wing white, the tips of the coverts black; under surface of the tail black; bill black, the basal portion of the under mandible yellowish white; tarsi black; inner side of the foot and back of the tarsi ash-grey; irides very dark brown.

The general colours of the female are similar to those of the male, but she differs from her mate in being entirely destitute of the white collar at the back of the neck, which part is deep prussian blue, thus uniting the blue of the occiput and of the back; in the tints being much less brilliant in the back, being of a dull brownish verditer green, and in the upper tail-coverts pale verditer green instead of blue; upper mandible black; lower mandible halfway from the tip and along the whole of the cutting edges black, the remainder being fleshy white tinged with blue where it joins the black; legs and feet greenish grey.

The young male resembles the female in colour, but is still less brilliant; has the back of a purer green; the under surface tinged with buff; the spot on the lores deep buff; and the collar at the back of a deep buff, interrupted by some of the feathers of the occiput.

Genus SYMA, *Lesson*.

The *S. Torotoro* of New Guinea and the *S. flavirostris* of Northern Australia are the only species of this form that have yet been discovered. The serrated edges of the mandibles indicate that they feed on some peculiar kind of food, and it would be interesting to know what special service the serrations are intended to perform.

Sp. 67. SYMA FLAVIROSTRIS, *Gould*.

YELLOW-BILLED KINGFISHER.

Halcyon (Syma?) flavirostris, Gould in Proc. of Zool. Soc., part xviii. 1850, p. 200.

Halcyon flavirostris, Gould, Birds of Australia, fol., Supplement, pl. .

This species might easily be mistaken for the *Syma Torotoro*; but there can be little doubt of its being distinct and new to science: its lesser size, less brilliant colouring, the yellow instead of orange hue of the bill, and the smaller size of the serrations of the mandibles, are some of the characters by which it may be distinguished from the New Guinea species.

It was in that rich district the peninsula of Cape York, which appears to have a fauna peculiar to itself, that the present bird was first procured; the following notes by Mr. Macgillivray comprise all the information I have been able to obtain respecting it:—

"The *Poditti*, as it is called by the aborigines, appears to be a rare bird; for although it was much sought for, not more than four or five examples were obtained during our stay. Like the *Tanysiptera Silvia*, it is an inhabitant of the brushes, while the *S. Torotoro* of New Guinea is a mangrove bird. I myself saw it alive only once, in a belt of tall trees, thick underwood, and clumps of the *Seaforthia* palm fringing a small stream about three miles from the sea. Attracted by the call of the bird, which was recognized by the accompanying natives as that of the much-prized *Poditti*, three or four of us remained for about ten minutes almost under the very tree in which it was perched, intently looking out for the chance of a shot, before I discovered it on a bare transverse branch, so high up as scarcely to be within range of small shot; however, it fell, but our work was only half over, as the wounded bird eluded our search for a long time; at length, one of our sable allies—his eyes brightened, I dare say, by visions of a promised axe—found it lying dead in a corner to which it had retreated. The more intelligent natives whom I questioned separately agreed in stating that its mode of nidification is similar to that of the *Tanysiptera Sylvia*, and that, like that species, it lays several white eggs."

The male has the crown of the head, back of the neck, ear-coverts, and flanks cinnamon-red; at the back of the neck a narrow broken collar of black; throat and lower part of the abdomen tawny white; back and wings sordid green; rump and tail greenish blue; bill pale orange, the apical two-thirds of the ridge of the upper mandible dark brown.

The female differs in being less brightly coloured, and in having an oblong patch of black on the centre of the head extending a little way down the occiput.

Genus TANYSIPTERA, *Vigors*.

The species of this genus are some of the most beautiful of the *Alcedinidæ*; for, independently of the pleasing contrasts of their colourings of red, blue, and white, their lengthened tail-plumes not only impart a peculiar elegance to their appearance, but render them very conspicuous objects.

The well-known *Tanysiptera Dea*, the type of this form, is a native of New Guinea and the neighbouring islands, where several others are also found; and we now know that one species is a native of Australia.

Sp. 68.   TANYSIPTERA SYLVIA, *Gould*.

White-tailed Kingfisher.

Tanysiptera Sylvia, Gould in Proc. of Zool. Soc., part xviii. 1850, p. 200.
Quaierur, of the Aborigines at Cape York.

Tanysiptera Sylvia, Gould, Birds of Australia, fol., Supplement, pl. .

Hitherto this species has only been found on the northern coast, Cape York being the sole locality it is at present known to inhabit; and where, judging from the numerous specimens lately sent to this country, it appears to be by no means scarce.

As is the case with the *Alcedinidæ* generally, the sexes appear to present but little difference in size and colouring, but the female may be distinguished from the male by being somewhat less brilliant in colour and in the lesser development of the central tail-feathers.

Mr. Macgillivray informed me that "this pretty *Tanysiptera* is rather plentiful in the neighbourhood of Cape York, where it frequents the dense brushes, and is especially fond of resorting to the small sunny openings in the woods, attracted probably by the greater abundance of insect food found in such places than elsewhere: I never saw it on the ground, and usually was first made aware of its presence by the glancing of its bright colours as it darted past with a rapid, arrow-like flight, and disappeared in an instant among the dense foliage. Its cry, which may be represented by 'whee-whee-whee' and 'wheet-wheet-wheet,' is usually uttered while the bird is perched on a bare transverse branch or woody rope-like climber, which it uses as a look-out station, and whence it makes short dashes at any passing insect or small lizard, generally returning to the same spot. It is a shy suspicious bird, and one well calculated to try the patience of the shooter, who may follow it in a small brush for an hour without getting a shot, unless he has as keen an eye as the native to whom I was indebted for first pointing it out to me. According to the natives, who know it by the name of 'Qualdicur,' it lays three white eggs in a hole dug by itself in one of the large ant-hills of red clay which form so remarkable a feature in the neighbourhood, some of them being as much as ten feet in height, with numerous buttresses and pinnacles. I believe that the bird also inhabits New Guinea; for at Redscar Bay, on the south-east side of that great island, in long. 140° 50′ E., a head strung upon a necklace was procured from the natives."

Crown of the head, wings, and five lateral tail-feathers on each side blue; ear-coverts, back of the neck, and mantle black; in the centre of the latter a triangular mark of white; rump and two middle tail-feathers pure white; under surface cinnamon-red; bill and feet sealing-wax red.

Genus ALCYONE, *Swainson*.

Of this genus several species are known, all of which are inhabitants of the Old World, principally Oceania, and Australia. These singular-footed birds very closely resemble the typical Kingfishers in their habits and manners; for, like them, they frequent rivers, brooks, and ponds, and plunge beneath the surface for the fish, crustaceans, and insects upon which they principally subsist.

So much difference exists in the species of this form inhabiting Australia, that I have been obliged to characterize two of them as distinct from *A. azurea*.

Sp. 69.   ALCYONE AZUREA.

AZURE KINGFISHER.

Alcedo azurea, Lath. Ind. Orn. Supp., p. xxxii.
—— *tribrachys*, Shaw, Nat. Misc., pl. 681.
Tri-digitated Kingfisher, Shaw, Gen. Zool., vol. viii. p. 105.
Azure Kingfisher, Lath. Gen. Syn. Supp., vol. ii. Add., p. 372.
Ceyx azurea, Jard. and Selb. Ill. Orn., vol. ii. pl. 55. fig. 1.
Alcyone Australis, Swains. Class. of Birds, vol. ii. p. 336.
Ceyx cyanea, Less. Traité d'Orn., p. 241.
Alcyone azurea, G. R. Gray, List of Gen. of Birds, 2nd edit. p. 14.

Alcyone azurea, Gould, Birds of Australia, fol., vol. ii. pl. 25.

With the exception of Swan River, every colony of Australia, from Port Essington on the north-west to Tasmania in the extreme south, is inhabited by Azure Kingfishers; but as they, although closely allied, constitute at least three species, the present page must necessarily treat exclusively of the one that inhabits New South Wales and South Australia, over the whole of which countries it is dispersed, wherever brooks, ponds and other waters occur suitable to its habits and mode of life. In size and in brilliancy of its plumage, the Azure Kingfisher is intermediate between the species inhabiting the north coast and that found in Tasmania; although generically distinct from the Kingfisher of Europe (*Alcedo Ispida*), it has many characters in common with that bird. It subsists almost exclusively on small fish and aquatic insects, which it captures in the water by darting down from some bare branch overhanging the stream, and to which it generally returns to kill and devour its prey, which is swallowed entire and head foremost, after the manner of the little favourite of our own island. It is a solitary bird, a pair, and frequently only one, being found at the same spot. During the breeding-season it becomes querulous and active, and even pugnacious if any intruder of the same species should venture within the precincts of its abode. The males at this season chase each other up and down the stream with arrow-like quickness, when, the rich azure-blue of the back glittering in the sun, they appear more like meteors, as they dart by the spectator, than birds. The task of incubation commences in August and terminates in January, during which period two broods are frequently brought forth. The eggs, which are of a beautiful pearly or pinkish white and rather round in form, are deposited at the extremity of a hole, in a perpendicular or shelving bank bordering the stream, without any nest being made for their reception; they are from five to seven in number, three quarters of an inch broad by seven-eighths of an inch long. The young at the first moult assume the plumage of the adult, which is never afterwards changed. The hole occupied by the bird is frequently almost filled up with the bones of small fish, which are discharged from the throat and piled up round the young in the form of a nest. Immediately on leaving their holes the young follow the parents from one part of the brook to another, and are fed by them while resting on some stone or branch near the water's edge; they soon, however, become able to obtain their own food, and may be observed at a very early age plunging into the water to a considerable depth to capture small fish and insects.

The sexes are precisely similar in the colouring of their plumage, neither do they differ in size. The young are very clamorous, frequently uttering their twittering cry as their parents pass and repass the branch on which they are sitting.

All the upper surface and a patch on each side of the chest fine ultramarine blue, becoming more vivid on the rump and upper tail-coverts; on each side of the neck behind the ear-coverts a tuft of yellowish-white feathers; wings black; throat white, slightly washed with buff; all the under surface, including the under side of the wing, ferruginous orange, the flanks tinged with bluish lilac, giving them a rich purple hue; line from the bill to the eye reddish orange; irides and bill black; feet orange.

Sp. 70.   ALCYONE DIEMENENSIS, *Gould.*

Alcyone Diemenensis, Gould in Proc. of Zool. Soc., part xiv. p. 19.

Alcyone Diemenensis, Gould, Birds of Australia, fol., vol. i., Introd. p. xxxi.

This, the most southern member of the genus, differs from the *A. azurea* both in colour and size. It is a native of Tasmania.

All the upper surface deep blue, becoming more vivid on the rump and upper tail-coverts; wings black, washed with blue; throat buff; under surface of the body and wings ferruginous orange; on each side of the chest a patch of bluish black; lores and a small patch behind the ears buff; crown of the head indistinctly barred with black; irides and bill black; feet orange.

Sp. 71.   ALCYONE PULCHRA, *Gould.*

Alcyone pulchra, Gould in Proc. of Zool. Soc., part xiv. p. 19.

Alcyone pulchra, Gould, Birds of Australia, fol., vol. i., Introd. p. xxxii.

As its name implies, this is a very beautiful species, and exceeds in richness of colouring both the *A. azurea* and *A. Diemenensis*. The portion of the country it inhabits is the neighbourhood of Port Essington and perhaps the north coast of Australia generally.

All the upper surface shining purplish blue; wings

brownish black; lores, tuft behind the ear, and throat buff; under surface deep ferruginous orange; sides of the chest fine purplish blue, passing into a rich vinous tint on the flanks; irides and bill black; feet orange.

Inhabits the north coast of Australia.

Sp. 72.   ALCYONE PUSILLA.

LITTLE KINGFISHER.

Ceyx pusilla, Temm. Pl. Col., 505. fig. 3.
Nu-rea-bin-mo, Aborigines of the Cobourg Peninsula.

Alcyone pusilla, Gould, Birds of Australia, fol., vol. ii., pl. 26.

This lovely little Kingfisher is a native of the northern portions of Australia; the specimens in my collection were all procured at Port Essington, where it is a rare bird; and from its always inhabiting the densest mangroves, is not only seldom seen, but is extremely difficult to procure; in general habits and manners it very much resembles the *Alcyone azurea*, but its note is somewhat more shrill and piping, and its flight more unsteady. Specimens of this species from New Guinea, which I have had opportunities of examining in the noble collection at Leyden, present no difference whatever from those found in Australia.

The food of the *Alcyone pusilla* consists of small fish, which are taken precisely after the manner of the Common Kingfisher of our own island.

The sexes are alike in size and colour.

Lores, a tuft behind the ear-coverts and under surface silky white; forehead, sides of the neck, wing-coverts and the margins of the secondaries green; primaries brownish black; all the upper surface and a large patch on each side on the chest brilliant intense blue; tail dull deep blue; irides dark blackish brown; bill black; legs and feet greenish grey.

Family ARTAMIDÆ.

The proper position of the members of this isolated form has been a stumbling-block to every ornithologist. Those who have had opportunities of observing them in a state of nature cannot have failed to notice how closely they resemble the Swallows in their actions and general mode of life; to a certain extent, they offer an alliance to some of the *Laniadæ*, as *Gymnorhina*, and Mr. Jerdon has applied to the Indian members the trivial name of Swallow-Shrikes; I shall, however, retain that of Wood Swallow, which is equally descriptive, and had been applied long before.

Australia is undoubtedly the head-quarters of these pretty birds; but other species are found throughout the Indian Islands to the continent of India. They are all insectivorous, and must perform a most important part in checking an undue increase of those creatures.

Sp. 73.   ARTAMUS SORDIDUS.
WOOD SWALLOW.

Turdus sordidus, Lath. Ind. Orn. Supp., p. xliii.
Sordid Thrush, Lath. Gen. Syn. Supp., vol. ii. p. 186.
Ocypterus albovittatus, Cuv. Règn. Anim., tom. iv. t. 3. f. 6.
Artamus lineatus, Vieill. 2nde Edit. du Nouv. Dict. d'Hist. Nat., tom. xvii. p. 297.
—— *albovittatus*, Vig. and Horsf. in Linn. Trans., vol. xv. p. 210.
Leptopteryx albovittata, Wagl. Syst. Av., sp. 5.
Ba-wo-wen, Aborigines of Western Australia.
Werle, Aborigines of King George's Sound.
Wood Swallow of the Colonists.

Artamus sordidus, Gould, Birds of Australia, fol., vol. ii. pl. 27.

No species of the Australian *Artami* with which I am acquainted possesses so wide a range as the present; the whole of the southern portion of the continent, as well as the island of Tasmania, being alike favoured with its presence.

The extent of its range northward has not yet been satisfactorily ascertained, beyond the certainty that it has not hitherto been received in any collection from the north coast.

It may be regarded as strictly migratory in Tasmania, where it arrives in October, and after rearing at least two broods departs again in a northern direction. On the continent of Australia it arrives rather earlier, and departs later; but a scattered few remain throughout the year in all the localities favourable to their habits, the number being regulated by the supply of insect food necessary for their subsistence. I may here observe, that specimens from Swan River, South Australia, and New South Wales present no difference either in size or colouring, while those from Tasmania are invariably larger in all their admeasurements, and are also of a deeper colour.

This Wood Swallow must, I think, ever be a general favourite with the Australians, not only from its singular and pleasing actions, but from its often taking up its abode and incubating near the houses, particularly such as are surrounded by paddocks and open pasture-lands skirted by large trees. It was in such situations in Tasmania that, at the commencement of spring, I first had an opportunity of observing this species; it was then very numerous on all the cleared estates on the north side of the Derwent, about eight or ten being seen on a single tree, and half as many crowding one against another on the same dead branch, but never in such numbers as to deserve the appellation of flocks: each bird appeared to act independently of the other; each, as the desire for food prompted it, sallying forth from the branch to capture a passing insect, or to soar round the tree and return again to the same spot; on alighting it repeatedly throws up one of its wings, and obliquely spreads its tail. At other times a few were seen perched on the fence surrounding the paddocks, on which they frequently descended, like Starlings, in search of coleoptera and other insects. The form of the wing of the *Artamus sordidus* at once indicates that the air is its peculiar province:

hence it is, that when engaged in pursuit of the insects which the serenity and warmth of the weather have enticed from their lurking-places among the foliage, to sport in higher regions, this species displays itself to the greatest advantage. But the greatest peculiarity in the habits of this bird is its manner of hanging together in clusters from the branch of a tree, like a swarm of bees.

The season of incubation is from September to December. The situation of the nest is much varied; I have seen one placed in a thickly-foliaged bough near the ground, while others were in a naked fork, on the side of the bole of a tree, in a niche formed by a portion of the bark having been separated from the trunk, &c. The nest is rather shallow, of a rounded form, about five inches in diameter, and composed of fine twigs neatly lined with fibrous roots. I observed that the nests found in Tasmania were larger, more compact, and more neatly formed than those on the continent of Australia.

The eggs are generally four in number; they differ much in the disposition of their markings; their ground-colour is dull white, spotted and dashed with dark umber-brown: in some a second series of greyish spots appear as if beneath the surface of the shell; their medium length is eleven lines, and breadth eight lines.

Head, neck, and the whole of the body fuliginous grey; wings dark bluish black, the external edges of the second, third, and fourth primaries white; tail bluish black, all the feathers, except the two middle ones, largely tipped with white; irides dark brown; bill blue, with a black tip; feet mealy lead-colour.

The sexes are alike in the colouring of their plumage, and are only to be distinguished by the female being somewhat smaller in size.

The young have an irregular stripe of dirty white down the centre of each feather of the upper surface, and are mottled with the same on the under part of the body.

L

Sp. 74.   ARTAMUS MINOR, *Vieillot.*

LITTLE WOOD SWALLOW.

Artamus minor, Vieill. Nouv. Dict. d'Hist. Nat., tom. xvii. p. 298.
Ocypterus fuscatus, Valenc. Mém. du Mus. d'Hist. Nat., tom. vi. p. 24,
   t. 9. fig. 1.
Leptopteryx minor, Wagl. Syst. Av., sp. 6.
Ocypterus minor, Gould, Syn. Birds of Australia.

Artamus minor, Gould, Birds of Australia, fol., vol. ii. pl. 28.

In its structure and in the disposition of the markings of its plumage, this species offers a greater resemblance to the *Artamus sordidus* than to any other member of the group; the habits of the two species are also very similar; if any difference exists, it is that the present bird is still more aërial, a circumstance indicated by the more feeble form of the foot, and the equal, if not greater, development of the wing. During fine weather, and even in the hottest part of the day, it floats about in the air in the most easy and graceful manner, performing in the course of its evolutions many beautiful curves and circles, without the least apparent motion of the wings, the silvery whiteness of which, as seen from beneath, and the snowy tips of its wide-spread tail strongly contrast with the dark colouring of the other parts of its plumage.

I found the *Artamus minor* abundant on the Lower Namoi, particularly on the plains thinly studded with the *Acacia pendula* and other low trees in the neighbourhood of Gummel-Gummel, where it had evidently been breeding, as I observed numerous young ones, whose primaries were not sufficiently developed to admit of their performing a migration of any distance; besides which, they were constantly being fed by the parents, who were hawking about in the air over and around the trees, while the young were quietly perched close to each other on a dead twig.

I have received two specimens from Port Essington, and

there are examples in the Paris Museum from, I believe, Timor; it is evident, therefore, that this bird has a wide range.

The sexes are alike in plumage, but the young differ considerably.

The whole of the head, back, and abdomen chocolate-brown; wings, rump, and under tail-coverts bluish black; tail deep bluish black, all the feathers except the two outer and two middle ones tipped with white; bill beautiful violet-blue at the base, darker at the tip; irides and feet nearly black.

Sp. 75.   ARTAMUS CINEREUS, *Vieillot.*

GREY-BREASTED WOOD SWALLOW.

Artamus cinereus, Vieill. Nouv. Dict. d'Hist. Nat., tom. xvii. p. 297.
Ocypterus cinereus, Valenc. Mém. du Mus. d'Hist. Nat., tom. vi. p. 22, t. 9. fig. 1.
Be-wol-wen, Aborigines of the lowland and mountain districts of Western Australia.
Wood Swallow of the Colonists of ditto.

Artamus cinereus, Gould, Birds of Australia, fol., vol. ii. pl. 29.

This bird exceeds in size all the other Australian Wood Swallows. Its large tail, most of the feathers of which are broadly tipped with white, as well as the colouring of its plumage, at once point out its close affinity to the *Artamus sordidus* and *A. minor.*

In Western Australia it is a very local but by no means an uncommon species, particularly at Swan River, where it inhabits the limestone hills near the coast, and the "Clear Hills" of the interior, assembling in small families, and feeding upon the seeds of the *Xanthorrhœa*, which proves that insects do not form the sole diet of this species; with such avidity in fact does it devour the ripe seeds of this grass-tree, that several birds may frequently be seen crowded together on the perpendicular seed-stalks of this plant busily engaged in extracting them; at other times, particularly among the limestone hills, where there are but few trees, it descends to the broken rocky ground in search of insects and their larvæ.

It breeds in October and November, making a round compact nest, in some instances of fibrous roots, lined with fine hair-like grasses, in others of the stems of grasses and small plants; it is built either in a scrubby bush or among the grass-like leaves of the *Xanthorrhœa*, and is deeper and more cup-shaped than those of the other members of the group. The eggs are subject to considerable variation in colour and in the character of their markings; they are usually bluish white, spotted and blotched with lively reddish brown, intermingled with obscure spots and dashes of purplish grey, all the markings being most numerous towards the larger end; they are about eleven lines long by eight lines broad.

The sexes are alike in colour, and can only be distinguished from each other with certainty by dissection. I have remarked that specimens from Timor rather exceed in size those collected on the Australian continent, and are somewhat lighter in colour; but these variations are too slight to be regarded as specific.

Crown of the head, neck, throat, and chest grey, passing into sooty grey on the abdomen; space between the bill and the eye, the fore part of the cheek, the chin, the upper and under tail-coverts jet-black; two middle tail-feathers black; the remainder black, largely tipped with white, with the exception of the outer feather on each side, in which the black colouring extends on the outer web nearly to the tip; wings deep grey; primaries bluish grey; under surface of the shoulder white, passing into grey on the under side of the primaries; irides dark blackish brown; bill light greyish blue at the base, black at the tip; legs and feet greenish grey.

Sp. 76.   ARTAMUS ALBIVENTRIS, *Gould.*

WHITE-VENTED WOOD SWALLOW.

Artamus albiventris, Gould in Proc. of Zool. Soc., part xv. 1847, p. 31.

Artamus albiventris, Gould, Birds of Australia, fol., vol. ii. pl. 30.

Two examples of this species are all that have come under my notice; one of these was killed on the Darling Downs in New South Wales, and the other some distance to the northward of that locality, it being one of the birds procured during Dr. Leichardt's expedition to Port Essington. Its nearest ally is the *Artamus cinereus*, a species inhabiting the opposite side of the continent; but it is somewhat smaller, and may moreover be distinguished from that bird by the white under tail-coverts, and the lighter colour of the lower part of the abdomen.

Lores, space beneath the eye, and the chin deep black; head, neck, and upper part of the back brownish grey; lower part of the back and the wings dark grey, becoming gradually deeper towards the tips of the feathers; primaries and secondaries narrowly edged with white at the tip; under surface of the wing white; ear-coverts, chest, and abdomen pale grey, passing into white on the under tail-coverts; upper tail-coverts and tail black; the apical third of all but the two middle ones white; irides dark brown; bill yellowish horn-colour, becoming black at the tip; feet blackish brown.

Sp. 77.   ARTAMUS MELANOPS, *Gould.*

BLACK-FACED WOOD SWALLOW.

Artamus melanops, Gould in Proc. of Zool. Soc. 1865, p. 198.

This fine species is unlike every other known member of the genus. It is most nearly allied to *A. albiventris*, but differs from that bird in the jet-black colouring of its under tail-coverts, and from *A. cinereus* in its smaller size and the greater extent of the black on the face. The specimen from which the above description was taken has been kindly sent to me by Mr. S. White, of the Reed Beds, near Adelaide, South Australia, who informs me that it was shot by him at St. à Becket's Pool, lat. 28° 30', on the 23rd of August 1863, and who, in the notes accompanying it, says, " I have never seen this bird south. It collects at night like *A. sordidus*, and utters the same kind of call. It seems to be plentiful all over the north country, and particularly about Chamber's Creek and Mount Margaret. It feeds on the ground, soars high in the air, and clings in bunches like the others. The two sexes appeared to be very similar in outward appearance; but the young are much speckled with dusky brown, particularly on the back."

Lores, face, rump, and under tail-coverts black; stripe over the eye, ear-coverts, sides of the face, and throat greyish buff, increasing in depth on the chest so as to form a well-marked band; under surface delicate vinous grey; two middle tail-feathers black, the remainder black, largely tipped with white; upper surface of the wings grey, their under surface white; bill leaden grey, darkest at the tip; feet blackish brown.

Total length $6\frac{3}{4}$ inches; bill $\frac{3}{4}$; wing $4\frac{3}{4}$; tail 3; tarsi $\frac{3}{4}$.

Sp. 78.   ARTAMUS PERSONATUS, *Gould*.

MASKED WOOD SWALLOW.

Ocypterus personatus, Gould in Proc. of Zool. Soc., part viii. p. 149.
Jil-bung, Aborigines of the mountain districts of Western Australia.

Artamus personatus, Gould, Birds of Australia, fol., vol. II. pl. 31.

My knowledge of the range of this species is very limited; a single specimen was sent me from South Australia, while fine examples were killed by Gilbert in the colony of Swan River. Its richly coloured black face and throat, separated from the delicate grey of the breast by a narrow line of snowy white, at once distinguishes it from every other species, while the strong contrast of these colours renders it a conspicuous object among the trees.

In size and structure it more nearly resembles the *Artamus superciliosus* than any other, and the two species form beautiful analogues of each other, one being in all probability confined to the eastern portion of the country, and the other to the western.

"I have only met," says Gilbert, "with this species in the York and Toodyay districts. It is very like *Artamus sordidus* in its habits, but is more shy and retired, never being seen but in the most secluded parts of the bush. It is merely a summer visitant here, generally making its appearance in the latter part of October, and immediately commencing the task of incubation. Its voice very much resembles the chirping of the English Sparrow.

"Its nest is placed in the upright fork of a dead tree, or in the hollow part of the stump of a grass-tree; it is neither so well nor so neatly formed as those of the other species of the group, being a frail structure externally composed of a very few extremely small twigs, above which is a layer of fine dried grasses. The eggs also differ as remarkably as the nest, their ground-colour being light greenish grey, dashed and speckled with hair-brown principally at the larger end, and slightly spotted with grey, appearing as if beneath the surface of the shell; they are ten and a half lines long by eight and a half lines broad. I found two nests in a York Gum Forest, about five miles to the east of the Avon River; each of those contained two eggs, which I believe is the usual number. Mr. Angas informs me that in South Australia this bird makes no nest, but places the eggs on a few bent stalks of grass in the bend of a small branch.

"Its food consists of insects generally and their larvæ."

The male has the face, ear-coverts, and throat jet-black, bounded below with a narrow line of white; crown of the head sooty black, gradually passing into the deep grey which covers the whole of the upper surface, wings, and tail; the latter tipped with white; all the under surface very delicate grey; thighs dark grey; irides blackish brown; bill blue at the base, becoming black at the tip; legs and feet mealy bluish grey.

The female differs in having the colouring of the bill and the black mask on the face much paler.

Sp 70.   ARTAMUS SUPERCILIOSUS, *Gould*.

WHITE-EYEBROWED WOOD SWALLOW.

Ocypterus superciliosus, Gould in Proc. of Zool. Soc., part iv. 1836, p. 142.

Artamus superciliosus, Gould, Birds of Australia, fol., vol. ii. pl. 32.

There is no species of *Artamus* yet discovered to which the present yields the palm, either for elegance of form or for the beauty and variety of its plumage; the only one known with which it could be confounded is the *Artamus rufiventer*, an Indian bird with the breast similarly coloured, but which is entirely destitute of the superciliary stripe of white, which suggested the specific name; in this character and in the rich chestnut colouring of the breast, it differs from every member of its genus. I am unable to say what is the extent of its range, but I am induced to believe that it is confined to Australia, and that in all probability it seldom leaves the interior of the country; the extreme limits of the colony of New South Wales, particularly those which border the extensive plains, being the only parts where it has yet been observed. I first met with it at Yarrundi on the Dartbrook, a tributary of the Hunter, where it was thinly dispersed among the trees growing on the stony ridges bordering the flats.

From this locality to as far as I penetrated northwards on the Namoi, as well as in the direction of the River Peel, it was distributed in similar numbers, intermingled with the *Artamus sordidus*, at about the ratio of one hundred pairs to the square mile, the two species appearing to live and perform the task of incubation in perfect harmony, both being frequently observed on the same tree. In their dispositions, however, and in many of their actions, they are somewhat dissimilar, the *A. superciliosus* being much more shy and difficult of approach than the *A. sordidus*, which is at all times very tame; it also gives a preference to the topmost branches of the highest trees, from which it sallies forth for the capture of insects, and to which it again returns, in the usual manner of the tribe. In every part where I have observed it, it is strictly migratory, arriving in summer, and departing northwards after the breeding-season.

The nest is most difficult of detection, being generally placed either in a fork of the branches or in a niche near the bole of the tree, whence the bark has been partially stripped. It is a round, very shallow, and frail structure, composed of small twigs and lined with fibrous roots; those I discovered contained two eggs, but I had not sufficient opportunities for ascertaining if this number was constant. Their ground-colour is dull buffy white, spotted with umber-brown, forming a zone near the larger end; in some these spots are sparingly sprinkled over the whole surface; they have also the obscure grey spotting like those of *A. sordidus*; the eggs are rather more than eleven lines long by eight and a half lines broad.

The male has the lores, space surrounding the eye, and the ear-coverts deep black; chin greyish black, passing into blackish grey on the chest; crown of the head greyish black; over each eye a pure white stripe commencing in a point, and gradually becoming wider or spatulate in form as it proceeds towards the occiput; all the upper surface, wings, and tail fuliginous grey, which is lightest on the rump and tail; all the tail-feathers tipped with white, except the outer web of the lateral feather, which is grey; under surface of the wing pure white; all the under surface rich deep chestnut; irides nearly black; bill light blue at the base, black at the tip; feet dark lead-colour.

The female has a similar distribution of colouring, but differs from her mate in the following particulars:—lores and a ring surrounding the eye jet-black; only an indication of the superciliary stripe; throat grey; tail not so distinctly tipped with white; under surface light chestnut-red.

Sp. 80. ARTAMUS LEUCOPYGIALIS, *Gould.*

WHITE-RUMPED WOOD SWALLOW.

Artamus leucopygialis, Gould in Proc. of Zool. Soc., part x. 1842, p. 17.

Artamus leucopygialis, Gould, Birds of Australia, fol., vol. ii. pl. 33.

On a careful comparison of specimens of the White-rumped *Artami* from India and the Indian Archipelago with those killed in Australia, I cannot but consider that at least two, if not three, species have been confounded under one name, and that the Australian bird had remained undescribed until characterized by me in the 'Proceedings of the Zoological Society' above quoted. The present species is most nearly allied to the *Artamus leucorhynchus*, but is readily distinguished from it by the blue colour of the bill; and I may here remark, that the Australian birds are also considerably smaller in all their admeasurements than those of the islands to the northwards.

Tasmania and Western Australia are the only colonies in which this bird has not been observed; its range, therefore, over the continent may be considered as very general: in South Australia and New South Wales it would appear to be migratory, visiting these parts in summer for the purpose of breeding. Among other places where I observed it in considerable abundance was Mosquito, and the other small islands near the mouth of the Hunter, and on the borders of the rivers Mokai and Namoi, situated to the northward of Liverpool Plains; in these last-mentioned localities it was breeding among the large flooded gum-trees bordering the rivers.

The breeding-season commences in September and continues until January, during which period at least two broods are reared. In the Christmas week of 1839, at which time I was on the plains of the interior, in the direction of the Namoi, the young progeny of the second brood were perched in pairs or threes together, on a dead twig near their nest. They were constantly visited and fed by the adults, who were hawking about for insects in great numbers, some performing their evolutions above the tops and among the branches of the trees, while others were sweeping over the open plain with great rapidity of flight, making in their progress through the air the most rapid and abrupt turns; at one moment rising to a considerable altitude, and the next descending to within a few feet of the ground, as the insects of which they were in pursuit arrested their attention. In the brushes, on the contrary, the flight of this bird is more soaring and of a much shorter duration, particularly when hawking in the open glades, which frequently teem with insect life. When flying near the ground, the white mark on the rump shows very conspicuously, and strikingly reminds one of the House Martin of our own country.

Two nests, taken by Gilbert on a small island in Coral Bay, near the entrance of the harbour at Port Essington, were compactly formed of dried wiry grass and the fine plants growing on the beach; they were placed in a fork of a slender mangrove-tree, within fifteen feet of the water, in which they were growing; but, like several other Australian birds, the *Artamus leucopygialis* often avails itself of the deserted nests of other species instead of building one of its own. Most of those I found breeding on the Mokai had possessed themselves of the forsaken nest of the *Grallina melanoleuca*, which they had rendered warm and of the proper size by slightly lining it with grasses, fibrous roots, and the narrow leaves of the *Eucalypti*. The eggs are generally three in number, and much lighter in colour and more minutely spotted than those of any other species of the genus I have seen; their ground-colour is flesh-white, finely freckled and spotted with faint markings of reddish brown and grey, in some instances forming a zone at the larger end; their medium length is ten lines, and breadth seven lines and a half.

The sexes are only to be distinguished by dissection, and may be described thus: head, throat, and back sooty grey; primaries and tail brownish black, washed with grey; chest, all the under surface, and rump pure white; irides brown; bill light bluish grey at the base, black at the tip; legs and feet mealy greenish grey.

Family AMPELIDÆ?

Genus PARDALOTUS, *Vieillot*.

This form is peculiar to Australia, in every portion of which great country, including Tasmania, one or other of the species are to be found; some of them associated in the same district, and even inhabiting the same trees, while in other parts only a single species exists: for instance, the *P. punctatus*, *P. quadragintus*, and *P. affinis* inhabit Tasmania; on the whole of the southern coast of the continent from east to west *P. punctatus* and *P. striatus* are associated; the north coast is the cradle of the species I have called *uropygialis*, and the east coast that of *melanocephalus*, from both of which countries the others appear to be excluded; the true habitat of the beautiful species I have described as *P. rubricatus* is the basin of the interior.

Sp. 81. PARDALOTUS PUNCTATUS, *Temm.*
Spotted Diamond-bird.

Pardalotus punctatus, Temm. Man., tom. i. p. lxv.
Pipra punctata, Lath. Ind. Orn. Supp., p. lvi. no. 1.
Speckled *Manikin*, Lath. Gen. Syn. Supp., vol. ii. p. 253.
Wĕ-dup-wĕ-dup, Aborigines of the lowland districts of Western Australia.
Diamond Bird, Colonists of New South Wales.

Pardalotus punctatus, Gould, Birds of Australia, fol., vol. ii. pl. 35.

No species of the genus *Pardalotus* is more widely and generally distributed than the Spotted Diamond-bird; for it inhabits the whole of the southern parts of the Australian continent from the western to the eastern extremities of the country, and is very common in Tasmania. It is incessantly engaged in searching for insects among the foliage, both of trees of the highest growth and of the lowest shrubs; it frequents gardens and enclosures as well as the open forest; and is exceedingly active in its actions, clinging and moving about in every variety of position both above and beneath the leaves with equal facility.

With regard to the nidification of this species, it is a singular circumstance that, in the choice of situation for the reception of its nest, it differs from every other known member of the genus; for while they always nidify in the holes of trees, this species descends to the ground, and availing itself of any little shelving bank, excavates a hole just large enough to admit of the passage of its body, in a nearly horizontal direction to the depth of two or three feet, at the end of which a chamber is formed in which the nest is deposited. The nest itself is a neat and beautifully built structure, formed of strips of the inner bark of the *Eucalypti*, and lined with finer strips of the same or similar materials; it is of a spherical contour, about four inches in diameter, with a small hole in the side for an entrance. The chamber is generally somewhat higher than the mouth of the hole, by which means the risk of its being inundated upon the occurrence of rain is obviated. I have been fortunate enough to discover many of the nests of this species, but they are most difficult to detect, and are only to be found by watching for the egress or ingress of the parent birds from or into its hole or entrance, which is frequently formed in a part of the bank overhung with herbage, or beneath the overhanging roots of a tree. How so neat a structure as is the nest of the Spotted Diamond-bird should be constructed at the end of a hole where no light can possibly enter is beyond our comprehension. The eggs are four or five in number, rather round in form, of a beautiful polished fleshy white, seven and a half lines long by six and a half lines broad.

The song of the Spotted Diamond-bird is a rather harsh piping note of two syllables often repeated.

The male has the crown of the head, wings, and tail black, each feather having a round spot of white near the tip; a stripe of white commences at the nostrils and passes over the eye; ear-coverts and sides of the neck grey; feathers of the back grey at the base, succeeded by a triangular-shaped spot of fawn-colour, and edged with black; rump rufous brown; upper tail-coverts crimson; throat, chest, and under tail-coverts yellow; abdomen and flanks tawny; irides dark brown; bill brownish black; feet brown.

The female may be distinguished by the less strongly contrasted tints of her colouring, and by the absence of the bright yellow on the throat.

Sp. 82. PARDALOTUS RUBRICATUS, *Gould*.

RED-LORED DIAMOND-BIRD.

Pardalotus rubricatus, Gould in Proc. of Zool. Soc., part v. p. 149.

Pardalotus rubricatus, Gould, Birds of Australia, fol., vol. ii. pl. 36.

The Red-lored Diamond-bird belongs to the same section of

the *Pardaloti* as the *P. punctatus* and *P. quadragintus*, and like them is distinguished from the other members of the group by the absence of the scaling-wax-like tips of the spurious wing-feathers. It is the largest species of the genus yet discovered, and is readily distinguished from its near allies, the *P. punctatus* and *P. quadragintus*, by the great size of the spots on the crown.

When I published my plate and description in the folio edition, only a single specimen of this bird had been discovered, and I was unaware in what part of Australia it had been obtained. I have, however, lately seen other specimens collected by Mr. Waterhouse during the overland expedition to the Victoria River under Mr. Stuart. Mr. White of Adelaide also informs me in a letter that he "saw this bird in considerable numbers about the lat. 27° or 28°."

The more I study the birds of Australia and witness the gradual discovery of new species of various forms, the more I am convinced that many still remain to be discovered, and that years must elapse before our knowledge of the entire avifauna of Australia can be considered complete.

Forehead crossed by a narrow band of dirty white; crown and back of the head deep black, each feather having a spot of white near its extremity; back of the neck, back, wing-coverts, and rump brownish grey; wings dark brown, margined with pale brown, the spurious wing, a small portion of the base of the primaries, and the outer margins of the secondaries fine golden orange; immediately before the eye a spot of bright, fiery orange; above and behind the eye a stripe of buff; upper tail-coverts bright olive-green; tail deep blackish brown, the extreme tips of the feathers being white; throat and abdomen greyish white; chest bright yellow; upper mandible and legs brown, under mandible greyish white.

Sp. 83. PARDALOTUS QUADRAGINTUS, *Gould.*
Forty-spotted Diamond-bird.

Pardalotus quadragintus, Gould in Proc. of Zool. Soc., part v. p. 148.
Forty-spot of the Tasmanian Colonists.

Pardalotus quadragintus, Gould, Birds of Australia, fol., vol. ii. pl. 37.

This species is, I believe, peculiar to Tasmania, where it inhabits the almost impenetrable forests which cover that island, particularly those of its southern portion. It is I think less numerous than either of its congeners, the *Pardalotus affinis* and *P. punctatus*, and appears to confine itself more exclusively to the highest gum-trees than those species. I found it very abundant in the gulleys under Mount Wellington, and observed it breeding in a hole in one of the loftiest trees, at about forty feet from the ground; I afterwards took a perfectly developed white egg from the body of a female killed on the 5th of October. The weight of this little bird was rather more than a quarter of an ounce; the stomach was muscular, and contained the remains of the larvæ of lepidoptera, which with coleoptera and other insects constitute its food.

It has a simple piping kind of note of two syllables.

In its actions it much resembles the Tits, creeping and clinging among the branches in every direction.

The eggs are white and nearly round in form, being seven lines and a half long and six broad.

The sexes are so much alike in colour, that a separate description is unnecessary.

Crown of the head and all the upper surface bright olive-green, each feather obscurely margined with brown; wings brownish black, all the feathers, except the first and second primaries, having a conspicuous spot of pure white near their extremities; tail blackish grey, the extreme tips of the feathers being white; cheeks and under tail-coverts yellowish olive; throat and under surface greyish white, passing into olive on the flanks; irides dark brown; bill brownish black; feet brown.

Sp. 84.   PARDALOTUS STRIATUS, *Temm.*

STRIATED DIAMOND-BIRD.

Pardalotus striatus, Temm. Man., part i. p. lxv.
Pipra striata, Lath. Ind. Orn., p. 558, No. 18.
Striped-headed Manakin, Lat. Gen. Syn., vol. iv. p. 525, pl. 54.
Pardalotus ornatus, Temm. Pl. Col. 394. fig. 1.
Wä-dup-wäl-dup, Aborigines of the lowland, and
Wä-dee-we-das, Aborigines of the mountain districts of Western Australia.

Pardalotus striatus, Gould, Birds of Australia, fol., vol. ii. pl. 38.

This beautiful species, like the *P. punctatus*, enjoys an extensive range of habitat, being found in all parts of the southern portion of the Australian continent; it has not as yet been discovered in Tasmania, its place in that island being apparently occupied by the *P. affinis*. I have carefully examined specimens killed at Swan River with others from New South Wales, and I cannot find any difference either in their size or markings. It will be interesting to know how far this species and the *P. punctatus* extend their range northwards, a point which can only be ascertained when the country has been fully explored. This active little bird is generally seen seeking insects among the leaves, for which purpose it frequents trees of every description, but gives a decided preference to the *Eucalypti*. Its flight is rapid and darting, hence it passes from tree to tree, or from one part of the forest to another with the greatest ease. Its voice is a double note several times repeated.

The nest, which is a very neat structure of dried soft grasses and the bark of the tea-tree, lined with feathers, is usually placed in a hole of a dead branch, but sometimes in the boll of the tree. It breeds in September, October, and

November, and lays three or four fleshy-white eggs, which are nine lines long by seven lines broad.

The sexes so closely assimilate in colour and markings that they are only to be distinguished with certainty by dissection.

The young assume the adult colouring from the nest, but have the tips of the spurious wing orange instead of red.

The validity of this latter passage has been questioned by Mr. Ramsay, no mean authority respecting Australian birds. I believe it is his opinion that the bird which I regarded as the young of *P. striatus* may prove to be a distinct species, intermediate between *P. striatus* and *P. affinis*. I have not, however, as yet seen any examples which would lead to confirm his view, and, without undervaluing his opinion, I leave the subject undetermined. Mr. Ramsay has, I think, found birds with this character of plumage breeding; but that is no proof, for I have positive evidence that some of the Australian species reproduce their kind before they have attained their adult livery.

Forehead and crown of the head black, the feathers of the latter having a stripe of white down the centre; a stripe of deep orange-yellow commences at the base of the upper mandible and runs above the eye, where it is joined by a stripe of white which leads to the occiput; back of the neck and back brownish olive-grey; rump and upper tail-coverts yellowish brown; wings black, the external edges of the third, fourth, fifth, sixth, and seventh primaries white at their base and tipped with white; secondaries margined with white and reddish brown; tail black, each feather tipped with white; sides of the face and neck grey; throat and upper part of the chest yellow; centre of the abdomen white; flanks and under tail-coverts brownish buff, the former tinged with yellow; irides brownish red; bill at the tip and along the culmen dark brown tinged with blue, the remainder yellowish white; legs and feet greenish grey.

Sp. 85.   PARDALOTUS AFFINIS, *Gould.*
           ALLIED DIAMOND-BIRD.

Pipra striata?, Gmel. et Auct.
Striped-headed Manakin, Shaw, Gen. Zool., vol. x. p. 29, pl. 4.
Pardalotus affinis Gould, in Proc. of Zool. Soc., part v. 1837, p. 25.

Pardalotus affinis, Gould, Birds of Australia, fol., vol. ii. pl. 39.

The *Pardalotus affinis* is distinguished by the yellow tips of its spurious wings, and by the margin of the third primary only being white. The bird figured by Shaw and Latham, as quoted above, has, in all probability, reference to the present species, but not, in my opinion, to the *Pipra striata* of Gmelin, whose description does not agree with the Tasmanian bird, or with any of those from New South Wales; he distinctly states that the tips of some of the wing-coverts are yellow, and that the spurious wing is tipped with white, and, moreover, adds that it is a native of South America.

The Allied Diamond-bird is distributed over every part of Tasmania, and may be regarded as the commonest bird of the island: wherever the gum and wattle exist, there also may the bird as certainly be found; giving no decided preference to trees of a high or low growth, but inhabiting alike the sapling and those which have attained their greatest altitude. It displays great activity among the branches, clinging and creeping about in the most easy and elegant manner, examining both the upper and under sides of the leaves with the utmost care in search of insects. It is equally common in all the gardens and shrubberies, even those in the midst of the towns, forming a familiar and pleasing object, and enlivening the scenery with its sprightly actions and piping though somewhat monotonous note. Its food consists of seeds, buds, and insects, in procuring which, its most elegant actions are brought into play.

I was formerly led to believe that the Allied Diamond-bird

was strictly confined to Tasmania and the islands in Bass's Straits; but I have lately seen specimens from Victoria and New South Wales.

The season of nidification occupies three or four months of the year, during which two or more broods are reared. Eggs may be found in September; and on reference to my journal I find that near George Town, on the 8th of January, I took from a nest in the hole of a tree five fully-fledged young. The nest in this instance was of a large size, and of a round domed form like that of the Wren, with a small hole for an entrance; it was outwardly composed of grasses and warmly lined with feathers. The eggs vary from three to five in number, and are of a beautiful white, nine lines long by seven lines in diameter.

The holes selected for the nest are sometimes high up in the loftiest trees, at others within a few feet of the ground. The young birds have the tips of the spurious wing orange instead of yellow; and although the whole plumage possesses the same character as that of the adults, the markings are less brilliant and well-defined. The sexes offer no observable difference in their colouring by which they can be distinguished.

Forehead and crown of the head black, the latter with a stripe of white down the centre of each feather; a stripe of yellow commences at the base of the upper mandible, and runs above the eye, where it is joined by a stripe of white, which proceeds nearly to the occiput; back of the neck and back greyish olive-brown; rump and upper tail-coverts olive-brown; wings black, each of the primaries slightly tipped with white, and the third externally edged with white; the secondaries edged with white and rufous, and the tips of the spurious wing yellow; tail blackish brown, each feather having a transverse mark of white at the tip; ear-coverts and cheeks grey; throat yellow, passing into lighter yellow on the flanks; centre of the abdomen white; irides olive-brown; bill black; feet brown.

Sp. 86. PARDALOTUS MELANOCEPHALUS, *Gould*.

BLACK-HEADED DIAMOND-BIRD.

Pardalotus melanocephalus, Gould in Proc. of Zool. Soc., part v. p. 149.

Pardalotus melanocephalus, Gould, Birds of Australia, fol., vol. ii. pl. 40.

I have received numerous examples of this beautiful and well-defined species from Moreton Bay, where it probably takes the place of the *P. striatus*, from which it is distinguished by the black colouring of its head and by its thicker bill, but to which it is very nearly allied, as well as to the *P. uropygialis*; it is, in fact, directly intermediate between the two, having the black head of the latter without the yellow colouring of the rump. There appears to be no external difference in the sexes.

Crown of the head, lores, and ear-coverts black; over each eye a stripe commencing at the nostrils, the anterior half of which is orange, and the posterior white; sides of the face and neck white; back of the neck and back olive-grey; upper tail-coverts brownish buff; tail black, each feather tipped with white; wings blackish brown, the third, fourth, fifth, sixth, and seventh primaries white; secondaries edged and tipped with white; one of the wing-coverts broadly margined on the inner web with white, forming an oblique line across the shoulder; spurious wing tipped with crimson; line down the centre of the throat, the breast, and middle of the abdomen bright yellow; vent and under tail-coverts buff; bill black; feet brown.

At present we are unaware whether this bird nidifies in the holes of trees like the other *Pardaloti* or not; the colour and number of the eggs are also unknown; they are probably pinky white; but these particulars must be left for the investigation of those who may be favourably situated for ascertaining them.

Sp. 87. PARDALOTUS UROPYGIALIS, *Gould*.

YELLOW-RUMPED DIAMOND-BIRD.

Pardalotus uropygialis, Gould in Proc. of Zool. Soc., part vii. 1839, p. 143.

Pardalotus uropygialis, Gould, Birds of Australia, fol., vol. ii. pl. 41.

For this very beautiful Diamond-bird, and several other interesting species from the north-west coast of Australia, I am indebted to the kindness of the late Benjamin Bynoe, Esq., Surgeon of Her Majesty's Surveying Ship the Beagle; to Captain Wickham and the other officers of which vessel my thanks are also due for their polite attention to my wishes, and the promise of communicating to me any novelties they might procure during their survey of the north-west coast.

The Yellow-rumped Diamond-bird is easily distinguished from every other species of the group with which I am acquainted by the bright yellow colouring of the lower part of the back, by the rich spot of orange before the eye, by having a shorter wing, and by being more diminutive in size than any of the others, with the exception of *Pardalotus punctatus*.

I am unable to give any account of its habits and manners; but in these respects it doubtless closely assimilates to the other members of its group.

Crown of the head, stripe before and behind the eye black; lores rich orange; a mark from above the eye to the occiput, chest, and centre of the abdomen white; throat and cheeks delicate crocus-yellow; rump and upper tail-coverts sulphur-yellow; back of the neck and back olive-grey; wings black, the external webs of the second and five following primaries white at the base; tips of the spurious wing scarlet; tail black; the three outer feathers tipped with white, the white spreading largely over the inner web of the outer feathers; bill black; feet lead-colour.

The sexes do not seem to differ in size or in the colour of their plumage.

Family LANIADÆ.

Genus STREPERA, *Lesson*.

On a careful examination of the members of this genus, it will be perceived that their relationship to the *Corvidæ*, to which they have been usually assigned, is very remote, their size and colour being, in fact, the only features of resemblance; their whole structure and economy are indeed very different from those of every other known bird, except those of *Gymnorhina* and *Cracticus*, with which genera, in my opinion, they form a very distinct group, the natural situation of which is among the *Laniadæ* or Shrikes.

Most of the species are peculiar to Australia, and strictly confined to the southern portion of that continent, their range being limited to the country comprised within the 25th and 40th degrees of south latitude; future research may, however, add both to the number of species and perhaps to the extent of their range; still their great stronghold is undoubtedly the most southern portion of the Australian continent, the islands of Bass's Straits, and Tasmania. I have, however, seen a species of this genus from Lord Howe's Island which is very similar to, if not the same as *S. graculina*.

These birds seek their food on or near the ground, sometimes in swampy situations and even on the sea-shore, at others on the most sterile plains far distant from water; grasshoppers and insects of every order are eaten by them with avidity, and to these grain seeds and fruits are frequently added; they hop with remarkable agility over the broken surface of the ground, and leap from branch to branch with great alacrity: their flight is feeble and protracted, and they seldom mount high in the air, except for the purpose of crossing a gully, or for passing from one part of the forest to another, and then merely over the tops of the trees; during flight they usually utter a peculiar shrill cry, which is frequently repeated and answered by other birds of the same troop, for they mostly flit about in small companies of from four to six in number, apparently the parents and their offspring of the year. All the species occasionally descend to the cultivated grounds, orchards, and gardens of the settlers, and commit considerable havoc among their fruits and grain; in many parts of Australia, and particularly in Tasmania, they form an article of food, and are considered good and even delicate eating. They usually build open cup-shaped nests as large as that of the Crow, composed of sticks and other coarse materials, lined with grasses or any other suitable substance that may be at hand; the eggs are generally three, but are sometimes four, in number. The sexes are similar in plumage, and the young assume the livery of the adult from the time they leave the nest.

Sp. 88. STREPERA GRACULINA.

PIED CROW-SHRIKE.

Réveilleur de l'Isle de Norfolk?, Daud. tom. ii. p. 267.
Corvus graculinus, White's Bot. Bay, pl. in p. 251.
Coracias strepera, Lath. Ind. Orn., vol. i. p. 173.
Corvus streperus, Leach, Zool. Misc., vol. ii. pl. 86.
Noisy Roller, Lat. Gen. Syn., Supp., vol. ii. p. 121.
Le Grand Calibé, Le Vaill. Ois. de Par., &c., pl. 24.
Cracticus streperus, Vieill. Gal. des Ois., pl. 109.
Gracula strepera, Shaw, Gen. Zool., vol. vii. p. 462.
Barita strepera, Temm. Man., part i. p. li.
Coronica strepera, Gould, Syn. Birds of Australia, part i.
Strepera, Less. Traité d'Orn., p. 329.
Strepera graculina, G. R. Gray, Gen. of Birds, 2nd edit., p. 50.

Strepera graculina, Gould, Birds of Australia, fol., vol. ii. pl. 49.

This species was originally described and figured in White's 'Voyage to New South Wales'; it is consequently the oldest and most familiarly known member of the group to which it belongs. It is very generally distributed over the colony of

New South Wales, inhabiting alike the brushes near the coast, those of the mountain ranges, and also the forests of *Eucalypti* which clothe the plains and more open country. As a great part of its food consists of seeds, berries, and fruits, it is more arboreal in its habits than some of the other species of its group, whose structure better adapts them for progression on the ground, and whose food principally consists of insects and their larvæ. Like the other members of the genus, it is mostly seen in small companies, varying from four to six in number, seldom either singly or in pairs: I am not, however, inclined to consider them as gregarious birds in the strict sense of the word, believing as I do that each of these small companies is composed of a pair and their progeny, which appear to keep together from the birth of the latter until the natural impulse for pairing prompts them to separate.

It is during flight that the markings of this bird are displayed to the greatest advantage, and render it a conspicuous object in the bush: while on the wing it utters a peculiar noisy cry, by which its presence is often indicated.

The nest, which is usually constructed on the branches of low trees, sometimes even on those of the *Casuarinæ*, is of a large size, round, open, and cup-shaped, built of sticks and lined with moss and grasses; the eggs, which I was not so fortunate as to procure, are said to be three or four in number.

The plumage of both sexes at all ages is so precisely similar, that by dissection alone can we distinguish the male from his mate, or the young from the adult; the female is, however, always a trifle less in all her admeasurements, and the young birds have the corners of the mouth more fleshy and of a brighter yellow than the adults.

Their general colour is fine bluish black; the basal half of the primaries, the basal half and the tips of the tail-feathers, including those portions of their shafts and the under tail-coverts, snow-white; irides beautiful yellow; bill and feet black.

Sp. 69.   STREPERA FULIGINOSA, *Gould.*
   Sooty Crow-Shrike.

Cracticus fuliginosus, Gould in Proc. of Zool. Soc., part iv. p. 100.
Coronica fuliginosa, Gould in Syn. Birds of Australia, part i.
Black Magpie of the Colonists.

Strepera fuliginosa, Gould, Birds of Australia, fol., vol. ii. pl. 48.

This species is a permanent resident in Tasmania; its range also extends to the islands in Bass's Straits, and a few individuals have been found in South Australia. Its browner colouring, more arched and gibbose bill, its smaller size, and the absence of the white colouring of the under tail-coverts and of the base of the primaries, are characters by which it may at once be distinguished from most of the other members of the group. The localities it frequents are also of a different description, those preferred being low swampy grounds in the neighbourhood of the sea and woods bordering rivers. Like the other species of the genus, it subsists on insects and grubs of various kinds, to which pulpy seeds and berries are frequently added.

It is very active on the ground, passing over the surface with great rapidity.

It breeds in the low trees, constructing a large and deep nest very similar to that of the European Crow, and lays three eggs, of a pale vinous brown marked all over with large irregular blotches of brown, one inch and five-eighths long by one inch and a quarter broad.

I have seen this bird in a state of captivity, and it appeared to bear confinement remarkably well.

The sexes present no visible difference except in size, the female being smaller than the male; they may be thus described:—

All the plumage sooty black, with the exception of the ends of the primaries and all but the two middle tail-feathers, which are white; irides bright yellow; bill and feet black.

Sp. 90. STREPERA ARGUTA, *Gould.*

HILL CROW-SHRIKE.

Strepera arguta, Gould in Proc. of Zool. Soc., part xiv. p. 19.
——— *melanoptera*, Gould, Id. p. 20.

Strepera arguta, Gould, Birds of Australia, fol., vol. ii. pl. 44.

The *Strepera arguta* is abundantly dispersed over Tasmania, but is more numerous in the central parts of the island than in the districts adjacent to the coast; it also inhabits South Australia, in which country it is more scarce, and all the specimens I have seen are rather smaller in size. I have never seen it in any part of New South Wales that I have visited, neither have specimens occurred in the numerous collections from the west coast that have come under my notice. It is the largest, the boldest, and the most animated species of the genus yet discovered. If not strictly gregarious, it is often seen in small companies of from four to ten, and during the months of winter even a greater number are to be seen congregated together. The districts most suited to its habits are open-glades in the forest and thinly-timbered hills; although it readily perches on the trees, its natural resort is the ground, for which its form is admirably adapted, and over which it passes with amazing rapidity, either in a succession of leaps or by running. Fruits being but sparingly diffused over Australia, insects necessarily constitute almost its sole food, and of these nearly every order inhabiting the surface of the ground forms part of its diet; grasshoppers are devoured with great avidity.

Its note is a loud ringing and very peculiar sound, somewhat resembling the words *clink, clink,* several times repeated, and strongly reminded me of the distant sound of the strokes on a blacksmith's anvil; and hence the term *arguta* appeared to me to be an appropriate specific appellation for this new species.

All the nests I found of this species either contained young birds or were without eggs; I am consequently unable to give their size and colour. The nest, which is of a large size, is generally placed on a horizontal branch of a low tree; it is round, deep, and cup-shaped, outwardly formed of sticks and lined with fibrous roots and other fine materials.

The sexes present no external difference whatever, neither is there much difference in size; the young are black from the nest, except that the tertiary feathers are strongly tipped with white, a character which is rarely I believe thrown off in adult age.

All the plumage brownish black, becoming much browner on the tips of the wing-feathers, and of a grey tint on the abdomen; base of the inner webs of the primaries and secondaries, the under tail-coverts and the apical third of the inner webs of the tail-feathers white; irides orange-yellow; bill and feet black; corner of the mouth yellow.

Upon a careful examination of the numerous specimens of this bird contained in my collection, I find among them two very singular varieties; one with the base of the primaries of a nearly uniform black and the tips white, and another in which the base of the primaries is white and the tips black. It is evident, therefore, that the markings of this species are not constant, and this induces me to believe that the bird I characterized as *S. melanoptera* is nothing more than one of the varieties above mentioned. I do not, however, venture to affirm that the birds received from South Australia with wholly black wings may not prove to be distinct from those from Tasmania; this is a matter for investigation of future Australian naturalists. For the present I sink the appellation *melanoptera* into a synonym.

Sp. 01.   STREPERA ANAPHONENSIS.

Grey Crow-Shrike.

Barita Anaphonensis, Temm. Pl. Col.
Cracticus cuneicaudatus, Vieill. 2° Edit. du Nouv. Dict. d'Hist. Nat. tom. v. p. 356.
Strepera plumbea, Gould in Proc. of Zool. Soc., part xiv. p. 20.
Corvus versicolor, Lath. ?
Strepera versicolor, Gray, Gen. of Birds, vol. ii. p. 302, *Strepera*, sp. 3.
Gymnorhina Anaphonensis, Gray, Gen. of Birds, vol. ii. p. 302, *Gymnorhina*, sp. 8.
Dje-Bak, Aborigines of Western Australia.
Squeaker of the Colonists.

Strepera Anaphonensis, Gould, Birds of Australia, fol., vol. ii. pl. 45.

Having formerly considered the Grey Crow-Shrikes of New South Wales and Western Australia as distinct species, I assigned to the Swan River bird the specific appellation of *plumbea*; subsequent research has, however, induced me to believe them identical; and if this be really the case, no one species of the genus has so wide a range as the present, extending as it does from New South Wales on the east to Swan River on the west coast. It is, however, more local in its habitat than any of them, at least such is the case in New South Wales; for although it is tolerably abundant at Illawarra, at Camden, and at Bong-bong, it was not seen in any other district that I visited. Gilbert states that in Western Australia he mostly met with it in the thickly wooded forests, singly or in pairs, feeding on the ground with a gait and manners very much resembling the Common Crow. Its flight is easy and long-sustained, and it occasionally mounts to a considerable height in the air.

The stomach is very muscular, and the food consists of coleoptera and the larvæ of insects of various kinds.

It breeds in the latter part of September and the beginning

of October, forming a nest of dried sticks in the thickest part of the foliage of a gum- or mahogany-tree and laying three eggs, the ground-colour of which is either reddish buff or wood-brown, marked over nearly the whole of the surface with blotches of a darker tint; their medium length is one inch and nine lines by one inch and two and a half lines broad.

The sexes resemble each other so closely in colour, that it is impossible to distinguish the one from the other, except by dissection.

All the upper surface leaden grey, becoming much darker on the forehead and lores; wings black; secondaries margined with grey and tipped with white; basal half of the inner webs of the primaries white, of the outer webs grey; the remainder of their length black, slightly tipped with white; tail black, margined with grey and largely tipped with white; all the under surface greyish brown; under tail-coverts white; irides orange; bill and feet black.

Genus GYMNORHINA.

Like *Strepera*, this is strictly an Australian form, the structure of which is a mere modification of that of the members of the last genus adapted to a somewhat different mode of life and habits. The species, being more pastoral in their habits than the *Streperæ*, frequent the open plains and grassy downs, over which they hop with great facility. Their chief food consists of grasshoppers and other insects, to which berries and fruits are added, when procurable. Few birds are more ornamental, or give a more animated appearance to the country, than the members of this genus, either when passing over the surface of the ground, or when pouring forth their singular choral-like notes while perched together on the bare branches of a fallen *Eucalyptus*. The form and situation of their nests are the same as those of the *Streperæ*.

Sp. 92.   GYMNORHINA TIBICEN.
              Piping Crow-Shrike.

Coracias Tibicen, Lath. Ind. Orn. Supp., p. xxvii.
Barita Tibicen, Temm. Man. d'Orn., part i. p. li.
Piping Roller, Lath. Gen. Hist., vol. iii. p. 86, no. 23.
Cracticus Tibicen, Vig. and Horsf. in Linn. Trans., vol. xv. p. 260.
Gymnorhina Tibicen, G. R. Gray, List of Gen. of Birds, 2nd edit. p. 50.
Ca-ruck, Aborigines of New South Wales.

Gymnorhina tibicen, Gould, Birds of Australia, fol., vol. ii. pl. 48.

This species is universally diffused over the colony of New South Wales, to which part of the Australian continent I believe it to be confined. It is true that a bird of this genus inhabits the neighbourhood of Swan River, whose size and style of plumage are very similar, but which I have little doubt will prove to be distinct; I shall therefore consider the habitat of the present bird to be restricted to New South Wales until I have further proofs to the contrary.

The *Gymnorhina Tibicen* is a bold and showy bird, which greatly enlivens and ornaments the lawns and gardens of the colonists by its presence, and with the slightest protection from molestation becomes so tame and familiar that it approaches close to their dwellings, and perches round them and the stock yards in small families of from six to ten in number. Nor is its morning carol less amusing and attractive than its pied and strongly contrasted plumage is pleasing to the eye. To describe the notes of this bird is beyond the power of my pen, and it is a source of regret to myself that my readers cannot, as I have done, listen to them in their native wilds, or that the bird is not introduced more frequently into this country; for a more amusing and easily-kept denizen for the aviary could not be selected. It lives almost entirely on insects, which are generally procured on the ground, and the number of locusts and grasshoppers it devours is immense. In captivity it subsists upon animal food of almost every kind, and that berries and fruits would be equally acceptable I have but little doubt.

Cleared lands, open flats, and plains skirted by belts of trees are its favourite localities; hence the interior of the country is more favourable to its habits than the neighbourhood of the coast.

The breeding-season commences in August and lasts until January, during which period two broods are generally reared by each pair of birds. The nest is round, deep, and open, composed outwardly of sticks, leaves, wool, &c., and lined with any finer materials that may be at hand. The eggs are either three or four in number; their colour and size I regret to say I cannot give, having unfortunately neglected to procure them while in New South Wales.

Crown of the head, cheeks, throat, back, all the under surface, scapularies, secondaries, primaries, and tips of the tail-feathers black; wing-coverts, nape of the neck, upper and under tail-coverts, and base of the tail-feathers white; bill bluish ash-colour at the base, passing into black at the tip; irides rich reddish hazel; legs black.

Sp. 93. GYMNORHINA LEUCONOTA, *Gould*.

WHITE-BACKED CROW-SHRIKE.

Barita Tibicen, Quoy et Gaim. Voy. de la Coq., pl. 20.
Gool'r-bat, Aborigines of the lowland districts of Western Australia.

Gymnorhina leuconota, Gould, Birds of Australia, fol., vol. ii. pl. 47.

This fine species of *Gymnorhina*, which has been confounded by the French writers with the *Coracias Tibicen* of Latham, inhabits South Australia, Victoria, and New South Wales. It is said to be tolerably abundant at Port Phillip, and that it is sometimes seen on the plains near Yass. For my own part I have never met with it in New South Wales, but observed it to be rather abundant in South Australia. In the extreme shyness of its disposition it presents a remarkable contrast to the *G. Tibicen*; it was indeed so wary and so difficult to approach, that it required the utmost ingenuity to obtain a sufficient number of specimens necessary for my purpose. Plain and open hilly parts of the country are the localities it prefers, where it dwells much on the ground, feeding upon locusts and other insects. In size it is fully as large as any species of the genus yet discovered; it runs over the ground with great facility, and frequently takes long flights across the plains from one belt of trees to another; in other parts of its economy it so nearly resembles the *G. Tibicen*, that it would be useless to repeat a description of them. The same single clear note and early carol of small companies perched on some leafless branch of a *Eucalyptus* appears characteristic of all the members of the genus.

It breeds in September and October, constructing a nest of dried sticks in an upright fork of a gum- or mahogany-tree. A nest taken in Angas Park, South Australia, Oct. 5, 1861, and presented to me by G. French Angas, Esq., measures about a foot across, and is constructed of coarse roots and twigs, with its shallow interior lined with coarse dried grasses; and Mr. Angas tells me that it is built in September, and always placed at a great height in red gum-trees. The eggs are three in number, very long in form, and of a dull bluish white, in some instances tinged with red, marked with large zigzag streakings of brownish red; the average length of the eggs is one inch and eight lines, and breadth one inch and one line. Occasionally eggs are met with which are spotted with black or umber-brown.

Immature birds of both sexes have the whole of the back clouded with grey, and the bill of a less pure ash-colour.

Back of the neck, back, upper and under coverts of the wings, basal portion of the spurious wing, upper and under tail-coverts, and base of the tail-feathers white; remainder of the plumage and the shafts of the white portion of the tail-feathers glossy black; irides light hazel; bill bluish-lilac, passing into black at the tip; legs and feet blackish grey.

Sp. 94. GYMNORHINA ORGANICUM, *Gould*.

TASMANIAN CROW-SHRIKE.

Cracticus hypoleucus, Gould in Proc. of Zool. Soc., part iv. p. 106.
Gymnorhina hypoleuca, Cab. Mus. Hein., Theil i. p. 226.
Organ-Bird and *White Magpie* of the Colonists.

Gymnorhina organicum, Gould, Birds of Australia, fol., vol. ii. pl. 48.

This animated and elegant bird is a native of Tasmania, and appears to be very local in its habitat, for while it is never found below Austin's Ferry on the southern bank of the river Derwent, it is very plentiful on the opposite side, and it is also to be met with in small troops in all the open parts of the country; but I did not observe it on the banks of the Tamar. When perched on the dead branches of the trees soon after day-break, it pours forth a succession of notes of the strangest description that can be imagined, much resembling the sounds of a hand-organ out of tune, which has obtained for it the colonial name of the Organ-Bird. It is very easily tamed; and as it possesses the power of imitation in an extraordinary degree, it may be readily taught to whistle various tunes as well as to articulate words; it consequently soon becomes a most amusing as well as an ornamental bird for the aviary or cage. The stomach is very muscular, and the food consists of insects of various kinds, grubs, caterpillars, &c., which are procured on the ground.

A nest I found among the topmost branches of a high gum-tree was round, and outwardly constructed of sticks interspersed with strips of bark, short grasses, and tufts of a species of swamp grass, to which succeeded an internal lining of coarse grass, which again was lined with the inner bark of the stringy bark-tree, sheep's wool, and a few feathers, felted together, and forming a dense and warm receptacle for the eggs; it was about ten inches in diameter, and about four or five inches in depth.

The eggs were four in number, of a lengthened form, with a ground-colour of greenish ashy grey, spotted and blotched, particularly at the larger end, with umber-brown and bluish grey, the latter colour appearing as if beneath the surface of the shell; they were one inch and five lines long by one inch broad. The young assume the adult livery from the nest, and appear to keep in company of the parent birds during the first ten months of their existence.

The male has the crown of the head, cheeks, throat, all the under surface, scapularies, primaries, and tips of the tail jet-black; nape of the neck, back, upper and under tail-coverts, and base of the tail-feathers white; bill dark lead-colour at the base, passing into black at the tip; legs black; irides bright hazel.

The female differs in having the nape of the neck and back grey, and the primaries and tips of the tail-feathers brownish black.

Genus CRACTICUS, *Vieillot*.

The members of this genus are universally dispersed over Australia, where they prey upon small quadrupeds, birds, lizards, and insects, which they frequently impale after the manner of the ordinary Shrikes. Their mode of nidification resembles that of the species belonging to the genera *Strepera* and *Gymnorhina*, the nest being a large round structure placed among the branches of the trees, and the eggs four in number. A great similarity exists between the species inhabiting New South Wales, Tasmania, and Western Australia, but the annexed descriptions, with a due attention to the localities, will obviate all difficulty in determining the species.

Sp. 95. CRACTICUS NIGROGULARIS, *Gould*.

BLACK-THROATED CROW-SHRIKE.

Vanga nigrogularis, Gould in Proc. of Zool. Soc., part x. p. 143.
Cracticus rarius, Vig. and Horsf. in Linn. Trans., vol. xv. p. 261.
—— *robustus*, Bonap. Conspectus Gen. Av., tom. i. p. 307, *Cracticus*, sp. 2.

Cracticus nigrogularis, Gould, Birds of Australia, fol., vol. ii. pl. 49.

The Black-throated Crow-Shrike finds a natural asylum in New South Wales, the only one of the Australian colonies in which it has yet been found, and where it is by no means rare, although the situations it affects render it somewhat local; it is a stationary species, breeding in all parts of the country suitable to its habits and mode of life; districts of rich land known as apple-tree flats, and low open undulating hills studded with large trees, are the kind of districts to which it peculiarly resorts: hence the cow-pastures at Camden, the fine park-like estate of Charles Throsby, Esq., at Bong-bong, and the entire district of the Upper Hunter are among the localities in which it may always be found.

It is usually seen in pairs, and, from its active habits and pied plumage, forms a conspicuous object among the trees, the lower and outspreading branches of which are much more frequented by it than the higher ones; from these lower branches it often descends to the ground in search of insects and small lizards, which however form but a portion of its food, for, as its powerful and strongly-hooked bill would lead us to infer, prey of a more formidable kind is often resorted to; its sanguinary disposition, in fact, leads it to feed on young birds, mice, and other small quadrupeds, which it tears piece-meal and devours on the spot.

The nest, which is rather large and round, is very similar to that of the European Jay; those I examined were outwardly composed of sticks, neatly lined with fine fibrous roots, and were generally placed on a low horizontal branch among the thick foliage.

The eggs are dark yellowish brown, spotted and clouded with markings of a darker hue, and in some instances with a few minute spots of black; their medium length is one inch and three lines by eleven lines in breadth.

The breeding-season commences in August, and continues during the four following months.

The sexes are so precisely alike in colouring, that they can only be distinguished with certainty by dissection.

Head, neck, and chest black; hinder part of the neck, shoulders, centre of the wing, rump, and under surface white; two middle tail-feathers entirely black, the remainder black largely tipped with white; bill lead-colour at the base, black at the tip; legs black; irides brown.

The young during the first autumn are very different from the adult, particularly in the colouring of the head and chest, which is light brown instead of black; the bill, as in most youthful birds, is also very different, the basal portion being dark fleshy brown instead of lead-colour.

Sp. 90.   CRACTICUS PICATUS, *Gould*.

PIED CROW-SHRIKE.

Cracticus picatus, Gould in Proc. of Zool. Soc., 1848, p. 40.
Ka-ra-a-ra, Aborigines of Port Essington.
Magpie of the Colonists.

Cracticus picatus, Gould, Birds of Australia, fol., vol. ii. pl. 50.

This is in every respect a miniature representative of the *Cracticus nigrogularis* of New South Wales; it must, however, be regarded as a distinct species; for its much more diminutive size will warrant such a conclusion by every ornithologist who compares them.

Gilbert, who found it at Port Essington in considerable abundance, states that it is an extremely shy and wary bird, inhabiting the most secluded parts of the forest, and is as frequently seen searching for its food on the ground as among the topmost branches of the highest trees. In its habits, manners, mode of flight, and in its loud, discordant, organ-pipe-like voice, it closely resembles the other members of the genus. It is usually seen in pairs, or in small families of four or five. Its nest is built of sticks in the upright fork of a thickly-foliaged tree, at about thirty or forty feet from the ground.

The stomach is muscular, and the food consists of insects of various kinds, but principally of coleoptera.

The sexes are not distinguished by any difference in the markings of the plumage, but the young are dressed in a brown colouring like those of the other members of the genus.

Collar at the back of neck, centre and edge of the wing, rump, abdomen, under tail-coverts, and tips of all but the centre tail-feathers white, remainder of the plumage deep black; irides dark reddish brown; bill ash-grey, the tip black; legs and feet dark greenish grey.

Sp. 07.   CRACTICUS ARGENTEUS, *Gould*.
SILVERY-BACKED CROW-SHRIKE.

Cracticus argenteus, Gould in Proc. of Zool. Soc., part viii. p. 120.

Cracticus argenteus, Gould, Birds of Australia, fol., vol. ii. p. 51.

Examples of this species were discovered on the north coast of Australia, both by Sir George Grey and B. Bynoe, Esq., to the latter of whom I am indebted for one of the specimens from which my description was taken.

The *Cracticus argenteus* is directly intermediate in size between *C. torquatus* and *C. nigrogularis*, and moreover exhibits a remarkable participation in the colouring of those two species, having the white throat and chest of the former, and the parti-coloured wings, conspicuous white rump, and white-tipped tail of the latter; it differs, however, from both, as well as from all the other members of the genus, in the light or silvery-grey colouring of the back, and hence the term of *argenteus* has been applied to it.

No account of its habits has yet been received, but they doubtless resemble those of the other species of the genus.

Crown of the head, ear-coverts, shoulders, primaries, and all the tail-feathers for three-fourths of their length from the base black; back silvery grey; throat, all the under surface, sides of the neck, some of the wing-coverts and the margins of several of the secondaries, rump, and tips of the tail-feathers pure white; bill horn-colour; feet blackish brown.

Sp. 08.   CRACTICUS QUOYII.

Quoy's Crow-Shrike.

Barita Quoyi, Less. Zool. de la Coq., tom. i. p. 639, pl. 24.
Mul-gul-ga, Aborigines of Port Essington.

Cracticus Quoyii, Gould, Birds of Australia, fol., vol. ii. pl. 53.

We have abundant evidence that the zoology and botany of New Guinea and Australia are very similar. In some instances the same species are found in both countries, of which fact the present bird is an example. M. Temminck, to whom I showed specimens of this bird killed in Australia, assured me that they were identical with others from New Guinea. The northern coast is the only portion of Australia in which this bird has been observed. It is tolerably abundant at Port Essington, where it inhabits the mangrove-swamps generally, even those close to the settlement.

Gilbert states that it is extremely shy and wary, and that the nature of its usual haunts precludes in a great measure all chance of getting a sight of it. He never met with it in any other situation than the darkest and thickest parts of the mangroves, where there is a great depth of mud, and where the roots of the trees are very thickly intertwined; it is among these roots that it is constantly seen searching for crabs. Its note is short and monotonous, and very like the name given to it by the aborigines, *Mal-gŏl-ga*, the second syllable being prolonged and forming the highest note; it also utters other sounds, some of them resembling those of the *Gymnorhina leuconota*; at other times it frequently emits a note very similar to the cry of young birds for food.

The stomach is muscular, and the food consists of crabs, and occasionally of coleoptera, neuroptera, and the larvæ of insects of various kinds.

The entire plumage black, each feather of the upper and under surface broadly margined with deep glossy green; irides dark reddish brown; bill very light ash-grey, passing into leaden grey at the base, and dark bluish grey on the culmen near the tip; legs and feet greenish grey.

The bill appears to vary very much in colour; being in some instances entirely ash-grey, except at the tip, where it is black; while in others the basal two-thirds is black and the tip grey: whether this difference is occasioned by age or sex has not yet been ascertained.

Sp. 99. CRACTICUS TORQUATUS.

COLLARED CROW-SHRIKE.

Lanius torquatus, Lath. Ind. Orn., Supp., p. xviii.
Vanga destructor, Temm. Man., part i. p. lix.
Barita destructor, Temm. Pl. Col. 273.
Bulestes torquatus, Cab. Mus. Hein. Theil i. p. 66.
Wăd-do-wăd-ong, Aborigines of the lowland districts of Western Australia.
Butcher-Bird of the Colonists of Swan River.

Cracticus destructor, Gould, Birds of Australia, fol., vol. II. pl. 59.

This bird is a permanent resident in New South Wales and South Australia, where it inhabits the margins of the brushy lands near the coast, the sides of hills, and the belts of trees which occur in the more open parts of the country; in fact I scarcely know of any Australian bird so generally dispersed. Its presence is at all times betrayed by its extraordinary note, a jumble of discordant sounds impossible to be described. It is nearly always on the trees, where it sits motionless on some dead or exposed branch whence it can survey all around, and particularly the surface of the ground beneath, to which it makes perpendicular descents to secure any large insect or lizard that may attract its sharp and penetrating eye; it usually returns to the same branch to devour what it has captured, but at times will resort to other trees and impale its victim after the manner of the true Shrikes: mice, small birds, and large *Phasmiæ* come within the list of its ordinary diet. September and the three following months constitute the period of incubation. The nest, which is large and cup-shaped, is neatly formed of sticks, and in some instances beautifully lined with the shoots of the *Casuarina* and fibrous roots. Considerable difference is found to exist in the colour of the eggs, the ground colouring of some being dark yellowish brown, with obscure blotches and marks of a darker hue, and here and there a few black marks not unlike small blots of ink; while in others the ground colour is much lighter and the darker markings are more inclined to red, and to form a zone round the larger end; the eggs are generally three in number, one inch and three lines long by eleven lines broad.

Under ordinary circumstances this species is very shy and retiring, but at times is altogether as bold; as an evidence of which I may mention, that having caught a young *Eöpsaltria* and placed it in my pocket, the cries of the little captive attracted the attention of one of these birds, and it continued to follow me through the woods for more than an hour, when the little tenant, disliking its close quarters, effected its escape and flitted away before me: I immediately gave chase; but the Crow-Shrike, which had followed me, pounced down within two yards of my face and bore off the poor bird to a neighbouring tree, and although I ran to the rescue, it was of no avail, the prize being borne away from tree to tree until the tyrant paid the forfeit of his life by being shot for his temerity.

The male has the crown of the head, ear-coverts, and back of the neck black; a white mark from the base of the bill to the eye; back and rump dark greyish brown; upper tail-coverts white; wings blackish brown; the middle secondaries white along their outer edges; tail black, all the feathers except the two middle ones tipped with white on their inner webs; under surface greyish white; bill bluish lead-colour at the base, passing into black at the tip; feet blackish lead-colour; irides very dark reddish brown.

The female resembles the male, but is more obscure in all her markings; and the young differ in being clothed in a plumage of mottled tawny and brown.

Sp. 100.   CRACTICUS CINEREUS, *Gould.*
      CINEREOUS CROW-SHRIKE.

Vanga cinerea, Gould in Proc. of Zool. Soc., part iv. p. 143.—Syn. Birds of Australia, part i. fig. of head.
Bulastes cinereus, Cab. Mus. Hein. Theil i. p. 66, note.

Cracticus cinereus, Gould, Birds of Australia, vol. i. Introd. p. xxxv.

Inhabits Tasmania, and may be distinguished from *C. torquatus* by its much longer bill, and, when fully adult, by its grey back.

By some ornithologists this bird may be considered only a local variety of *C. torquatus,* but I did not fail to notice that the two birds appeared very different in their respective countries, and ornithologists will observe on examination that a marked difference occurs in individuals from Tasmania and New South Wales. I will not, however, affirm that this bird is confined to Tasmania, for I have lately received evidence of its also occurring on the shores of the opposite part of the continent.

The male has the crown of the head, ears, and back of the neck black; back, shoulders, and rump delicate grey; upper tail-coverts white; tail black, largely tipped with white on the inner webs, except the two middle feathers, which are wholly black; space between the bill and the eye, middle of the secondaries, greater wing-coverts, throat, and all the under surface white; primaries black; bill bluish lead-colour at the base, passing into black at the tip; legs black.

The female differs in being browner, and less distinct in all her markings.

Total length 12½ inches; bill 1⅜; wing 6; tail 5⅞; tarsi 1¼.

Sp. 101. CRACTICUS LEUCOPTERUS, *Gould.*

WHITE-WINGED CROW-SHRIKE.

Balaites leucopterus, Cab. Mus. Hein. Theil i. p. 67.

Cracticus leucopterus, Gould, Birds of Australia, fol., vol. i. Introd. p. xxxv.

This species, which inhabits Western Australia, is very closely allied to *C. torquatus* and *C. cinereus*; but differs from the former in the white mark on the wings being much more extensive, and from the latter in its smaller size.

Family —— ?

Genus GRALLINA, *Vieillot.*

The only known species of this form is one of the anomalies of the Australian avifauna; for its alliance to any group with which we are acquainted is very limited. Its colouring and general contour remind us of the *Motacillæ*; but its habits and mode of nidification clearly indicate that it must not be associated with those birds. Uncertain where to place it, I

shall assign it the same position in the present as in the folio
work ; not that it has any special affinity to the birds which
immediately precede or follow it. I find it impossible to
arrange the birds of a single country in a linear series without
numerous hiati.

Sp. 102.   GRALLINA PICATA.

PIED GRALLINA.

Gracula picata, Lath. Ind. Orn. Supp., p. 29.
Tanypus Australis, Oppel.
Grallina melanoleuca, Vieill. Anal. d'une Nouv. Orn., pp. 42 and 68.
Cracticus cyanoleucus, Vieill. 2^e Edit. du Nouv. Dict. d'Hist. Nat., tom. v. p. 356.
Grallina Australis, G. R. Gray, List of Gen. of Birds, 2nd Edit., p. 33.
—— *picata*, Strickl. in Mag. Nat. Hist., vol. xi. p. 335.
—— *cyanoleuca*, Gray and Mitch. Gen. of Birds, vol. i. p. 204.
Corvus cyanoleucos, Lath. Gen. Hist., vol. iii. p. 49.
Magpie Lark, Colonists of New South Wales.
Little Magpie, Colonists of Swan River.
Bÿ-yoo-gool-yee-de and *Dil-a-but*, Aborigines of Western Australia.

Grallina australis, Gould, Birds of Australia, fol., vol. ii. pl. 54.

Future research will, in all probability, ascertain that
this bird is universally dispersed over the greater portion
of Australia; I have specimens in my collection from New
South Wales, Swan River, and Port Essington, all of which
are so closely alike that no character of sufficient importance
to establish a second species can be detected. Those that
came under my observation in New South Wales frequented
alluvial flats, sides of creeks and rivulets.

Few of the Australian birds are more attractive or more
elegant and graceful in its actions, and these, combined
with its tame and familiar disposition, must ever obtain
for it the friendship and protection of the settlers, whose
verandahs and house-tops it constantly visits, running along

the latter like the Pied Wagtail of our own island. Gilbert states that in Western Australia he observed it congregated in large families on the banks and muddy flats of the lakes around Perth, while in the interior he only met with it in pairs, or at most in small groups of not more than four or five together; he further observes, that at Port Essington, on the north coast, it would seem to be only an occasional visitant, for on his arrival there in July it was tolerably abundant round the lakes and swamps, but from the setting in of the rainy season in November to his leaving that part of the country in the following March not an individual was to be seen; it is evident therefore that the bird removes from one locality to another according to the season and the more or less abundance of its peculiar food. I believe it feeds solely upon insects and their larvæ, particularly grasshoppers and coleoptera.

The flight of the Pied Grallina is very peculiar—unlike that of any other Australian bird that came under my notice, and is performed in a straight line with a heavy flapping motion of the wings.

Its natural note is a peculiarly shrill whining whistle often repeated. It breeds in October and November.

The nest is from five to six inches in breadth and three in depth, and is formed of soft mud, which, soon becoming hard and solid upon exposure to the atmosphere, has precisely the appearance of a massive clay-coloured earthenware vessel; and as if to attract notice, this singular structure is generally placed on some bare horizontal branch, often on the one most exposed to view, sometimes overhanging water, and at others in the open forest. The colour of the nest varies with that of the material of which it is formed; sometimes the clay or mud is sufficiently tenacious to be used without any other material; in those situations where no mud or clay is to be obtained, it is constructed of black or brown mould; but the bird, appearing to be aware that this substance will not hold together for want of the adhesive quality of the clay, mixes with it a great quantity of dried grass, stalks, &c., and thus forms a firm and hard exterior, the inside of which is slightly lined with dried grasses and a few feathers. The eggs differ considerably in colour and in shape, some being extremely lengthened, while others bear a relative proportion; the ground-colour of some is a beautiful pearl white, of others a very pale buff; their markings also differ considerably in form and disposition, being in some instances wholly confined to the larger end, in others distributed over the whole of the surface, but always inclined to form a zone at the larger end; in some these markings are of a deep chestnut-red, in others light red, intermingled with large clouded spots of grey appearing as if beneath the surface of the shell. The eggs are generally four, but sometimes only two in number; their average length is one inch and three lines, and their breadth nine lines.

The sexes are very similar in size, but the female may at all times be distinguished from the male by her white forehead and throat, a fact I determined many times by actual dissection, thus showing the fallacy of the opinion entertained by some naturalists of there being two distinct species.

The male has a line over the eye, a patch on each side of the neck, a longitudinal stripe on the wing, tips of the secondaries, rump, upper tail-coverts, the basal two-thirds and the tips of the tail, under surface of the shoulder, breast, flanks, abdomen, and under tail-coverts white; the remainder of the plumage black, with a deep bluish tinge on the head, throat, chest, and back, and a green tinge on the primaries and tail; bill yellowish white; irides straw-yellow; feet black.

The female differs in having the forehead, lores, and chin white. The young on leaving the nest have the irides black; in other respects they resemble their parents, but are of course far less brilliant in colour.

In a note on the name of this species by the late Mr. Strickland, that gentleman says, "As this bird was very accurately described by Latham in his second 'Supplement' under the name of *Gracula picata*; and as the name *picata* is more correctly descriptive than *cyanoleuca*, which he had previously applied to it, I should prefer making the *permanent* designation of the bird *Grallina picata*, rather than *G. cyanoleuca*"; coinciding with Mr. Strickland's views, I have adopted his suggestion.

Family CAMPEPHAGINÆ.

The birds which I intend to keep under the above family name are very numerous in Australia, in the Indian Islands, and in the Peninsulas of India and Malacca. The Australian members appear to be naturally divided into two or three well-marked forms—*Graucalus*, *Pteropodocys*, and *Campephaga*. These three forms, however, constitute but a small portion of this extensive family, in which, perhaps, the beautifully coloured *Pericrocoti* should be comprised. All the species are individually very numerous, and, being truly insectivorous, must perform a most important part in the economy of nature.

Most of the members of this group build a flat slight nest of fine short dead twigs, curiously joined together with cobwebs, on which they lay two eggs.

Genus GRAUCALUS, *Cuvier*.

The infinite changes of plumage which some of the Australian members of this genus undergo from youth to maturity render their investigation very perplexing. I have done my best to define them correctly; if I have committed some errors, let us hope that a son of the great southern land may be imbued with a sufficient love for natural science to pay attention to the subject, and place it in a truer light.

All the members of the present genus are of large size compared with the other forms of the family.

Sp. 103.   GRAUCALUS MELANOPS.

BLACK-FACED GRAUCALUS.

Corvus melanops, Lath. Ind. Orn. Supp., p. xxiv. no. 1.
Ceblepyris melanops, Temm. Man., p. lxii.
Rollier à masque noir, Le Vaill, Ois. de Parad., pl. 30.
Black-faced Crow, Lath. Gen. Syn. Supp., vol. ii. p. 110.
Graucalus melanops, Vig. and Horsf. in Linn. Trans., vol. xv. p. 210.
——— *melanotis*, Gould in Proc. of Zool. Soc., part v. p. 143, and in Syn. Birds of Australia, part iv. Young.
Campephaga melanops, G. R. Gray, Cat. Mamm. and Birds of New Guinea in Brit. Mus., p. 32.
Kai-a-lora, Aborigines of New South Wales.
Nu-lar̈-go, Aborigines of the lowland districts of Western Australia.
Blue Pigeon of the Colonists.

Graucalus melanops, Gould, Birds of Australia, fol., vol. ii. pl. 55.

New South Wales, Tasmania, Swan River, and Port Essington are each inhabited by *Graucali* so nearly allied, that by many persons it would be considered questionable whether they were not referable to one and the same species; but as this is by no means certain, I shall confine my remarks to the bird inhabiting New South Wales, which is one of the largest of the genus yet discovered, which is distinguished from its near allies by the greater depth of the blue-grey colouring of the upper surface, and to which the synonyms given above refer.

The *Graucalus melanops*, then, is a very common bird in New South Wales, but is far less numerous in winter than in summer, when it is so generally dispersed over the colony, that to particularize situations in which it may be found is quite unnecessary; hills of moderate elevation, flats, and plains thinly covered with large trees being alike resorted to;

but I do not recollect meeting with it in the midst of the thick brushes,—situations which probably are uncongenial to its habits and mode of life. It is very abundantly dispersed over the plains of the interior, such as the Liverpool and those which stretch away to the northward and eastward of New South Wales.

Its flight is undulating and powerful, but is seldom exerted for any other purpose than that of conveying it from one part of the forest to another, or to sally forth in pursuit of an insect which may pass within range of its vision while perched upon some dead branch of a high tree, a habit common to this bird and the other members of the genus. On such an elevated perch it sometimes remains for hours together; but during the heat of the day seeks shelter from the rays of the sun by shrouding itself amidst the dense foliage of the trees. Its food consists of insects and their larvæ, and berries, but the former appear to be preferred, all kinds being acceptable, from the large Mantes to others of a minute size.

When the young, which are generally two in number, leave the nest, the feathers of the body are brown, margined with light grey; this colouring is soon exchanged for one of a uniform grey, except on the lower part of the abdomen and under tail-coverts, which are white, and a mark of black which surrounds the eye and spreads over the ears: the throat and forehead in this stage are lighter than the remainder of the plumage, which is somewhat singular, as in the next change that takes place those parts become of a jet-black; and this colour, I believe, is never afterwards thrown off, but remains a characteristic of the adult state of both sexes, which are at all times so similar in size and colour as not to be distinguished from each other.

It breeds in October and the three following months. The nest is often of a triangular form, in consequence of its being made to fit the angle of the fork of the horizontal branch in which it is placed; it is entirely composed of small dead twigs, firmly matted together with a very fine, white, downy substance like cobwebs and a species of *Lichen*, giving the nest the same appearance as the branch upon which it is placed, and rendering it most difficult of detection. In some instances I have found the nest ornamented with the broad, white, mouse-eared Lichen; it is extremely shallow in form, its depth and breadth depending entirely upon that of the fork in which it is built; the largest I have seen did not exceed six inches in diameter.

The ground-colour of the eggs, which are usually two in number, varies from wood-brown to asparagus-green, the blotches and spots, which are very generally dispersed over their surface, varying from dull chestnut-brown to light yellowish brown; in some instances they are also sparingly dotted with deep umber-brown; their medium length is thirteen lines, and breadth ten lines.

Its note, which is seldom uttered, is a peculiar single purring or jarring sound, repeated several times in succession.

The adults have the forehead, sides of the face, ear-coverts, and throat jet-black; crown of the head, all the upper surface, and wing-coverts delicate grey; primaries black, their outer edges and tips margined with grey; secondaries grey, with their inner webs black; tail grey at the base, gradually passing into black near the extremity, and broadly tipped with white; chest blackish grey, into which the black of the throat gradually passes; lower part of the abdomen pale grey; under tail-coverts white; irides, bill, and feet black.

Sp. 104. GRAUCALUS PARVIROSTRIS, *Gould.*

Graucalus parvirostris, Gould in Proc. of Zool. Soc., part v. p. 143.

Graucalus parvirostris, Gould, Birds of Australia, vol. i. Introd. p. xxxv.

In my description of *Graucalus melanops*, I have stated that New South Wales, Tasmania, Swan River, and Port Essington are each inhabited by *Graucali* so nearly allied to each other that it was questionable whether they were not one and the same species, and that the slight differences they present were attributable to some peculiarity in the districts they inhabit; after much attention to the subject, I have been induced to regard the Tasmanian bird as distinct, and I have therefore assigned it a name, *parvirostris*.

Forehead, sides of the face, and the throat jet-black; crown of the head, all the upper surface and centre of the wings delicate grey; primaries and the inner webs of the secondaries deep brownish black, the former narrowly and the latter broadly margined with greyish white; tail grey at the base, passing into deep brownish black, and largely tipped with white, the grey colouring predominating on the two centre feathers, which are destitute of the white tips; chest grey, into which the black of the throat gradually passes; lower part of the abdomen, under surface of the wing and under tail-coverts white; flanks and thighs grey; bill and feet brownish black.

Total length 12 inches; bill 1$\frac{1}{8}$; wing 7$\frac{1}{2}$; tail 6; tarsi 1.

Sp. 106. GRAUCALUS MENTALIS, *Vig. and Horsf.*

VARIED GRAUCALUS.

Graucalus mentalis, Vig. & Horsf. in Linn. Trans., vol. xv. p. 217.
Lanius robustus, Lath. Ind. Orn. Supp., p. xviii ?
Robust Shrike, Lath. Gen. Syn. Supp., vol. ii. p. 74 ?

Graucalus mentalis, Gould, Birds of Australia, fol. vol. ii. pl. 56.

New South Wales, or the south-eastern division of Australia, is the native habitat of the present species; it is by no means a rare bird in the Upper Hunter and all similar districts, yet I did not succeed in finding its nest and eggs; they are therefore desiderata with me.

There is no one member of the family to which it belongs which undergoes so many changes of plumage as the present species, and it is consequently very puzzling to the ornithologist. In extreme youth, or during the first few months after it has left the nest, the throat, chest, and back of the neck are jet-black, while the breast and abdomen are rayed with obscure arrow-shaped markings of the same colour on a greyish white ground; from this state individuals in every variety of change, to the uniform grey throat and head, with black lores and mark under the eye, are to be met with. Independently of a difference in its markings, its much smaller size will at all times serve to distinguish it from *Graucalus melanops*, which inhabits the same districts. Insects of various orders and caterpillars, which are either captured on the wing or taken from the branches, form its diet.

In the adult the upper surface and wings are dark slate-grey, passing into paler grey on the forehead and on the rump and upper tail-coverts; primaries and secondaries slaty black, narrowly edged with greyish white; outer webs of the three secondaries nearest the body grey; tail black, the lateral feathers largely tipped with white; lores deep velvety black, which colour is continued above and below the eye; throat and breast grey; insertion of the wing, under surface of the wing, abdomen, and under tail-coverts white; bill black; irides and feet dark brown.

Sp. 106. GRAUCALUS HYPOLEUCUS, *Gould*.

WHITE-BELLIED GRAUCALUS.

Graucalus hypoleucus, Gould in Proc. of Zool. Soc. 1848, p. 38.
Campephaga hypoleuca, G. R. Gray, Cat. of Mam. and Birds of New Guinea in Brit. Mus., p. 82.

Graucalus hypoleucus, Gould, Birds of Australia, fol., vol. ii. pl. 57.

This species inhabits the neighbourhood of Port Essington, where it is a very familiar bird, constantly flitting about the branches overhanging the houses of the settlement. In its general habits, manners, and note it closely assimilates to the

Graucalus melanops. It is abundant in every part of the Cobourg Peninsula, and is generally seen in small families of from four to ten or twelve in number.

The whiteness of the under surface serves to distinguish this from all the other species of the genus yet discovered in Australia.

The stomach is muscular, and the food consists of insects of various genera, which are generally taken from the leafy branches of the highest trees.

The sexes assimilate very closely in colouring, and only differ in the females and young males having the lores of a dull brown instead of black.

Lores black; crown of the head and all the upper surface dark grey; wings and tail black; chin, under surface of the wings, abdomen, and under tail-coverts white; breast pale greyish white; irides brownish black; bill blackish brown; legs and feet black; insides of the feet and spaces between the scales of the tarsi mealy grey.

Sp. 107.   GRAUCALUS SWAINSONII, *Gould.*

SWAINSON'S GRAUCALUS.

Coblepyris lineatus, Swains. in Zool. Journ., vol. i. p. 466.
Graucalus Swainsonii, Gould in Syn. Birds of Australia, part iv.

Graucalus Swainsonii, Gould, Birds of Australia, fol., vol. ii. pl. 58.

This species of *Graucalus,* which is distinguished from all the other Australian members of the genus by the beautiful barring of the breast, was originally described by Swainson under the specific appellation of *lineatus*; but that term having been previously applied to another species of the group, it became necessary to change it; and in substituting that of *Swainsonii,* I was desirous of paying a just tribute to the talents of a gentleman who has laboured most zealously in the cause of natural science, and whose researches and writings are so well known to all ornithologists.

Examples of this species occur in almost every collection sent from Moreton Bay; I regret to add that it is one of the few birds I had no opportunities of observing in a state of nature, and that little is at present known of its habits and economy, beyond what is stated in the following note which was sent to me by the late F. Strange:—"During the summer months this species feeds exclusively on the wild figs, in company with members of the following genera, *Ptilonorhynchus*, both species, *Sericulus*, *Scythrops*, and *Carpophaga*, with all of which it seems to be quite familiar, but does not appear to mix with the other species of its own genus, all of which are strictly insectivorous. A female shot on the 24th of November contained a fully developed egg." Judging from the specimens I have examined, I believe that the sexes are alike in plumage.

Lores black; head, all the upper surface, wing-coverts, throat, and breast grey; primaries and secondaries black, the former narrowly, and the latter broadly, margined on their external edges with grey; tail grey at the base, black for the remainder of its length; abdomen, under surface of the shoulder, and under tail-coverts white, crossed by numerous decided narrow bars of black; irides straw-colour; bill and feet black.

Genus PTEROPODOCYS, *Gould*.

To say that this is the terrestrial form of the Australian *Campephaginæ* will, I think, be consistent with truth; for while all the others affect the branches, and either sally thence to capture their insect food or search for them and their larvæ among the leafy tops of trees, the only known member of the present genus looks for them on the ground. Its lengthened tarsi would indicate that this was its habit, and in accordance with this inference it is most frequently found thereon. The increased length of the tarsi and tail, and the narrow form of the bill, are the most striking of the structural differences between *Pteropodocys* and *Graucalus*, and are so apparent as to be perceptible at a single glance. Only a single species of this form has yet been discovered.

Sp. 108. PTEROPODOCYS PHASIANELLA, *Gould*.
GROUND GRAUCALUS.

Graucalus Phasianellus, Gould in Proc. of Zool. Soc., part vii. p. 142.
Ceblepyris maxima, Rüpp. Mon. in Mus. Senckenbergianum, 1839, p. 28. taf. iii.
Goo-rá-ling, Aborigines of York, Western Australia.

Pteropodocys phasianella, Gould, Birds of Australia, fol., vol. ii. pl. 59.

The rarity of this species in our collections is sufficient evidence that it is a bird inhabiting the interior of the country, and that its native localities have been seldom visited by the explorer; hence it was a source of no ordinary gratification to me when I first encountered it on the plains bordering the River Nomoi in New South Wales, and perceived that no very lengthened study of its habits and mode of life was requisite to ascertain that its structure is as beautifully adapted for terrestrial progression and for a residence on the ground, as the structure of the *Graucali* fits them to inhabit the branches of the trees; more beautiful modifications of form, in fact, can scarcely be seen than occur among the members of this group, which now comprehends a considerable number of species; the present bird, however, is the only terrestrial one that has yet come under my notice, either from Australia or the great nursery of these birds—India and the Indian islands. Plains and open glades skirted by belts of high trees are the localities in which I generally met with this bird, either in pairs or small parties of from three to six or eight in number. Its actions are very animated; at the same time it is cautious and shy.

Its powers of progression on the ground are considerable;

when disturbed it flies across the plain to the belts of lofty trees, when the white mark on the rump shows very conspicuously, and may be seen at a considerable distance.

Its range extends over the whole of the interior of Southern Australia from east to west; how far it proceeds northwards has not yet been ascertained.

Of its nidification I regret to say nothing is at present known.

The sexes, which exhibit no external differences, may be thus described:—

Head, neck, chest, and back delicate grey, becoming darker on the ear-coverts; rump and abdomen white, crossed by narrow irregular bars of black; under tail-coverts white; wings and tail black, the latter having the tips of the outer and the basal portion of all the feathers white; bill and feet black, tinged with olive; irides buffy white.

Genus CAMPEPHAGA, *Vieillot*.

Several species of this form are found in the Indian Islands and Africa; and three or four in Australia; some of these have been separated and placed in the genus *Lalage*, but I do not perceive the necessity of such a measure.

The *Campephagæ* are allied to the *Graucali*; but are much smaller in size, and more active among the branches.

The sexes are generally very dissimilar in colour and markings, while in *Graucalus* they are alike. The nidification and the form of the nests of the members of the two genera are very similar.

Sp. 109. CAMPEPHAGA JARDINII, *Rüppell*.

JARDINE'S CAMPEPHAGA.

Graucalus tenuirostris, Jard. and Selb. Ill. Orn., vol. ii. pl. 114.
Ceblepyris Jardinii, Rüpp. Mon. in Orn. Misc. 1839, p. 80.

Campephaga Jardinii, Gould, Birds of Australia, fol., vol. ii. pl. 60.

The only parts of Australia wherein this species has been

observed are Moreton Bay and the Liverpool Range in New South Wales, and the Cobourg Peninsula: it is likely that it ranges over the whole of the intermediate country, but this can only be determined by future research. Its smaller size, the more attenuated form of its bill, and the great difference in the colouring of the sexes, point out most clearly that it is a member of the genus *Campephaga*, and not of *Graucalus*, to which it was first assigned. It is far less common in New South Wales than it is at Port Essington, where Gilbert collected the following particulars respecting it:—

"This bird is extremely shy and retiring in its habits. It generally inhabits the topmost branches of the loftiest and most thickly-foliaged trees growing in the immediate vicinity of swamps. Its note is altogether different from that of any other species of the genus, being a harsh, grating, buzzing tone, repeated rather rapidly about a dozen times in succession, followed by a lengthened interval. It appears to be a solitary species, as I never saw more than one at a time."

The stomach is muscular, and the food consists of insects of many kinds, but principally coleoptera.

The adult male has the lores black; all the upper and under surface, wing-coverts, edges of the primaries and secondaries, basal three-fourths of the two central and the tips of the outer tail-feathers deep blue-grey; primaries, secondaries, and the other parts of the tail black; irides dark brown; bill blackish brown; legs and feet very dark greenish grey.

The female has the whole of the upper surface, wings, and tail brown, the two latter edged with buff; line over the eye and all the under surface buff, the feathers of the side of the neck, the breast, and the flanks with an arrow-head-shaped mark of brown in the centre.

The young male is bluish brown above; wings and tail as in the female; under surface buff, crossed with numerous transverse narrow irregular bars of black.

Sp. 110.   CAMPEPHAGA KARU.
NORTHERN CAMPEPHAGA.

Lanius Karu, Less. Zool. de la Coq., pl. 12.
Notodela Karu, Less. Traité d'Orn., p. 374.
Lalage Karu, Cab. Mus. Hein., Theil i. p. 60.
Campephaga (Lalage) Karu, G. R. Gray, Cat. of Birds of Trop. Islands of Pacific Ocean in Coll. Brit. Mus., p. 23.

Campephaga Karu, Gould, Birds of Australia, fol., vol. ii. pl. 61.

Gilbert, who met with this species at Port Essington, states that it is a very shy and timid bird, that it is generally seen creeping about in pairs among the thickets and clumps of mangroves, that its note is a somewhat shrill piping call, that its stomach is tolerably muscular, and that it feeds upon insects of various kinds: this, I regret to say, is all that is known respecting it.

In referring this species to the *Lanius Karu* of Lesson, I am rather influenced by a desire not to add to the number of useless synonyms, than from any positive conviction of their being identical; for although, with only Lesson's figure to refer to, I am unable to detect any difference of sufficient importance to be considered specific, it is possible that the two birds are really distinct.

The male has the head, all the upper surface, wings, and tail black; the wing-coverts largely tipped, primaries narrowly edged and tipped, secondaries broadly margined on their external webs, rump and upper tail-coverts slightly, the external tail-feather largely, and the next on each side slightly tipped with white; line from the nostrils over each eye to the occiput buffy white; under surface pale grey, crossed on the breast and flanks with narrow irregular bars of slaty black, and washed with fulvous, gradually increasing in intensity until on the vent and under tail-coverts it becomes of a deep tawny buff; irides dark brown; bill black; feet blackish grey caternally, bluish grey internally; light mealy ashy grey between the scales and inside the feet.

The female differs in being somewhat smaller than the male; in having the upper surface and tail brown, instead of black; the upper tail-coverts tipped with buff instead of white, and the barrings of the under surface broader, darker, and more distinct.

Sp. 111. CAMPEPHAGA LEUCOMELA, *Vig. and Horsf.*
BLACK AND WHITE CAMPEPHAGA.

Campephaga leucomela, Vig. and Horsf. in Linn. Trans., vol. xv. p. 215.
Lalage leucomela, Cab. Mus. Hein., Theil i. p. 60, note.

Campephaga leucomela, Gould, Birds of Australia, fol., vol. ii. pl. 62.

This species, which frequents the brushes of the eastern parts of New South Wales between the river Hunter and Moreton Bay, differs from the *Campephaga Karu* in its much greater size, in the rufous colouring of the lower part of the abdomen and under tail-coverts, in the more uniform grey colouring of the breast, and in the barring of this part being much less conspicuous. I have had examples of this species in my collection for many years, but was not fortunate enough to see it alive during my visit to Australia. Strange also sent me a pair which he shot in the scrubs on the banks of the Clarence. Its nest and eggs, and any information of its habits, are desiderata to me.

The sexes, as is the case with the other species, differ very considerably from each other in their colouring; they may be thus described:—

The male has the head, back, wings, and tail deep glossy black; wing-coverts largely tipped and the secondaries broadly margined with white; the two outer tail-feathers tipped with white, the external one also narrowly margined on the outer web with the same hue; rump and upper tail-coverts very dark grey; line over the eye snow-white; under surface greyish white, gradually passing into rufous on the abdomen and under tail-coverts, and indistinctly rayed with dark grey; bill, feet, and irides black.

The young male is brown where the male is black; has the wings not so conspicuously marked with white; the under surface washed with rufous and conspicuously rayed with brown; and the under tail-coverts deep rufous.

Sp. 112. CAMPEPHAGA HUMERALIS, *Gould.*

WHITE-SHOULDERED CAMPEPHAGA.

Ceblepyris humeralis, Gould in Proc. of Zool. Soc., part v. p. 143.
Lalage humeralis, Cab. Mus. Hein., Theil i. p. 60.
Goo-mal-cut-long, Aborigines of the mountain districts of Western Australia.

Campephaga humeralis, Gould, Birds of Australia, fol., vol. ii. pl. 62.

This bird occurs in considerable numbers throughout the southern portion of Australia during the months of summer; it is strictly migratory, arriving in the month of September, and having performed the task of reproduction departs again northwards in the months of January and February. It is a most animated, lively, and spirited bird, constantly singing a loud and pretty song while actively engaged in pursuit of insects, which it captures on the wing, among the branches, or on the ground. It commences breeding soon after its arrival, constructing a shallow round nest of small pieces of bark, short dead twigs and grasses interwoven with fine vegetable fibres, cobwebs, white moss, &c., and sometimes a few grasses and fine fibrous roots by way of lining; it is usually placed in the fork of a horizontal dead branch of the *Angophoræ* and *Eucalypti*, and is not easily seen from below. During the early part of the breeding-season the male frequently chases the female from tree to tree, pouring forth his song all the while. The eggs, which are generally two, but sometimes three in number, differ very considerably in colour, some being of a light green blotched all over with wood-brown, while others have a lighter ground so largely blotched with chestnut-brown as nearly to cover the entire surface of the shell, and I have seen some of an almost uniform greyish green; their medium length is nine and a half lines, and breadth seven and a half lines.

In his Notes from Western Australia, Gilbert says, " This bird is a migratory summer visitant to this part of the country, where it arrives about the beginning of September, after which it is to be met with in considerable numbers among the mountains of the interior, but is very rarely seen in the lowland districts.

" Its powers of flight are considerable, and when excited during the breeding-season the males become very pugnacious, and not only attack each other in the most desperate manner, but also assault much larger birds that may approach the nest. Its usual flight is even, steady, and graceful, and while flying from tree to tree it gives utterance to its sweet and agreeable song, which at times is so like the full, swelling, shaking note of the Canary, that it might easily be mistaken for the song of that bird. It is a remarkably shy species, especially the females, which are so seldom seen that I was at first inclined to think they were much less numerous than the other sex, but this I afterwards found was not the case. Their favourite haunts are thickly-wooded places and the most secluded spots. The nest is so diminutive that it is very difficult to detect it, and so shallow in form that it is quite surprising the eggs do not roll out when the branch is shaken by the wind. The nests I discovered were placed on a horizontal dead branch of a Eucalyptus; they were formed of grasses and contained two eggs. It breeds in the latter part of September and the beginning of October."

Gilbert subsequently met with the bird at Port Essington,

where also it appears to be migratory, for not a single individual was to be seen from the early part of November to the month of March; females and young birds were very abundant on his arrival in July, but he only met with one old male during his residence in the colony, a period of eight months.

The stomach is muscular, and the food consists of insects of various kinds and their larvæ.

The sexes differ considerably in colour, as will be seen by the following description:—

The male has the forehead, crown of the head, back of the neck, and upper part of the back glossy greenish black; shoulders and upper wing-coverts pure white, forming an oblique line along the wing; the remainder of the wing dull black, with the secondaries slightly margined and tipped with white; lower part of the back and rump grey; tail dull black, the two outer feathers on each side largely tipped with white; throat, chest, and all the under surface white; bill and feet black; irides nearly black.

The female has all the upper surface, wings, and tail brown; wing-coverts and secondaries margined with buff; throat and all the under surface buffy white, with the sides and front of the breast speckled with brown; irides very dark brown; upper mandible and tip of the lower dark reddish brown; basal portion of the latter saffron-yellow; legs and feet dark greyish black, slightly tinged with lead-colour.

Genus PACHYCEPHALA, *Swainson*.

The *Pachycephala gutturalis* may be regarded as the type of this genus, the members of which are peculiar to Australia and the adjacent islands to the northward. Their habits differ from those of most other insectivorous birds, particularly in their quiet mode of hopping about and traversing the branches of the trees in search of insects and their larvæ: caterpillars constitute a great portion of their food; but coleoptera and other insects are not rejected. The more gaily-attired species, such as *P. gutturalis, P. glaucura, P. melanura,* and *P. pectoralis,* resort to the flowering *Acaciæ, Eucalypti,* and other stately trees, while the more dull-coloured frequent the ground: they all build a neat, round, cup-shaped nest, and the eggs are generally four in number. Their powers of flight are not great; some enjoy a wide range of habitat, while others are extremely local. The song of some is loud and rather pleasing, while others merely emit a whistling note, slowly but frequently repeated.

Sp. 113. PACHYCEPHALA GUTTURALIS.

WHITE-THROATED THICKHEAD.

Turdus gutturalis, Lath. Ind. Orn. Supp., p. xlii.
Muscicapa pectoralis, Lath. Ib., p. li.
Orange-breasted Thrush, Lewin, Birds of New Holl., pl. 7.
Black-crowned Thrush, Lewin, Ib., pl. 10.
Motacilla dubia, Shaw, Nat. Misc., vol. xxii. pl. 949.
Guttural Thrush, Lath. Gen. Syn. Supp., vol. ii. p. 162.
Black-breasted Flycatcher, Lath. Ib., vol. ii. p. 222.
Pachycephala gutturalis, Vig. and Horsf. in Linn. Trans., vol. xv. p. 289.
Turdus lunularis, Steph. Cont. of Shaw's Gen. Zool., vol. xiii. part ii. p. 200.
La Cravate blanche, Le Vaill. Ois. d'Afriq., tom. iii. pl. 115.
Pachycephala fusca, Vig. and Horsf. in Linn. Trans., vol. xv. p. 240.
—— *fuliginosa,* Vig. and Horsf. in Linn. Trans., vol. xv. p. 241, female or young.
Pe-dil-me-dong, Aborigines of Western Australia.
Thunder Bird, Colonists of New South Wales.

Pachycephala gutturalis, Gould, Birds of Australia, fol., vol. ii. pl. 64.

It would seem that the whole extent of the southern coast of Australia is inhabited by the present species, for on comparing adult males from New South Wales, South Australia, and Swan River, I find that they do not present any material differences; the apical half of the tail is blackish brown in all, and the colouring of the under surface of the richest yellow. It is rather abundantly dispersed over the forests of *Eucalypti* and the belts of *Acaciæ*, among the flowering branches of which latter trees the male displays himself to the greatest advantage, and shows off his rich yellow breast as if desirous of outvieing the beautiful blossoms with which he is surrounded.

The stomach is very muscular, and the principal food consists of insects of various genera, which are sought for and captured both among the flowers and leaves as well as on the ground.

It is generally met with in pairs, and the males are more shy than the females. It flies in short and sudden starts, and seldom mounts far above the tops of the trees.

The voice of the male is a single note, seven or eight times repeated, and terminating with a sharp higher note much resembling the smack of a whip.

Gilbert mentions that it is sparingly dispersed throughout the Swan River colony, but is more abundant in the best-watered districts, such as Perth and Freemantle.

I did not succeed in finding the nest of this species, but was informed that it breeds in September and October, and lays three or four eggs, ten and a half lines long by eight lines broad, with a ground-colour of brownish-buff, sparingly streaked and spotted with reddish brown and bluish grey, the latter colour appearing as if beneath the surface of the shell.

The male has the crown of the head, lores, line beneath the eye, ear-coverts, and a crescent-shaped mark from the latter across the breast deep black; throat, within the black, white; back of the neck, a narrow line down each side of the chest behind the black crescent, and all the under surface gamboge yellow; back and upper tail-coverts yellowish olive;

wing-coverts blackish brown, margined with yellowish olive; primaries and secondaries blackish brown, margined with greyish olive; basal half of the tail grey, apical half blackish brown, tipped with grey; irides dark brown; bill black; legs and feet blackish grey.

The female has the whole of the upper surface and tail greyish brown; primaries and secondaries brown, margined with grey; throat pale brown, freckled with white; remainder of the under surface pale brown, passing into deep buff on the abdomen.

Sp. 114. PACHYCEPHALA GLAUCURA, *Gould*.

GREY-TAILED THICKHEAD.

Pachycephala glaucura, Gould, in Proc. of Zool. Soc., part xiii. p. 19.

Pe-dil-me-dang, Aborigines of the lowland districts of Western Australia.

Pachycephala glaucura, Gould, Birds of Australia, fol., vol. ii. pl. 65.

Although the present bird is very nearly allied to the *P. gutturalis*, it may be readily distinguished from that species by its larger size, by its shorter and more robust bill, by the uniform grey colouring of its tail, and by the lighter and more washy tint of the yellow of the under surface. Tasmania and the islands in Bass's Straits are the only countries in which it has yet been discovered, and where it takes the place of the *P. gutturalis*, which latter species appears to be exclusively confined to the Australian continent.

The *P. glaucura* frequents the vast forests of *Eucalypti* that cover the greater part of Tasmania, and although it is rather thinly dispersed, is to be met with in every variety of situation, the crowns of the hills and the deep and most secluded gulleys being alike visited by it. It frequently descends to the ground in search of insects, but the leafy branches of the trees, particularly those of a low growth, are the situations to which it gives the preference.

The adult male, like most other birds of attractive plumage, is of a shy disposition; hence there is much more difficulty in obtaining a glimpse of that sex in the woods than of the sombre-coloured and comparatively tame female, or even of the young males of the year, which during this period wear a similar kind of livery to that of the latter.

The actions of this species are somewhat peculiar, and unlike those of most other insectivorous birds: it pries about the leafy branches of the trees, and leaps from twig to twig in the most agile manner possible, making all the while a most scrutinizing search for insects, particularly coleoptera. When the male exposes himself, as he occasionally does, on some bare twig, the rich yellow of his plumage, offering a strong contrast to the green of the surrounding foliage, renders him a conspicuous and doubtless highly attractive object to his sombre-coloured mate, who generally accompanies him. It sometimes resorts to the gardens and shrubberies of the settlers, but much less frequently than might be supposed, when we consider that the neighbouring forests are its natural place of abode.

The Grey-tailed Pachycephala utters a loud whistling call of a single note several times repeated, by which its presence is often detected. I was unsuccessful in my search for its nest, and the eggs are still desiderata to my collection. Soon after leaving the nest, the ground-colour of the entire plumage is grey, washed, both on the upper and under surface, with rusty or chestnut-red, which gradually gives place to a uniform olive-brown above and pale brown beneath.

The adult male has the crown of the head, lores, space beneath the eye, and a broad crescent-shaped mark from the latter across the breast deep black; throat, within the black, white; back of the neck, a narrow line down each side of the chest behind the black crescent, and the under surface yellow;

back and wing-coverts yellowish olive; wings dark slate-colour, margined with grey; tail entirely grey; under tail-coverts white, or very slightly washed with yellow; irides reddish brown; bill black; feet dark brown.

Total length 7 inches; bill ⅝; wing 4; tail 3¼; tarsi 1.

Sp. 115. PACHYCEPHALA MELANURA, *Gould.*
BLACK-TAILED THICKHEAD.

Pachycephala melanura, Gould in Proc. of Zool. Soc., part x. p. 134.

Pachycephala melanura, Gould, Birds of Australia, fol., vol. ii. pl. 66.

The *Pachycephala melanura* is a native of the northern coasts of Australia, where it was procured by B. Bynoe, Esq., during the surveying voyage of H.M.S. the Beagle. It may be readily distinguished from the *P. gutturalis* and *P. glaucura* by the jet-black colouring of the tail (which organ is also shorter and more square than that of any other species), by its much longer bill, and by the colouring of the back of the neck and the under surface being richer than that of either of those above named. I have not yet seen a female of this fine species. Whenever this sex is collected, it will be found to bear a very general resemblance to the females of *P. gutturalis* and *P. glaucura.*

Head, crescent commencing behind the eye and crossing the chest, and the tail black; throat pure white; collar round the back and sides of the neck, and all the under surface, very rich gamboge yellow; upper surface rich yellowish olive; wings black, the coverts margined with yellowish olive; the primaries narrowly, and the secondaries broadly margined with yellowish grey; bill and feet black; irides brown.

Total length 6 inches; bill ⅞; wing 3¼; tail 2½; tarsi ⅞.

Sp. 110. PACHYCEPHALA RUFIVENTRIS.

Rufous-breasted Thickhead.

Sylvia rufiventris, Lath. Ind. Orn. Supp., p. lix.
Rufous-vented Warbler, Lath. Gen. Syn. Supp., vol. ii. p. 248.
Orange-breasted Thrush, Lewin, Birds of New Holland, pl. 8.
Pachycephala pectoralis, Vig. and Horsf. in Linn. Trans., vol. xv. p. 239.
—— *striata*, Vig. and Horsf. in Linn. Trans., vol. xv. p. 240, female or young male.
—— *rufiventris*, G. R. Gray, Ann. & Mag. Nat. Hist., vol. xi. p. 198.
Lanius macularius, Quoy et Gaim., Voy. d'Astrolabe, p. 257, pl. 31. f. 1, young male.
Rufous-vented Honey-eater, Lath. Gen. Hist., vol. iv. p. 183.

Pachycephala pectoralis, Gould, Birds of Australia, fol., vol. ii. pl. 67.

This very common species ranges over the whole of the southern portion of the Australian continent, from Swan River on the west to Moreton Bay on the east; but the extent of its range northwards has not yet been determined. During the spring and the earlier months of summer there are few birds that give utterance to a more animated and lively song—a loud continuous ringing whistle, frequently terminating in a sharp smack, which latter note is peculiar to most members of the group. In New South Wales and South Australia it is abundantly dispersed over all the thinly-timbered forests, keeping among the leafy branches of the highest trees. I do not recollect having met with it in the cedar-brushes of New South Wales; in Western Australia the thick scrubs are said to be its favourite places of resort.

Although it does not migrate, it makes a slight change in the situations it frequents, according to the state of the seasons, or the more or less abundant supply of food, which consists of insects of various kinds, caterpillars, and berries: like the other members of the group, it creeps and hops about the branches in a gentle and quiet manner.

The breeding-season commences in August or September,

and continues during the three following months. The nest is cup-shaped, and is rather a frail structure, being often so slight that the eggs may be descried through the interstices of the fine twigs and fibrous roots of which it is composed. In New South Wales I found the nest upon the small horizontal branches of large trees, but at Swan River it is more frequently constructed in shrubs, particularly the *Melaleuca*: the eggs are generally three in number, of an olive tint, with a zone of indistinct spots and blotches at the larger end; they are eleven lines long by eight lines broad.

The sexes differ very considerably both in the arrangement of their markings and in the general colouring of their plumage, and it is not until the second year that the young males assume the band on the chest and the pure white throat of the adult.

Sp. 117. PACHYCEPHALA FALCATA, *Gould*.

LUNATED THICKHEAD.

Pachycephala falcata, Gould in Proc. of Zool. Soc., part x. p. 134.

Pachycephala falcata, Gould, Birds of Australia, fol., vol. ii. pl. 68.

We find in this species of *Pachycephala*, which inhabits the northern parts of Australia, a beautiful representative of the *P. pectoralis* of the southern parts of the continent; from which it differs in its much smaller size, and in the black crescent which bounds the white throat of the male not extending upwards to the ear-coverts, which with the lores are grey. All the specimens I possess were killed on the Cobourg Peninsula, near the settlement at Port Essington, where, as well as on the adjacent islands, it is a stationary species and very abundant. It breeds in September and the two following months, and lays two eggs. Its habits and manners are precisely similar to those of the other members of the family.

The adult male has the crown of the head, lores, ear-coverts, back, and upper tail-coverts grey; wings dark brown, all the feathers margined with grey; throat white, bounded below by a distinct crescent of black: abdomen, flanks, and under tail-coverts orange-brown; tail dark brown, the basal portion of the webs edged with grey; irides reddish brown; bill black; feet blackish brown.

The adult female has the crown of the head and all the upper surface grey; ear-coverts brownish grey; throat buffy white, passing into light buff or fawn-colour on the chest, flanks, abdomen, and under tail-coverts; the feathers of the throat and chest with a narrow dark line down the centre; wings and tail as in the male.

The young male is similar in colour to the female, but has the throat whiter, and the markings on the chest much more distinct and extending over the abdomen also.

In very young individuals a rich rufous or tawny tint pervades the greater part of the upper surface.

Sp. 118. PACHYCEPHALA LANOIDES, *Gould*.

SHRIKE-LIKE THICKHEAD.

Pachycephala lanoides, Gould in Proc. of Zool. Soc., part vii. p. 142.

Pachycephala lanoides, Gould, Birds of Australia, fol, vol. ii. pl. 69.

The single specimen of this species which has come under my notice was procured on the north-west coast of Australia, and is probably unique. It is a most robust and powerful bird, and may hereafter be made the type of a new genus; but until the female has been discovered, and more examples obtained, I retain it among the *Pachycephalæ*.

That it feeds on insects of a large size there can be little doubt, its whole structure indicating that it subsists upon this kind of food.

No information whatever has been obtained with respect to its habits and economy; this blank therefore remains to be filled up by those naturalists who may hereafter visit the part of the country of which it is a denizen.

Crown of the head, ear-coverts, and chest black, bounded posteriorly by a narrow band of chestnut; throat, centre of the abdomen, and under tail-coverts white; flanks, back, shoulders, and external webs of the primaries, secondaries, and wing-coverts grey; tail, bill, and feet black.

Sp. 110. PACHYCEPHALA RUFOGULARIS, *Gould*.

Red-throated Thickhead.

Pachycephala rufogularis, Gould in Proc. of Zool. Soc., part viii. p. 164.

Pachycephala rufogularis, Gould, Birds of Australia, fol, vol. ii. pl. 70.

All the examples of this species of *Pachycephala* I have yet seen, were obtained by myself during my explorations in South Australia, where I found it anything but abundant; in fact many days frequently elapsed without my procuring a specimen. Its stronghold, probably a part of the vast interior, has yet to be discovered. From the little I saw of it, I am induced to believe that it is a very solitary bird; for I usually met with only one at a time, hopping about on the ground in the thinly-timbered forest which surrounds the city of Adelaide; but its actions were so particularly quiet, and its plumage so unattractive, that it might easily be overlooked. I never heard it utter any note, nor did I observe anything in its habits and economy worthy of remark. It doubtless resorted to the ground for coleopterous and other insects, the remains of which formed the contents of the stomachs of those I procured.

The adult males and females differ considerably in the colouring of their plumage; the young males resemble the females. The rusty colouring of the throat and face distinguishes this species from every other member of the genus.

The male has the crown of the head and all the upper surface deep brownish grey; wings and tail dark brown, the feathers margined with greyish brown; lores, chin, throat, under surface of the shoulder and all the under surface reddish sandy brown, crossed on the breast by a broad irregular band of greyish brown; irides reddish brown; bill black; feet blackish brown.

The female differs from the male in having the throat and under surface greyish white, the chest being crossed by an obscure mark of greyish brown, and with a line down the centre of each feather.

Sp. 120. PACHYCEPHALA GILBERTI, *Gould*.

GILBERT'S THICKHEAD.

Pachycephala Gilbertii, Gould in Proc. of Zool. Soc., part xii. p. 107.
—— *inornata*, Gould, Ib., part viii. p. 164 (young).

Pachycephala Gilbertii, Gould, Birds of Australia, fol., vol. ii. pl. 71.

Although the practice of naming species after individuals is a means by which the names of men eminent for their scientific attainments may be perpetuated to after-ages, I have ever questioned its propriety, and have rarely resorted to it; but in assigning the name of *Gilberti* to this interesting bird, I feel that I only paid a just compliment to one who most assiduously assisted me in the laborious investigations required for the production of the 'Birds of Australia,' and who was the discoverer of the species. The specimens transmitted to me by Gilbert are, I believe, all that have yet been procured.

Although the *P. Gilberti* is nearly allied to the *P. rufogularis*, it may be readily distinguished by the rufous colouring being confined to the throat, and not ascending upon the forehead and occupying the space between the bill and the eyes as in that species; it is also a smaller bird in all its admeasurements.

The Red-throated Thickhead is an inhabitant of the interior

of Western Australia. The following notes, which are all that is known of its history, accompanied the specimens sent to me:—" This species inhabits the the thick brushes of the interior. It is an early breeder, as is proved by my finding a nest with three newly hatched young birds in the middle of August. The nest was built in the upright fork of a small shrub about four feet from the ground. It was deep, cup-shaped in form, and constructed of dried grasses, and, except that it was rather more compactly built, it was very similar to those of the other members of the genus."

The sexes of the present bird exhibit a similar difference in colour to those of *P. rufogularis*; the females of both species being very sombre and devoid of any rufous colouring on the throat and breast.

The male has the upper surface dark greyish olive-brown; head dark slate-grey; breast of a lighter grey; lores black; throat rust-red; under surface of the shoulder, centre of the abdomen, and under tail-coverts sandy buff; irides light brown; bill and feet black.

Sp. 121. PACHYCEPHALA SIMPLEX, *Gould*.

PLAIN-COLOURED THICKHEAD.

Pachycephala simplex, Gould in Proc. of Zool. Soc., part x. p. 135.

Pachycephala simplex, Gould, Birds of Australia, fol., vol. ii. pl. 72.

The *Pachycephala simplex* is a native of the north-western parts of Australia, but does not appear to be very numerous in any locality yet explored; Gilbert, who discovered it in the neighbourhood of Port Essington, states that it is of a very shy and retiring disposition, and that it is usually met with in pairs hopping and creeping about among the under-wood or very thickly-foliaged trees, but may be more frequently seen in thickets situated in the midst of swamps or among the mangroves. In its mode of feeding and in many of its actions it greatly resembles the Flycatchers, but does not, like them, shake the tail. Its voice is peculiarly soft and mournful, and its call consists of a single note four times repeated with rather lengthened intervals; at other times it utters a somewhat pleasing and lengthened song; "but," says Gilbert, "I never heard it emit that sharp terminating note, resembling the smack of a whip, which concludes the song of all the other species of the genus."

The stomach is muscular, and the food consists of insects and seeds of various kinds.

It appears to breed during the months of December, January, and February; for the ovarium of a female killed on the third of the last-mentioned month contained eggs very fully developed, and, from the bare state of the breast, it appeared to have been already engaged in the task of incubation.

All the upper surface brown; under surface brownish white, with a very faint stripe of brown down the centre of each feather; irides light brown; bill and feet black.

Sp. 122. PACHYCEPHALA OLIVACEA, *Vig. and Horsf.*

OLIVACEOUS THICKHEAD.

Pachycephala olivacea, Vig. and Horsf. in Linn. Trans., vol. xv. p. 241.
Native Thrush of the Tasmanians.

Pachycephala olivacea, Gould, Birds of Australia, fol, vol. ii. pl. 73.

This species, the largest of the genus yet discovered, is a native of Tasmania, where it inhabits forests and thick-scrubby situations, and is very generally dispersed over the island from north to south; I observed it also on Flinders Island in Bass's Straits, but no instance has come under my notice of its occurrence on the continent of Australia. It is rather recluse in its habits; and were it not for its oft-repeated, loud, sharp, liquid, whistling note, its presence would not always be detected. I usually met with it in the thickest parts of the forests, where it appeared to resort to the ground rather than to the branches, and to frequent gulleys and low swampy situations beneath the branches of the dwarf *Eucalypti* and other trees, with which its olive-brown colouring so closely assimilated, that it was very difficult to perceive it.

Although I felt assured that the bird was breeding in many parts of the country, and made repeated attempts to discover its nest, I could never succeed in so doing; the eggs are therefore among the desiderata of my cabinet.

But little outward difference is observable in the sexes; the male is rather the largest, and has the head of a sooty greyish brown, while the head of the female is olive-brown. The young resemble the female, and assume the adult colouring at an early age.

The stomachs of several specimens dissected were very muscular, and contained the remains of coleoptera and hemiptera mingled in some instances with small stones and seeds.

Crown of the head and ear-coverts dark brown; back, wings, and tail chestnut-olive; throat greyish white, each feather tipped with brown; chest, abdomen, and under tail-coverts reddish brown; bill black; irides reddish brown; feet mealy reddish brown.

Genus COLLURICINCLA, *Vigors and Horsfield.*

The members of the present genus are more strictly confined to Australia than those of the last mentioned. Each of the colonies, from north to south and from east to west, is inhabited by a species peculiarly and restrictedly its own. They have many characters which would appear to ally them to the *Pachycephalæ*, which they also somewhat resemble in their nidification. They are neither Shrikes nor Thrushes, but are most nearly allied to the former; and feed on insects to a very great extent, but occasionally partake of mollusks and berries. Some of them defend themselves vigorously with both bill and claws when attacked. Their voice is a loud whistle, some parts of which are not devoid of melody, particularly the loud swelling notes.

The nest is rather slightly built, cup-shaped in form, and is mostly placed in the hollow spout of a tree: the eggs are four in number.

It is somewhat singular that each of the great divisions of Australia should, as before mentioned, be tenanted by a different species of this genus, each possessing distinctive characters by which they may be readily recognized

Sp. 123. COLLURICINCLA HARMONICA.

HARMONIOUS SHRIKE-THRUSH.

Turdus harmonicus, Lath. Ind. Orn. Supp., p. xli.
Harmonic Thrush, Lath. Gen. Syn. Supp., vol. ii. p. 182.
Grey-headed Thrush, Lath. Gen. Hist., vol. v. p. 118.
Colluricincla cinerea, Vig. and Horsf. in Linn. Trans., vol. xv. p. 214.
Lanius saturninus, Nordm.
Turdus dilutus, Lath. Ind. Orn. Supp., p. xl?
Dilute Thrush, Lath. Gen. Syn. Supp., vol. ii. p. 182?
Turdus badius, Lath. Ind. Orn. Supp., p. xli?
Port Jackson Thrush, Lath. Gen. Syn., vol. ii. p. 183.
Austral Thrush, Lath. Gen. Hist., vol. v. p. 124?
Pinarocichla harmonica, Cab. Mus. Hein., Theil i. p. 66.
Certhia cinerea, Lath.?

Colluricincla harmonica, Gould, Birds of Australia, fol., vol. ii. pl. 74.

The *Colluricincla harmonica* is an inhabitant of New South Wales and South Australia, and is one of the oldest-known of the Australian birds, having been described in Latham's 'Index Ornithologicus,' figured in White's 'Voyage,' and included in the works of all subsequent writers.

So generally is it dispersed over the countries of which it is a native, that there are few localities in which it is not to be found, the brushes near the coast, as well as the plains of the interior, being equally frequented by it; it is a very active bird, living much among the branches, and feeding upon insects of various kinds, caterpillars, and their larvæ.

The term *harmonica* applied to this species is very appropriate; for although it does not give utterance to any continued song, it frequently pours forth a number of powerful swelling notes, louder but less varied than those of the Song-Thrush of Europe; and it is somewhat singular that these notes are emitted while in the act of feeding, and while engaged in search of its insect food.

The site of the nest is very varied: sometimes a hollow in the upright bole of a small tree is chosen; at others the ledge of a decayed branch, or a rock, or any similar situation. The nest is a cup-shaped and somewhat slight structure, externally composed of the outer and inner bark of trees, and leaves, and lined with fibrous roots; I have occasionally seen wool intermingled with the outer materials. The eggs, which are three in number, and one inch and two lines long by ten lines broad, are of a beautiful pearly white, thinly sprinkled with large blotches of light chestnut-brown and dull bluish grey, the latter colour appearing as if beneath the surface of the shell. In one instance I found a nest of eggs which were brownish white instead of pearly white.

The sexes are very nearly alike, the only difference being that the female has the bill browner and an indication of a white stripe over the eye.

Head brownish grey, with an indistinct line of brown down the centre of each feather; back of the neck, back, and shoulders olive-brown; wings slaty black, margined with grey; rump and tail grey, the latter with dark-brown shafts; under surface light brownish grey, fading into pure white on the vent and under tail-coverts, and greyish white on the throat, each of the throat- and breast-feathers with a fine line of brown down the centre; irides dark brown; bill blackish brown; feet dark greenish grey.

Sp. 124. COLLURICINCLA RUFIVENTRIS, *Gould*.

BUFF-BELLIED SHRIKE-THRUSH.

Colluricincla rufiventris, Gould in Proc. of Zool. Soc., part viii. p. 164.
Goŏ-de-lang, Aborigines of Western Australia.
Thrush of the Colonists.

Colluricincla rufiventris, Gould, Birds of Australia, fol., vol. ii. pl. 75.

This species is about the size of the *Colluricincla harmonica*, for which at a first glance it might be mistaken, but from which on comparison it will be found to differ in the following particulars:—the whole of the upper surface is pure grey instead of brown; the abdomen and under tail-coverts are deep buff instead of greyish white; and the lores are much more distinctly marked with white. It is a native of Western Australia, where it is to be found in all thickly-wooded places, feeding as much on the ground as upon the trees and scrubs.

It breeds in the latter part of September and the beginning of October, and the nest, which is generally placed in the hollow part of a high tree, is formed of dried strips of gum-tree bark very closely packed; it is deep, and is sometimes lined with soft grasses. The eggs, which are two or three in number, are of a beautiful bluish or pearly white, with large blotches of reddish olive-brown and dark grey, the latter appearing as if beneath the surface of the shell; the medium length of the eggs is one inch and one line, by ten lines in breadth.

Gilbert mentions that upon two occasions he found the eggs of this bird in old nests of *Pomatorhinus superciliosus*.

The stomach is muscular, and the food consists of insects, principally of the coleopterous order, and seeds.

Lores greyish white; crown of the head, and all the upper surface deep grey, slightly tinged with olive; primaries and tail dark brown, margined with brownish grey; throat and under surface darkish grey, passing into buff on the vent and under tail-coverts; all the feathers of the under surface have a narrow dark line down the centre; thighs grey; irides dark reddish brown; bill blackish brown; feet dark greenish leaden grey.

Sp. 125. COLLURICINCLA BRUNNEA, *Gould.*

BROWN SHRIKE-THRUSH.

Colluricincla brunnea, Gould in Proc. of Zool. Soc., part viii. p. 164.
Men-e-luö-roo, Aborigines of Port Essington.

Colluricincla brunnea, Gould, Birds of Australia, fol., vol. ii. pl. 76.

This bird is abundantly dispersed over the Cobourg Peninsula, and is to be met with in all the forests in the immediate neighbourhood of Port Essington and the north coast generally, in which distant localities it represents the *Colluricincla harmonica* of New South Wales, the *C. Selbii* of Tasmania, and the *C. rufiventris* of Western Australia. As might be expected, its habits, manners, and general economy are very similar to those of the other species of the genus; consequently the description of those of *C. harmonica* is equally descriptive of those of *C. brunnea.*

A nest of this bird found on the 2nd of February was built in the upper part of a hollow stump, and was outwardly formed of narrow strips of the bark of the *Melaleuca,* and lined with fine twigs. The eggs are of a pearly bluish white, spotted and blotched with markings of olive-brown and grey, the latter colour appearing as if beneath the surface of the shell; their medium length is one inch and two lines, by ten lines in breadth.

It is a larger and more robust species than either *C. harmonica* or *C. rufiventris,* the bill is shorter and much stouter, and the colouring is of a uniform light brown; even the primaries and tail-feathers are of the same hue.

All the upper surface pale brown; primaries and tail the same, but somewhat lighter; all the under surface brownish white, becoming almost pure white on the vent and under tail-coverts; thighs greyish brown; bill black; feet blackish brown.

Sp. 120.   COLLURICINLA SELBII, *Jardine.*

SELBY'S SHRIKE-THRUSH.

Colluricincla Selbii, Jard. in Jard. and Selby's Ill. Orn., vol. i. note to text of pl. 71.
—— *rectirostris*, Jard. in Jard. and Selby's Ill. Orn., vol. iv. pl. xxxi.
—— *strigata*, Swains. Anim. in Menag., &c., p. 283, female or young male.

Whistling Dick of the Colonists of Tasmania.

Colluricincla Selbii, Gould, Birds of Australia, fol., vol. ii. pl. 77.

The *Colluricincla Selbii* is a native of, and a permanent resident in, Tasmania and Flinders Island, over all parts of which it is very generally, but nowhere very abundantly, distributed; it appears to give a decided preference to the thick woods, wherein its presence may always be detected by its loud, clear, liquid, and melodious whistle. It does not appear to confine itself to any particular part of the forest; for it may sometimes be observed on the low scrub near the ground, and at others on the topmost branches of the highest trees. It is distinguished from all the other members of the genus by the greater length of the bill.

It feeds on caterpillars and insects of various kinds, which it often procures by tearing off the bark from the branches of the trees in the most dexterous manner with its powerful bill, and while thus employed frequently pours forth its remarkable note. In disposition it is lively and animated, confident and fearless, and might doubtless be easily tamed, when it would become a most interesting bird for the aviary.

The nest, although composed of coarse materials, is a remarkably neat structure, round, rather deep, and cup-shaped, outwardly formed of strips of the rind of the stringy-bark tree and lined with a few grasses; it is about five inches in diameter and four in height, the interior being three inches and a half in breadth by two and a half in depth. The sites usually selected for the nest are the hollow open stump of a tree, a cleft in a rock, &c.

The male has the general plumage dark slate-grey, deepening into brown on the back and wings, much paler on the under surface, and fading into white on the throat and breast; over the eye a faint stripe of greyish white; bill black; irides brown; feet light lead-colour.

The female or young male has all the upper surface, wings, and tail brown; upper tail-coverts slate-grey; over the eye a stripe of rust-red; under surface light grey tinged with brown on the throat and breast, and each feather with a stripe of dark brown down the centre; bill horn-colour at the base, black at the tip.

Sp. 127. COLLURICINCLA PARVULA, *Gould*.

LITTLE SHRIKE-THRUSH.

Colluricincla parvula, Gould in Proc. of Zool. Soc., part xiii. 1845, p. 62.

Colluricincla parvula, Gould, Birds of Australia, fol., vol. ii. pl. 78.

This species, to which I have given the name of *parvula*, from the circumstance of its being the smallest of the genus that has come under my notice, is a native of Port Essington and the neighbouring parts of the northern coast of Australia. Gilbert, to whose notes I must refer for all that is known about it, states that it is "an inhabitant of the thickets, is an extremely shy bird, and is generally seen on or near the ground. Its note is a fine thrush-like tone, very clear, loud, and melodious. The stomach is muscular, and the food consists of insects of various kinds, but principally of coleoptera. The nest and eggs were brought me by a native; they were taken from the hollow part of a tree, about four feet from the ground; the former, which was too much injured to be preserved, was formed of small twigs and narrow strips of the bark of a *Melaleuca*. The eggs were two in number, of a beautiful pearly flesh-white, regularly spotted all over with dull reddish orange and umber-brown; like the eggs of the other species of the genus, they are also sprinkled over with bluish markings, which appear as if beneath the surface of the shell; their medium length is one inch, and breadth nine lines."

The sexes are so nearly alike in plumage, that they are not readily distinguished from each other; but the male is somewhat larger than his mate.

All the upper surface, wings, and tail olive-brown; a faint line over the eye and the chin white; all the under surface pale buff, the feathers of the throat and breast with a broad stripe of brown down the centre; irides dark brownish red; bill blackish grey; tarsi bluish grey.

Sp. 128. COLLURICINCLA RUFIGASTER, *Gould*.

RUSTY-BREASTED SHRIKE-THRUSH.

Colluricincla rufogaster, Gould in Proc. of Zool. Soc., part xiii. 1845, p. 80.

Colluricincla rufogaster, Gould, Birds of Australia, fol., vol. I. Introd., p. xxxvii.

I assigned this name to a bird sent to me by the late F. Strange from the brushes of the Clarence in New South Wales; it may hereafter prove to be identical with the last-mentioned species, *C. parvula*, the form and admeasurements being precisely the same; but the bird from New South Wales has a lighter-coloured bill, and the whole of the under surface washed with deep rufous.

Strange informed me that the bird "is tolerably common in the brushes skirting the lower part of the Clarence and Richmond rivers; but I never saw it out of the brushes or on the ground, as you may *C. harmonica* and the other species of the genus. It imitates the note of *Ptilonorhynchus holosericeus* so exactly that I have often been deceived by it. You mostly meet with the bird amongst the vines and supplejacks trailing over a few stunted trees; here it will be seen hopping up the thick limbs in search of food, just after the manner of the members of the genus *Climacteris*; like them too, they are continually on the move."

All the upper surface, wings, and tail olive-brown, with the exception of the inner webs of the primaries, which are dark brown; throat pale buffy white, streaked with brown; all the under surface rusty red; irides black; bill and feet fleshy brown.

Total length 7½ inches; bill 1⅛; wing 3¾; tail 3½; tarsi 1¼.

Genus FALCUNCULUS, *Vieillot*.

The two species of this genus are not only strictly Australian, but are confined to the southern parts of the country; the *F. frontatus* inhabiting New South Wales and South Australia, and the *F. leucogaster* Western Australia. When attacked by other birds or by man, both species defend themselves with their powerful bill and claws with the utmost fury; they also use their strongly toothed bills for tearing off pieces of rotten wood and the thin scaly bark of the *Eucalypti* in search of insects. The large branches of trees are their usual place of resort, and in many of their actions and habits they closely resemble the Tits of Europe and India (genus *Parus*), while they also assimilate to the *Pachycephalæ*. They build a round, cup-shaped nest, and lay three or four eggs.

Sp. 129. FALCUNCULUS FRONTATUS, *Vieillot*.

FRONTAL SHRIKE-TIT.

Lanius frontatus, Lath. Ind. Orn., p. xviii.
Frontal Shrike, Lath. Gen. Syn. Supp., vol. ii. p. 75, pl. 122.
Falcunculus frontatus, Vieill. Gal. des Ois., tom. i. pl. 138.
—— *flavigulus*, Gould in Proc. of Zool. Soc., part v. p. 144, female.
—— *Gouldi*, Cab. Mus. Hein., Theil i. p. 66.

Falcunculus frontatus, Gould, Birds of Australia, fol., vol. ii. pl. 79.

I had many opportunities of observing this bird, both in New South Wales and South Australia, over both of which countries it is very generally although not numerously dispersed. It alike inhabits the thick brushes as well as the trees of the open plains. Its chief food is insects, which are either obtained among the foliage or under the bark of the larger branches and trunks of the tree; in procuring these it displays great dexterity, stripping off the bark in the most determined manner, for which purpose its powerful bill is admirably adapted.

It is very animated and sprightly in its actions, and in many of its habits bears a striking resemblance to the Tits, particularly in the manner in which it clings to and climbs among the branches in search of food. While thus employed it frequently erects its crest and assumes many pert and lively positions: no bird of its size with which I am acquainted possesses greater strength in its mandibles, or is capable of inflicting severer wounds, as I experienced on handling one I had previously winged, and which fastened on my hand in the most ferocious manner.

As far as I am aware, the *Falcunculus frontatus* is not distinguished by any powers of song, for I only heard it utter a few low piping notes.

I could neither succeed in procuring the nest of this species nor obtain any authentic information respecting its nidification.

The stomachs of the specimens I dissected were filled with the larvæ of insects and berries.

The male has immediately above the bill a narrow band of white, from which, down the centre of the head, is a broad stripe of black feathers forming a crest; sides of the face and head white, divided by a line of black which passes through the eye to the nape; back, shoulders, and wing-coverts olive; primaries and secondaries blackish brown, broadly margined with grey; tail blackish brown, broadly margined with grey, especially on the two centre feathers; two outer tail-feathers and tips of the remainder white, the white diminishing on each feather as it approaches the centre of the tail; throat black; all the under surface bright yellow; irides reddish brown; bill black; legs and feet bluish grey.

The sexes may at all times be distinguished from each other by the smaller size of the female, and by the colouring of the throat being green instead of black; by the irides being darker, and the feet bluish lead-colour.

Sp. 130. FALCUNCULUS LEUCOGASTER, *Gould.*

WHITE-BELLIED SHRIKE-TIT.

Falcunculus leucogaster, Gould in Proc. of Zool. Soc., part v. p. 144.
Goore-beet-goore-beet, Aborigines of the lowland districts of Western Australia.
Jil-le-ē-lee, Aborigines of the mountain districts of ditto.
Djoon-dool-goo-roon, Aborigines of the Murray in ditto.

Falcunculus leucogaster, Gould, Birds of Australia, fol., vol. ii. pl. 80.

This species is an inhabitant of the western portions of Australia, where it represents the *Falcunculus frontatus* of the eastern coast, from which it may be readily distinguished by its white abdomen; it is very generally dispersed over the colony of Swan River, although, like its near ally, it is not to be met with in great abundance. It is usually seen in pairs among the thickly-foliaged trees, particularly such as grow in quiet secluded places, and is a most active little bird, running over the trunks and branches of the trees with the greatest facility, and tearing off the bark in its progress in search of insects: the habits in fact of the present and Frontal Shrike-Tit are so closely similar that a further description is unnecessary. Its flight is of short duration, and is seldom employed for any other purpose than that of flitting from branch to branch, or from one tree to another. Its note is a series of mournful sounds, the last of which is drawn out to a great length.

Gilbert, while staying in the Toodyay district in the month of October, found the nest of this species among the topmost and weakest perpendicular branches of a *Eucalyptus*, at a height of fifty feet: it was of a deep cup-shaped form, composed of the stringy bark of the gum-tree, and lined with fine grasses, the whole matted together externally with cobwebs; the eggs, which are three or four in number, are of a glossy white with numerous minute speckles of dark olive most thickly disposed at the larger end; they are seven-eighths of an inch long by five-eighths of an inch in breadth. It is a shy bird, but when breeding becomes more bold and familiar.

The stomach is extremely muscular, and its food consists principally of coleoptera.

The male has immediately above the bill a narrow band of white, from which, down the centre of the head, is a broad stripe of black feathers forming a crest; sides of the face and head white, divided by a line of black, which passes through the eye to the nape; back, rump, shoulders, and wing-coverts bright yellowish olive; primaries and secondaries blackish brown, margined with olive-yellow; tail-feathers blackish brown, margined with olive-yellow, except the two outer, which are grey, broadly margined with white; all the tail-feathers tipped with white, the white diminishing on each feather as it approaches the centre of the tail; throat black; chest, upper part of the breast, and under tail-coverts bright yellow; abdomen and thighs white; irides wood-brown; bill dark brown, becoming lighter at the edges of the mandibles; legs and feet greenish blue.

The female differs from the male in being somewhat smaller in size, and in having the throat green instead of black.

Genus OREOÏCA, *Gould*.

The only species known of this form is strictly Australian, and is a sprightly animated bird frequenting the sterile districts studded with large trees, where it hops about on the ground in search of insects. Notwithstanding the singularly lengthened form of its scapularies and its terrestrial habits, it appears to me to partake of the characters of the *Colluricinclæ* and the *Pachycephalæ*; its loud piping note and mode of nidification also favour this opinion. It lays three or four eggs in a round cup-shaped nest, placed either in a *Xanthorrhœa* or in a hole in the stump of a tree.

Sp. 131. OREOÏCA CRISTATA.
CRESTED OREOÏCA.

Turdus cristatus, Lewin, Birds of New Holl., pl. 9. fem.
Falcunculus gutturalis, Vig. and Horsf. in Linn. Trans., vol. xv. p. 212.
Oreoica gutturalis, Gould in Proc. of Zool. Soc., part v. p. 151.
Oreica cristata, G. R. Gray, Ann. and Mag. Nat. Hist., vol. xi. p. 190, note.
Ba-ḧorn-bo-ḧorn, Aborigines of the mountain districts of Western Australia.
Bell-bird, Colonists of Swan River.

Oreoica gutturalis, Gould, Birds of Australia, fol., vol. ii. pl. 81.

This very singular bird possesses an extremely wide range of habitat, being dispersed over the whole of the southern portion of Australia from east to west. It has not yet been discovered in Tasmania or in any of the islands in

Bass's Straits, neither has the extent of its range northwards been ascertained. It is, I believe, everywhere a stationary species, but although its distribution is so general, it is nowhere very plentiful. From what I observed of it, it appeared to give a decided preference to the naked sterile crowns of hills and open bare glades in the forests, and I should say that its presence is indicative of a poor and bad land. It resorts much to the ground, over the surface of which it hops with great quickness, often in small companies of from three to six in number. When flushed it flies but a short distance, generally to a large horizontal branch of a neighbouring *Eucalyptus*, along which it passes in a succession of quick hops, similar to those of the Common Sparrow of Europe. It is very animated in many of its actions, particularly the male, whose erected crest and white face, relieved by the beautiful orange-colour of the eye, give it a very sprightly appearance. The female, on the other hand, being nearly uniform in colour, having the eye hazel and the crest less developed, is by no means so attractive. I regret much that it is not in my power to convey an idea of the sounds uttered by this bird, for they are singular in the extreme; besides which, it is a perfect ventriloquist, its peculiar, mournful, piping whistle appearing to be at a considerable distance, while the bird is perched on a large branch of a neighbouring tree. Gilbert having described to the best of his power the singular note of this species, I give his own words; but no description can convey anything like an accurate idea of it; notes of birds, in fact, are not to be described,—they must be heard to be understood. "The most singular feature," says Gilbert, "connected with this bird is, that it is a perfect ventriloquist. At first its note commences in so low a tone that it sounds as if at a considerable distance, and then gradually increases in volume until it appears over the head of the wondering hearer, the bird that utters it being all the while on the dead part of a tree, perhaps not more than a few yards distant; its motionless attitude rendering its discovery very difficult. It has two kinds of song, the most usual of which is a running succession of notes, or two notes repeated together rather slowly, followed by a repetition three times rather quickly, the last note resembling the sound of a bell from its ringing tone; the other song is pretty nearly the same, only that it concludes with a sudden and peculiar fall of two notes."

In Western Australia its nest is formed of strings of bark lined with a few fine dried grasses, and is generally placed in a *Xanthorrhæa* or grass-tree, either in the upper part of the grass or rush above, or in the fork of the trunk, and is of a deep, cup-shaped form. It breeds in October, and generally lays three eggs, which vary much in colour; the ground-tint being bluish white, in some instances marked all over with minute spots of ink-black, in others with long zigzag lines and blotches of the same hue. In some these markings are confined to the larger end, where they form a zone; in others they are equally spread all over the surface, intermingled with the black markings; also blotches of grey appear as if beneath the surface of the shell, and some eggs have been found with the ground-colour of the larger end of a beautiful bluish green.

In its nidification and in many of its actions it offers considerable resemblance to the members of the genus *Colluricincla*.

It has a thick muscular gizzard, and its food consists of seeds, grain, coleoptera, and the larvæ of all kinds of insects. In Western Australia it often resorts to newly ploughed land, as it there finds an abundance of grubs and caterpillars, its most favourite food.

The male has the face white; feathers on the fore part of the head, along the centre of the crest, line from the eye bounding the white of the face, and a large gorget-shaped mark on the breast deep black; sides of the head and crest grey; all the upper surface and flanks light brown; wings brown, margined with lighter brown; tail dark brown; centre of the abdomen brownish white; vent and under tail-coverts buff; irides beautiful orange, surrounded by a narrow black lash; bill black; legs and feet blackish brown.

The female resembles the male, but differs in having the face and forehead grey, only a line of black down the centre of the crest, the chin dull white, in having a mere indication of the black gorget, the irides hazel, and the feet olive- or dark brown.

Family DICRURIDÆ.

"The family of Drongo-Shrikes," says Mr. Jerdon, "comprises a small number of birds found in Africa, India, and Malayana, and extending in fewer numbers to Australia and the neighbouring islands. They have almost always black plumage and longish forked tails of only ten feathers, being one of the very few groups in which there are fewer than the normal number of twelve. The bill varies much, being short and depressed in some, lengthened and curved in others. They are capable of strong, rapid, and vigorous, but not of sustained flight; and they feed almost entirely on insects, which they capture on the wing, or on the ground, or occasionally on leaves or flowers; their legs are short, and their feet are only fitted for grasping. Some live in the open country, in gardens, and fields, others occur only in the forests, and they are found from the level of the sea to an altitude of 8000 feet and upwards. They form a most characteristic feature in Indian ornithology, for, go where you will in India, you are sure to see one or more of the genus. They build rather loosely constructed nests, and lay three or four eggs, which are usually white with a reddish tinge, and marked with spots and blotches of various shades of red or purple."

Genus CHIBIA, *Hodgson.*

The following is the only species of this form that has yet been found in Australia.

Sp. 132.   CHIBIA BRACTEATA, *Gould.*

SPANGLED DRONGO-SHRIKE.

Dicrurus balicassius, Vig. and Horsf. in Linn. Trans., vol. xv. p. 211.
—— *bracteatus,* Gould in Proc. of Zool. Soc., part x. p. 132.

Dicrurus bracteatus, Gould, Birds of Australia, fol., vol. ii. pl. 82.

Having carefully compared the bird here represented with the other species of the genus inhabiting the Indian islands and the continent of India, I find it to be quite distinct from the whole of them; I have therefore assigned to it a separate specific title, and selected that of *bracteatus* as expressive of its beautifully spangled appearance. Its range is very extensive, the bird being equally abundant in all parts of the northern and eastern portions of Australia; it was found by Sir George Grey on the north-west coast, by Gilbert at Port Essington, and it has also been observed in the neighbourhood of Moreton Bay. I did not encounter it myself during my rambles in Australia; we are therefore indebted to Gilbert's notes for all that is known of its history. "This species is one of the commonest birds of the Cobourg Peninsula, where it is generally seen in pairs and may be met with in every variety of situation, but more frequently among the thickets and mangroves than elsewhere. It is at all times exceedingly active, and its food consists entirely of insects of various kinds, particularly those belonging to the orders *Coleoptera* and *Neuroptera.* Its usual note is a loud, disagreeably harsh, cackling or creaking whistle, so totally different from that of any other bird, that having been once heard it is readily recognised.

"I found five nests on the 16th of November, all of which

contained young birds, some of them nearly able to fly, and
others apparently but just emerged from the egg. The whole of
these nests were exactly alike and formed of the same material, the dry wiry climbing stalk of a common parasitic plant,
without any kind of lining; they were exceedingly difficult
to examine from their being placed on the weakest part of
the extremities of the horizontal branches of a thickly-foliaged
tree at an altitude of not less than thirty feet from the ground;
they were of a very shallow form, about five inches and a half
in diameter; the eggs would seem to be three or four in
number, as three of the nests contained three, and the other
two four young birds in each."

The head and the body both above and below are deep
black, the feathers of the head with a crescent, and those of
the breast with a spot of deep metallic green at the tip;
wings and tail deep glossy green; under wing-coverts black,
tipped with white; irides brownish red; bill and feet blackish
brown.

Genus MANUCODIA, *Boddaert*.

Of this genus only a single species is found in Australia,
the exact position of which in the natural system has not, in
my opinion, been satisfactorily determined. I think it is as
well placed here as elsewhere.

Sp. 133. MANUCODIA GOULDII, *G. R. Gray*.

GOULD'S MANUCODE.

Manucodia gouldii, G. R. Gray in Proc. Zool. Soc., part xxvii. p. 156, note.

Manucodia Keraudreni, Gould, Birds of Australia, fol., Supplement pl.

New Guinea, owing to the hostile character of its native
population, is a sealed country to the collector, and we really
know but little of its natural productions. There are doubtless many fine birds in its mountain districts which never
quit their own forests, while others are from time to time

found on the Cape York Peninsula and other northern promontories of Australia, and this is probably one of them.

I have seen two or three specimens of this bird, all of which were collected during Captain Stanley's Expedition. A fine example in the British Museum, obtained at Cape York, is stated by Mr. Macgillivray to be a male, and is the one from which my description was taken.

Centre of the crown, the lengthened ear-plumes, the lanceolate feathers on the sides of the neck, back, rump, and breast green; shoulders, primaries, and tail purplish-black, as are also the thighs, lower part of the abdomen, and under tail-coverts; bill and legs black.'

When I published this species I believed it to be identical with the *Manucodia keraudreni*; but in his 'List of Birds sent by Mr. Wallace from New Guinea,' Mr. G. R. Gray says, "The specimen figured by Mr. Gould, in his 'Birds of Australia,' as from Cape York, is of a uniform glossy golden-green, with the feathers of the neck of a less pointed form than those of the Dorey examples. It is certainly distinct from the *M. keraudreni* of Dorey, and therefore will warrant a new specific name being given to it; and I now propose that of *Manucodia gouldii*."

Family MUSCICAPIDÆ.

Birds pertaining to this family are found in nearly every part of the globe. As their name implies, they live almost solely on insects, and must perform a most important office in keeping those creatures in check.

Genus RHIPIDURA, *Vigors* and *Horsfield*.

Many species of this genus occur in India, the Indian islands, New Guinea, and Polynesia; and several are comprised in the fauna of Australia, in every part of which country, including Tasmania, one or other member of the group is found.

Sp. 134. RHIPIDURA ALBISCAPA, *Gould.*

WHITE-SHAFTED FANTAIL.

Rhipidura flabellifera, Vig. and Horsf. in Linn. Trans., vol. xv. p. 247.
Rhipidura albiscapa, Gould in Proc. of Zool. Soc., part viii. 1840, p. 113.

Rhipidura albiscapa, Gould, Birds of Australia, fol., vol. ii. pl. 83.

Specimens of this bird from Tasmania are always much darker than those of the continent, and have the tail-feathers less marked with white; others from Western Australia, again, are somewhat lighter in colour, and have the white markings of the tail more extensive than in those I collected in South Australia or New South Wales; the bird from Western Australia has been characterized as distinct, and named *R. Preissi* by M. Cabanis.

In Tasmania I have seen the White-shafted Fantail in the depth of winter in the gullies on the sunny sides of Mount Wellington; and it is my opinion that it only retires at this season to such localities as are sheltered from the bleak south-westerly winds which then so generally prevail, and where insects are still to be found. The bird is also subject to the same law on the continent of Australia; but as the temperature of that country is more equable, its effects are not so decided; and in support of this opinion I may adduce the remark of Caley, who says, "The species is very common about Paramatta; and I do not recollect having missed it at any period of the year."

It is generally found in pairs, but I have occasionally seen as many as four or five together. It inhabits alike the topmost branches of the highest trees, those of a more moderate growth, and the shrouded and gloomy foliaged dells in the neighbourhood of rivulets: from these retreats it darts out a short distance to capture insects, and in most instances returns again to the same branch it had left. While in the air it often assumes a number of lively and beautiful positions, at one moment mounting almost perpendicularly, constantly spreading out its tail to the full extent, and frequently tumbling completely over in the descent; at another it may be seen flitting through the branches, and seeking for insects among the flowers and leaves, repeatedly uttering a sweet twittering song.

This Fantail is rather a late breeder, scarcely ever commencing before October, during which and the three following months it rears two and often three broods. Its elegant little nest, closely resembling a wine-glass in shape, is woven together with exquisite skill, and is generally composed of the inner bark of a species of *Eucalyptus*, neatly lined with the down of the tree-fern intermingled with flowering stalks of moss, and outwardly matted together with the webs of spiders, which not only serve to envelope the nest, but are also employed to strengthen its attachment to the branch on which it is constructed. The situation of the nest is much varied: I have observed it in the midst of dense brushes, in the more open forest, and placed on a branch overhanging a mountain rivulet, but at all times within a few feet of the ground. The eggs are invariably two in number, seven lines long; their ground-colour white, blotched all over, but particularly at the larger end, with brown slightly tinged with olive: the young from the nest assume so closely the colour and appearance of the adults, that they are only to be distinguished by the secondaries and wing-coverts being margined with brown, a feature lost after the first moult. The adults are so precisely alike, that actual dissection is necessary to determine the sexes.

In its disposition this little bird is one of the tamest imaginable, allowing of a near approach without evincing the slightest timidity, and will even enter the houses of persons resident in the bush in pursuit of gnats and other insects. During the breeding-season, however, it exhibits extreme anxiety at the sight of an intruder in the vicinity of its nest.

All the upper surface, ear-coverts, and a band across the chest sooty black, slightly tinged with olive, the tail, crown of the head, and pectoral band being rather the darkest; stripe over the eye, lunar-shaped mark behind the eye, throat, tips of the wing-coverts, margins of the secondaries, shafts, outer webs, and tips of all but the two middle tail-feathers white; under surface buff; eyes black; bill and feet brownish black.

Sp. 135.   RHIPIDURA PREISSI, *Cabanis*.
Preiss's Fantail.

Rhipidura Preissi, Cab. Mus. Hein., Theil i. p. 57.

This is the bird I have alluded to in my account of *R. albiscapa*. As I have now no specimens in my collection, I am unable to institute a comparison and form an opinion as to its specific value; it is, therefore, given on the authority of the learned Berlin Professor, who has named it in honour of Dr. Preiss, an ardent collector of natural history, who spent some years in the neighbourhood of Swan River. If not identical, it is very closely allied to *R. albiscapa*. Its habitat is Western Australia.

Sp. 136.   RHIPIDURA RUFIFRONS.
Rufous-fronted Fantail.

Muscicapa rufifrons, Lath. Ind. Orn. Suppl., p. 1.
Orange-rumped Flycatcher, Lewin, Birds of New Holl., pl. 13.
Rufous-fronted Flycatcher, Lath. Gen. Syn. Supp., vol. ii. p. 220.
Rhipidura rufifrons, Vig. and Horsf. in Linn. Trans., vol. xv. p. 248.
Bur-ril, Aborigines of New South Wales.

Rhipidura rufifrons, Gould, Birds of Australia, fol., vol. ii. pl. 84.

The Rufous-fronted Fantail is one of the most beautiful and one of the oldest known members of the group to which it belongs, having been originally described by Latham in his 'Index Ornithologicus,' and included in the works of nearly every subsequent writer on ornithology. In Mr.

Caley's short but valuable 'Notes on the Birds of New South Wales,' he says, "This bird appears to me to be a rare one; at least I do not recollect having ever seen any other specimen than the present. I met with it on the 15th of October, 1807, in a thick brush or underwood, the resort of the *great Bat*," at Cardunny, a place about ten miles to the north-east of Paramatta. The fact of the colony having at that early date been but little explored will readily account for Caley's opinion of the rarity of this bird; but had he visited the dense brushes of Illawarra, the Liverpool range, and the Hunter, he would have found it in considerable numbers.

Although many of its habits closely resemble those of the *Rhipidura albiscapa*, they are, as the greater length of its legs would indicate, far more terrestrial. It runs over the ground and the fallen logs of trees with great facility. While thus engaged, and particularly when approached, it constantly spreads and displays its beautiful tail, and evinces a great degree of restlessness. It is always found in the most secluded parts of the forest, no portion of which appears to be too dense for its abode.

I never met with it in Tasmania or on the islands in Bass's Straits, neither do I recollect having seen it in South Australia; and it has not yet been found in Western Australia or on the north coast, in which latter locality it is represented by the *Rhipidura dryas*.

I had but little opportunity of observing it during the breeding-season, but frequently found its deserted wineglass-shaped nest, which bore a general resemblance to that of *R. albiscapa*. In one of them I found a single egg, which may be thus described:—Ground-colour stony-white, speckled all over with purple and yellowish-brown spots and markings, disposed so numerously as to form a zone at the larger end. It is about eight lines long and six broad.

The sexes are precisely alike in colour; and their only

outward difference consists in the somewhat smaller size of the female.

Forehead rusty red, continuing over the eye; crown of the head, back of the neck, upper part of the back, and wings olive-brown; lower part of the back, tail-coverts, and the basal portions of the tail rusty red; remainder of the tail blackish brown, obscurely tipped with light grey; the shafts of the tail-feathers, for nearly half their length from the base, light rusty red; throat and centre of the abdomen white; ear-coverts dark brown; chest black, the feathers of the lower part edged with white; flanks and under tail-coverts light fawn-colour; eyes, bill, and feet brown.

Sp. 137.   RHIPIDURA DRYAS, *Gould*.
   Wood Flycatcher.

Rhipidura dryas, Gould, Birds of Australia, fol. vol. i. Introd., p. xxxix.

This bird differs from *R. ruffrons* in being of a smaller size, in its dark-grey tail-feathers being more largely tipped with white, and merely fringed with rufous at the base only, in the breast being white, crossed by a distinct band of black, and devoid of the dark spotted markings seen on the chest of its ally.

Total length 5¾ inches; wing 2⅝; tail 3¼; tarsi ⅞.

The *R. dryas* inhabits the north-western portion of Australia, where it appears to be as common as the *R. ruffrons* is in the south-eastern. I have several specimens, all of which bear a general resemblance to each other.

Sp. 138.   RHIPIDURA ISURA, *Gould*.
   Northern Fantail.

Rhipidura isura, Gould in Proc. of Zool. Soc., part viii. p. 174.

Rhipidura isura, Gould, Birds of Australia, fol. vol. ii. pl. 85.

This species is an inhabitant of the north and north-west

coasts of Australia, in which localities specimens have been procured by Sir George Grey and by Gilbert, the latter of whom states that it is abundant in all parts of the Cobourg Peninsula, and that it is to be met with in every variety of situation, that it is usually seen in pairs, and that it secludes itself during the heat of the day amidst the dense thickets of mangroves.

A nest found by Gilbert in the early part of November appeared to have been recently inhabited by young birds; it was placed in the centre of three upright twigs of a species of *Banksia*, and was formed of narrow strips of bark, firmly bound together on the outside with cobwebs and vegetable fibres; it was very cup-like in shape, about two inches and a half in height, one inch and three-quarters in diameter, and three-quarters of an inch in depth.

The food consists of insects of various kinds and their larvæ.

All the upper surface dull brown; wings and tail darker brown, the outer feather of the latter on each side margined externally and largely tipped with white, the next having a large irregular spot of white at the tip, and the next with a minute line of white near the tip; chin and under surface buffy white, with an indication of a dark brown band across the chest; bill and feet black.

Total length 8 inches; bill ⅝; wing 3⅞; tail 3½; tarsi 1¹⁄₈.

Genus SAULOPROCTA, *Cabanis*.

M. Cabanis has considered it desirable to separate the *Rhipidura motacilloides* of Vigors and Horsfield and one or two other nearly allied birds from the smaller *Rhipiduræ* and to form them into a distinct genus, believing that their greater size, longer wings and legs, and different style of colouring justified his so doing.

Besides the two species found in Australia several others exist in the islands lying to the northward of that country, all of which bear a general resemblance to each other.

Sp. 139. SAULOPROCTA MOTACILLOIDES.

BLACK FANTAIL.

Rhipidura motacilloides, Vig. and Horsf. in Linn. Trans., vol. xv. p. 248.
Sauloprocta motacilloides, Cab. Mus. Hein. Theil i. p. 57.
Wīl-la-ring, Aborigines of the lowland, and
Jil-te-jil-te, Aborigines of the mountain districts of Western Australia.
Wagtail Flycatcher of the Colonists of Swan River.

Rhipidura motacilloides, Gould, Birds of Australia, fol., vol. ii. pl. 86.

With the exception of Tasmania, this bird has been found in every part of Southern Australia yet visited by Europeans.

At the same time that it is one of the most widely diffused, it is also one of the most tame and familiar of the Australian birds, and consequently a general favourite; it is constantly about the houses, gardens, and stock-yards of the settlers, often running along the backs and close to the noses of the cattle in order to secure the insects which are roused and attracted by the heat from their nostrils, along the roofs of the buildings, the tops of pailings, gates, &c.; constructing its pretty neat beneath the verandah, and even entering the rooms to capture its insect prey. It passes much of its time on the ground, over which it runs and darts with the utmost celerity, and when skirting the stream with tail erect and shaking from side to side, it presents an appearance very similar to that of the Pied Wagtails; the movements of the tails of the two birds, however, are very different, that of the European being perpendicular, while that of the Australian is a kind of lateral swing.

Its song, which consists of a few loud and shrill notes, is continually poured forth throughout the entire night, especially if it be moonlight.

Its flight is at times gracefully undulating; at others it consists of a series of sudden zigzag starts, but is always of very short duration; it never poises itself in the air, like the *Seisura volitans*, and never mounts higher than the tops of the trees.

It commences breeding in September, and generally rears two or three broods. Its beautiful deep, cup-shaped and compact nest is very often built on a branch overhanging water, or on the dead limb of a tree overshadowed by a living branch above it, but the usual and favourite site is the upper side of a fallen branch without the slightest shelter from the sun and rain, at about three or four feet from the ground; the nest itself is constructed of dried grasses, strips of bark, small clumps of grass, roots, &c., all bound and firmly matted together and covered over with cobwebs, the latter material being at times so similar in appearance to the bark of the branch, that the entire nest looks like an excrescence of the wood, when it is almost impossible to detect it; it is lined with a finer description of grass, small wiry fibrous roots, or feathers. The eggs are generally three in number, of a dull greenish white, banded round the centre or towards the larger end with blotches and spots of blackish and chestnut-brown, which in some instances are very minute; the medium length of the egg is nine lines and a half, by seven lines in breadth. On an intruder approaching the nest, the birds fly about and hover over his head, and will even sit on the same branch on which the nest is placed while the eggs are being taken; uttering all the time a peculiar cry which may be compared to the sound of a child's rattle, or the noise produced by the small cog-wheels of a steam-mill.

The sexes are alike in plumage, and may be thus described:—

Head, neck, throat, sides of the chest, upper surface, and tail glossy greenish black; over each eye a narrow line of white; wings brown; wing-coverts with a small triangular spot of white at the tip; under surface pale buffy white; irides, bill, and feet black.

Sp. 140.   SAULOPROCTA PICATA, *Gould.*
PIED FANTAIL.

Sauloprocta picata, Cab. Mus. Hein. Theil i. p. 57 (note).

Rhipidura picata, Gould, Birds of Australia, fol., vol. i. Introd. p. xxxix.

This northern species is a minute representative of the *S. motacilloides* of the south. It is a native of Port Essington and the surrounding country, and I have specimens brought by Mr. Wallace from the Aru Islands which, if not identical, are so similar that I have failed to detect any difference.

The colouring of the *S. picata* being the same as that of *S. motacilloides*, a description of it is unnecessary; the following are its admeasurements:—

Total length $6\frac{7}{8}$; wing $3\frac{1}{2}$; tail $3\frac{1}{2}$; tarsi $\frac{7}{8}$.

Genus SEISURA, *Vig. and Horsf.*

The present genus and *Rhipidura* are mere modifications of each other; a difference of structure, however, exists of sufficient importance to justify their separation, and, as is always the case, a corresponding difference is found in the habits and actions of the species.

Sp. 141.   SEISURA INQUIETA.
RESTLESS FLYCATCHER.

Turdus inquietus, Lath. Ind. Orn., Supp., p. xl.
—— *volitans*, Lath., ib., p. xli.
—— *muscicola*, Lath., ib., p. xli.
—— *dubius*, Lath., ib., p. xl.
Restless Thrush, Lath. Gen. Syn. Supp., vol. ii. p. 181.
Volatile Thrush, Lath., ib., p. 183.
Seisura volitans, Vig. and Horsf. in Linn. Trans., vol. xv. p. 250.
Jit-tee-gnul, Aborigines of Western Australia.
The Grinder of the Colonists of Swan River and New South Wales.

Seisura inquieta, Gould, Birds of Australia, fol., vol. ii. pl. 87.

This species ranges over the whole of the southern portions

of the Australian continent, and appears to be as numerous at Swan River as it is in New South Wales, where it may be said to be universally distributed; for I observed it in every part I visited, both among the brushes as well as in the more open portions of the country, in all of which it is apparently a stationary species. It is a bird possessing many peculiar and very singular habits. It not only captures its prey after the usual manner of the other Flycatchers, but it frequently sallies forth into the open glades of the forest and the cleared lands, and procures it by poising itself in the air with a remarkably quick motion of the wings, precisely after the manner of the English Kestrel (*Tinnunculus alaudarius*), every now and then making sudden perpendicular descents to the ground to capture any insect that may attract its notice. It is while performing these singular movements that it produces the remarkable sound, which has procured for it from the colonists of New South Wales the appellation of "The Grinder." The singular habits of this species appear to have attracted the notice of all who have paid any attention to the natural history of New South Wales: Mr. Caley observes, "It is very curious in its actions. In alighting on the stump of a tree it makes several semicircular motions, spreading out its tail at the time, and making a loud noise somewhat like that caused by a razor-grinder at work. I have seen it frequently alight on the ridge of my house, and perform the same evolutions." To this I may add the following account of the actions and manners of this species as observed by Gilbert in Western Australia:—

"This bird is found in pairs in every variety of situation. Its general note is a loud harsh cry several times repeated; it also utters a loud clear whistle; but its most singular note is that from which it has obtained its colonial name, and which is only emitted while the bird is in a hovering position at a few feet above the ground; this noise so exactly resembles a grinder at work, that a person unaware of its being produced by a bird might easily be misled. Its mode of flight is one of the most graceful and easy imaginable; it rarely mounts high in flying from tree to tree, but moves horizontally with its tail but little spread, and with a very slight motion of the wings; it is during this kind of flight that it utters the harsh note above-mentioned—the grinding note being only emitted during the graceful hovering motion, the object of which appears to be to attract the notice of the insects beneath, for it invariably terminates in the bird descending to the ground, picking up something, flying into a tree close by, and uttering its shrill and distinct whistle."

The months of September, October, and November constitute the breeding-season. The nests observed by me in New South Wales were rather neatly made, very similar to those of *Sauloprocta motacilloides*, cup-shaped, and composed of fine grasses matted together on the outside with cobwebs, and lined with very fine fibrous roots and a few feathers; they were placed on horizontal branches frequently overhanging water. The eggs, which are sometimes only two, but mostly three in number, are dull white, distinctly zoned round the centre with spots of chestnut and greyish brown, the latter colour appearing as if beneath the surface of the shell; their medium length is nine lines and a half by seven lines in breadth. The nests found by Gilbert in Western Australia were remarkably neat and pretty, and were formed of cobwebs, dried soft grasses, narrow strips of gum-tree bark, the soft paper-like bark of the *Melaleucæ*, &c., and were usually lined with feathers or a fine wiry grass, and in some instances horse-hair. The situations chosen for their erection are the most difficult of access, being the upper side, the extreme end, and the dead portion of a horizontal branch. The bird is very reluctant to leave the nest, and will almost suffer itself to be handled rather than desert its eggs.

The sexes are very similar in plumage, but the female and

young males have the lores or space between the bill and the eye not so deep a black as in the male.

Head and all the upper surface shining bluish black; wings dark brown; tail brownish black; lores deep velvety black; under surface silky white, with the exception of the sides of the chest, which are dull black; irides dark brown; basal half of the sides of the upper mandible and the basal two-thirds of the lower mandible greenish blue; the remainder of the bill bluish black; legs and feet dark bluish brown.

Genus PIEZORHYNCHUS, *Gould.*

The only species of this genus yet discovered in Australia is a native of the northern parts of that country, from Cape York to Port Essington, where it frequents the dense beds of mangroves.

Mr. G. R. Gray, in his Catalogue of the Birds of the Tropical Islands of the Pacific Ocean in the collection of the British Museum, enumerates two species of this form from New Ireland, and in his Catalogue of the Birds of New Guinea one from the Aru Islands.

Sp. 142. PIEZORHYNCHUS NITIDUS, *Gould.*

SHINING FLYCATCHER.

Piezorhynchus nitidus, Gould in Proc. of Zool. Soc., part viii. p. 171.
Ur̈g-bur-ka, Aborigines of Port Essington.

Piezorhynchus nitidus, Gould, Birds of Australia, fol., vol. ii. pl. 89.

This Flycatcher is by no means scarce at Port Essington, but, from the extreme shyness of its disposition and the situations it inhabits, it is seldom seen; specimens in fact are not procured without considerable trouble and difficulty. As

I have not myself seen the bird in its native haunts, I transcribe Gilbert's notes respecting it:—"Inhabits the densest mangroves and thickets, and is usually seen creeping about close to the ground among the fallen trees in the swamps, at which time it utters a note so closely resembling the croak of a frog, that it might easily be mistaken for the voice of that animal; this peculiar note would seem to be only emitted while the bird is feeding on the ground; for when it occasionally mounts to the higher branches of the trees it utters rather a pleasing succession of sounds resembling *twit-te-twite*; on the slightest disturbance it immediately descends again to the underwood and recommences its frog-like note. The nest is either built among the mangroves, or on the verge of a thicket near an open spot. One that I found among the mangroves was built on a seedling-tree not more than three feet from the ground; another was on a branch overhanging a small running stream within reach of the hand; while a third, constructed on the branches of the trees bordering a clear space in the centre of a dense thicket, was at least twenty feet high. The nest at all times so closely resembles the surrounding branches, that it is very difficult to detect unless the birds are very closely watched; in some instances it looks so like an excrescence of the tree, and in others is so deeply seated in the fork whereon it is placed, that it can hardly be discovered when the bird is sitting upon it. The nest is about two inches and a half in height and three and a quarter in diameter, is of a cup-shaped form, with the rim brought to a sharp edge, and is outwardly composed of the stringy bark of a *Eucalyptus* bound together on the outside with vegetable fibres, among which in some instances cobwebs are mixed: all over the outside of the nest small pieces of bark resembling portions of lichens are attached, some of them hanging by a single thread and moving about with every breath of air; the internal surface is lined with a strong thread-like fibrous root, whereby the whole structure is rendered nearly as firm as if it were bound with wire."

The eggs, which are two in number, are ten lines long and seven lines broad, of a bluish white, blotched and spotted all over with olive and greyish brown, the spots of the latter hue being less numerous and more obscure; the spots inclining towards the form of a zone at the larger end.

The male has the whole of the plumage rich deep glossy greenish black; irides red; bill greyish blue at the base, black at the tip; tarsi greenish grey.

The female has the top and sides of the head and the back of neck rich deep glossy greenish black; the remainder of the upper surface, wings, and tail rusty brown; and the whole of the under surface white.

Total length $7\frac{1}{4}$ inches; bill $1\frac{1}{8}$; wing $3\frac{1}{4}$; tail $3\frac{1}{4}$; tarsi $\frac{3}{4}$.

Genus ARSES, *Lesson.*

The members of this form are allied to those of *Monarcha*, and should be placed between them and the *Tchitreæ*. One species inhabits Australia, and others are found in the Aru and neighbouring islands.

Sp. 143.   ARSES KAUPI, *Gould.*

KAUP'S FLYCATCHER.

Arses Kaupi, Gould in Proc. of Zool. Soc., part xviii. p. 278.

Arses Kaupi, Gould, Birds of Australia, fol., Supplement, pl.

I have some little doubt as to the propriety of placing this bird in the genus *Arses*, but rather than multiply the number of genera, perhaps unnecessarily, I have assigned it a place therein, as it accords more nearly with that form than with *Monarcha*, the only other genus to which it presents alliance. I am happy to have this opportunity of paying a just compliment to my friend Dr. Kaup of Darmstadt, an ornithologist of vast acumen and research, and whose philosophical labours are well known to all naturalists.

The specimen described was killed on the north coast of Australia.

Small spot on the chin, crown of the head, lores, line beneath the eye, ear-coverts, broad crescentic band across the back, and a broad band across the breast, deep shining bluish black; wings and tail brownish black; throat and a broad band across the back of the neck white; lower part of the back and abdomen white, the base of the feathers black, which occasionally showing through give those parts a mottled appearance; bill bluish horn-colour, becoming lighter at the tip; feet black.

Total length 6½ inches; bill ½; wing 5⅛; tail 3¼; tarsi ⅞.

Genus MYIAGRA, *Vig. and Horsf.*

A group of insectivorous birds, the greater number of which inhabit the Indian Islands and Polynesia, and of which four species are found in Australia.

Sp. 144. MYIAGRA PLUMBEA, *Vig. and Horsf.*

LEADEN-COLOURED FLYCATCHER.

Myiagra plumbea, Vig. and Horsf. in Linn. Trans., vol. xv. p. 254.

Myiagra plumbea, Gould, Birds of Australia, fol., vol. ii. pl. 89.

A summer visitant to New South Wales, where it takes up its abode on high trees bordering creeks and low valleys, and captures its insect food under the shady branches, the *Myiagra plumbea* is mostly seen in pairs, which are rather thinly dispersed over the districts forming its usual place of resort. A low whistling note, frequently uttered by the males, is, in all probability, indicative of the season of love. On the approach of winter it retires northwards, and returns again the following August or September, the months in which spring commences in Australia.

It is a most active bird; in fact all its actions are characterized by great liveliness; for even while in a state of comparative repose, or when not actually in pursuit of insects, it displays a constant tremulous motion of the tail, by which means its presence is often betrayed when it would otherwise remain unnoticed.

As is the case with all the other members of the genus, the sexes present considerable difference in their plumage, the female having the throat of a bright rusty red, while the throat of the male is of a rich greenish lead-colour, like the upper surface,—a style of colouring which has suggested the specific name of *plumbea*. The young males during the first year so closely assimilate in plumage to the female, that by dissection alone can they be distinguished with certainty.

The nest is cup-shaped, rather deep, formed of moss and lichens, and neatly lined with feathers, and is generally placed on the horizontal branch of a tree. I did not succeed in procuring the eggs.

The male has the whole of the upper surface, wings, tail, and breast lead-colour, glossed with green on the head, neck, and breast, and becoming gradually paler towards the extremity of the body and on the wings and tail; primaries slaty black; secondaries faintly margined with white; under surface of the wing, abdomen, and under tail-coverts white; bill leaden blue, except at the extreme tip, which is black; irides and feet black.

The female has the head and back lead-colour, without the greenish gloss; wings and tail brown, fringed with bluish grey, particularly the secondaries; throat and breast rich rusty red, gradually fading into the white of the lower part of the abdomen and under tail-coverts; upper mandible black; under mandible pale blue, except at the extremity, which is black.

Sp. 145.   MYIAGRA CONCINNA, *Gould*.

PRETTY FLYCATCHER.

Myiagra concinna, Gould, Birds of Australia, fol., vol. ii. pl. 90.

This species is a native of the north-western portion of Australia, where it inhabits the dense mangroves and thickets adjacent to swamps. It is very shy and retiring in its disposition, but may occasionally be seen on the topmost branches of the highest trees of the forest. Like the other Flycatchers, it has the habit of sitting for a long time on a branch, watching the various insects as they pass, now and then darting forth and capturing one on the wing, and then returning again to the branch from which it had flown.

When among the low mangroves it utters a rather agreeable twittering song; but on high trees it emits a loud and shrill whistle, drawn out at times to a considerable length.

The stomach is muscular, and the food consists of insects of various kinds and their larvæ.

Like the other members of the genus, the sexes differ considerably in colour; they may be thus described:—

The male has the whole of the upper surface, wings, tail, and breast lead-colour, glossed with green on the head, neck, and breast, and becoming gradually paler towards the extremity of the body and on the wings and tail; primaries slaty black; secondaries faintly margined with white; under surface of the wing, abdomen, and under tail-coverts white; bill leaden blue, except at the extreme tip, which is black; irides brown; feet blackish grey.

The female has the head and back lead-colour, without the greenish gloss; wings and tail brown, fringed with bluish grey, particularly the secondaries; throat and breast rich rusty red; abdomen and under tail-coverts white, which colour does not gradually blend with the rusty red of the breast as in the female of *Myiagra plumbea*; upper mandible black; under mandible pale blue, except at the tip, which is black.

Sp. 140.   MYIAGRA NITIDA, *Gould.*

SHINING FLYCATCHER.

Todus rubecula, Lath. Ind. Orn. Supp., p. xxii, female.
Red-breasted Tody, Lat. Gen. Syn. Supp., vol. ii. p. 147.
Platyrhynchus rubecula, Vieill., 2nde édit. du Nouv. Dict. d'Hist. Nat. tom. xxvii. p. 16.
Myiagra rubeculoides, Vig. and Horsf. in Linn. Trans., vol. xv. p. 253, female.
—— *nitida*, Gould in Proc. of Zool. Soc., part v. p. 142.
Satin Sparrow of the Colonists of Tasmania.

Myiagra nitida, Gould, Birds of Australia, fol., vol. ii. pl. 91.

The *Myiagra nitida* appears in Tasmania about the end of September, commences breeding soon after its arrival, rears a somewhat numerous progeny during the months of summer, and departs again in February. In performing these migrations it necessarily passes directly over the colonies of South Australia and New South Wales, yet it seldom occurs in collections from those countries. It is a most lively, showy, and active bird, darting about from branch to branch and sallying forth in the air in pursuit of its insect prey with a singular, quick, oscillating or trembling motion of the tail.

I experienced but little difficulty in obtaining several of its nests and eggs among the gullies and forest land on the north side of Mount Wellington, particularly those immediately in the rear of New Town, near the residence of the Rev. Thomas J. Ewing, who frequently accompanied and aided me in my search. The nest is usually placed at the extreme tip of a dead branch, at a height varying from twenty to forty feet from the ground. Some nests are formed of a minute species of light green moss; others are constructed of fine threads of stringy bark; all are rendered very warm by a dense lining of soft hair of the opossum, the flocculent fibres of the tree fern, and blossoms of many other kinds of plants; and the outsides of all are decorated with small pieces of lichen stuck on without any degree of regularity; these different materials are all felted together with cobwebs. The form of the nest appears to depend upon the nature of the site upon which it is built: if placed on a level part of the branch, the nest is large and high; if in a fork, then it is a more shallow structure; in each case the opening is as perfect a circle as the nature of the materials will admit: the height varies from two inches to three inches and a quarter, the average breadth of the opening is about one inch and three-quarters, and the depth one inch. The eggs are generally three in number, somewhat round in form, and of a greenish white spotted and blotched all over with umber brown, yellowish brown, and obscure markings of purplish grey; the medium length is nine lines, and breadth seven lines.

The note is a loud piping whistle frequently repeated.

The male has the lores deep velvety black; all the upper surface, wings, tail, and breast of a rich deep blackish green with a metallic lustre; primaries deep brown; under surface of the shoulder, abdomen, and under tail-coverts white; bill lead-colour at the base, passing into black at the tip; irides and feet black.

The female differs considerably from the male—the upper surface being much less brilliant, and the throat and breast of a rich rusty red—a style of colouring which is also characteristic of the young males during the first autumn of their existence.

Sp. 147.   MYIAGRA LATIROSTRIS, *Gould.*
BROAD-BILLED FLYCATCHER.

Myiagra latirostris, Gould in Proc. of Zool. Soc., part viii. p. 172.

Myiagra latirostris, Gould, Birds of Australia, fol., vol. ii. pl. 92.

This species was procured on the north coast by Mr. Dring,

and at Port Essington by Gilbert. It is in every respect a true *Myiagra*, and is rendered remarkably conspicuous by the great breadth or lateral dilatation of the bill. As no notes accompanied the specimens, I am unable to give any particulars as to its habits and economy; in all probability they are very similar to those of the other members of the genus.

All the upper surface, wings, and tail dark bluish grey, with a shining greenish lustre on the head and back of the neck; throat and chest sandy buff; under surface white; bill black; irides blackish brown; feet black.

Total length 6 inches; bill $\frac{3}{4}$; wing $2\frac{3}{4}$; tail $2\frac{3}{4}$; tarsi $\frac{1}{2}$.

Genus MACHÆRIRHYNCHUS, *Gould*.

This is a very singular and distinct form among the smaller Flycatchers. The bill is laterally developed to a greater extent than in any other bird of its size. At least two species are known, one of which inhabits Australia, the other, *M. xanthogenys*, the Aru Islands.

Sp. 148. MACHÆRIRHYNCHUS FLAVIVENTER, *Gould*.

YELLOW-BREASTED FLYCATCHER.

Machærirhynchus flaviventer, Gould in Proc. of Zool. Soc., part xviii. p. 277, Aves, pl. xxxiii.

Machærirhynchus flaviventer, Gould, Birds of Australia, Supplement, pl.

Mr. Macgillivray informed me that a single specimen of this Flycatcher was shot at Cape York by Mr. James Wilcox, who observed it on the skirts of one of the dense brushes or jungles, making short flights in the air, snapping at passing flies, and returning again to the same tree, the *Wormia alata* of botanists, distinguished by its red papery bark, large glossy leaves and handsome yellow flowers, which attract numbers of insects. The place was frequently visited afterwards, but no other example was seen.

Since Mr. Macgillivray's visit to Cape York other examples have been procured in that locality.

Crown of the head, lores, ear-coverts, wings, and tail black; wing-coverts tipped with white; secondaries margined with white; outer tail-feathers margined on the apical portion of the external web and largely tipped with white, the white becoming less and less, until only a slight trace of it is found on the central feathers; back olive-black; throat white; line from the nostrils over each eye and the breast, abdomen, and under tail-coverts bright yellow; bill black; feet bluish black.

Total length 5 inches; bill $\frac{1}{2}$; wing 2; tail $2\frac{1}{4}$; tarsi $\frac{1}{2}$.

Genus MICRŒCA, *Gould.*

Three species of this form inhabit Australia, to which country they are probably confined.

Sp. 149.   MICRŒCA FASCINANS.

BROWN FLYCATCHER.

Loxia fascinans, Lath. Ind. Orn. Supp., p. xlvi.
Fascinating Grosbeak, Lath. Gen. Syn. Supp., vol. ii. p. 197.
Myiagra macroptera, Vig. and Horsf. in Linn. Trans., vol. xv. p. 254.
Micrœca macroptera, Gould in Proc. of Zool. Soc., part viii. p. 172.
Sylvia leucophæa, Lath.
Brown Flycatcher of the Colonists.

Micrœca macroptera, Gould, Birds of Australia, fol., vol. ii. pl. 93.

This bird is generally dispersed over the colonies of New South Wales and South Australia, where it inhabits nearly every kind of situation, from the open forest lands of the interior to the brushes of thickly-grown trees near the sea-coast, shrubs not a yard high and the branches of the highest gum-trees being alike resorted to. It is certainly one of the least ornamental of the Australian birds; for it is neither gaily coloured, nor is it characterized by any conspicuous markings; these deficiencies, however, are, as is usually the case, amply compensated for by the little sombre tenant of the forest being endowed with a most cheerful and pleasing song, the notes of which much resemble, but are more clear and powerful than the spring notes of the Chaffinch (*Fringilla Cœlebs*), and which are poured forth at the dawn of day from the topmost dead branch of a lofty gum-tree, an elevated position which appears to be frequently resorted to for the purpose of serenading its mate, its usual place of abode being much nearer the ground. It is mostly met with in pairs, and may be frequently seen perched on the low bushy twigs of a thistle-like plant, occasionally on the gates and palings and in the gardens of the settlers. Mr. Caley states that "it has all the actions of the British *Robin Red-breast*, except coming inside houses. When a piece of ground was fresh dug it was always a constant attendant." It appeared to me that its actions resemble quite as much those of the Flycatchers as of the Robins, and at the same time are sufficiently distinct from either to justify the bird being made the type of a new genus; I may particularly mention a singular lateral movement of the tail, which it is continually moving from side to side.

Its food consists of insects, which it captures both among the foliage of the trees and on the wing, frequently flying forth in pursuit of passing flies, and returning again to the branch it had left.

It generally rears two broods in the course of the year.

The nest, which is built in October, is a slight, nearly flat, and very small structure, measuring only two inches and a half in diameter by half an inch in depth; it is formed of fine fibrous roots decorated externally with lichens and small flat pieces of bark, attached by means of fine vegetable fibres and cobwebs, and is most artfully placed in the fork of a dead horizontal branch, whereby it is rendered so nearly invisible from beneath, that it easily escapes detection from all but the scrutinizing eye of the aboriginal native. The eggs are generally two in number, of a pale greenish blue, strongly marked with dashes of chestnut-brown and indistinct blotches of grey; they are eight and a half lines long by five and a half lines broad.

The sexes are alike in colour; the young differs from the adult in being much paler, and in being spotted with white on the head and back, and with brown on the breast.

The adult has all the upper surface and wings pale brown; wing-coverts slightly tipped with white, and a wash of white on the margins of the tertiaries and tips of the upper tail-coverts; tail dark brown, the external feather white, and the next on each side with a large spot of white on the inner web at the tip; all the under surface pale brownish white, fading into nearly pure white on the chin and abdomen; bill, irides, and feet brown.

Sp. 150.   MICRŒCA ASSIMILIS, *Gould.*

ALLIED FLYCATCHER.

Micrœca assimilis, Gould in Proc. of Zool. Soc., part viii. p. 172.

Micrœca assimilis, Gould, Birds of Australia, fol., vol. i. Introd. p. xl.

This species inhabits Western Australia, and is nearly allied to the *Micrœca fascinans*, from which it only differs in being much less in size, and in having the base of the outer tail-feather brown instead of white.

All the upper surface brown; primaries dark brown; tail brownish black; the tips and the terminal half of the external margins of the two outer feathers white; the three next on each side also tipped with white, the extent of the white becoming less upon each feather as they approach the centre of the tail; the four middle feathers without the white lip; throat, centre of the abdomen, and under tail-coverts white, passing into pale brown on the sides of the chest and flanks; irides reddish brown; bill and feet blackish brown.

Total length $4\frac{3}{8}$ inches; bill $\frac{9}{16}$; wings $3\frac{5}{8}$; tail $2\frac{1}{4}$; tarsi $\frac{8}{16}$.

Sp. 151. MICRŒCA FLAVIGASTER, *Gould*.
Yellow-bellied Flycatcher.

Micrœca flavigaster, Gould in Proc. of Zool. Soc., part x. p. 132.

Micrœca flavigaster, Gould, Birds of Australia, fol., vol. ii. pl. 84.

This little Flycatcher is met with in the neighbourhood of Port Essington in every variety of situation, and is particularly abundant on all the islands in Van Diemen's Gulf. "It gives utterance," says Gilbert, "to many different notes, pouring forth at the dawn of day a strain much resembling that of some of the *Petroicæ*, and like them remaining stationary for a long time while singing its agreeable melody. In the middle of the day, when the sun is nearly vertical, it leaves the trees and soars upward in circles, like the Skylark, until it arrives at so great a height as to be scarcely perceptible; it then descends perpendicularly until it nearly reaches the trees, when it closes its wings and apparently falls upon the branch on which it alights. During the whole of this movement it pours forth a song, some parts of which are very soft and melodious, but quite different from that of the morning; in the evening the song is again varied, and then so much resembles the unconnected notes of the *Gerygonæ*, that I have frequently been misled by it. The *Micrœca flavigaster* is a very familiar species, inhabiting the trees and bushes close around the houses, and is little alarmed or disturbed at the approach of man. At times it is extremely pugnacious; I have seen a pair attack a Crow and beat it until it was obliged to seek safety by flight, all the while calling out most lustily. Notwithstanding it is so abundant everywhere, and it must have been breeding during my stay here, as is proved by my killing young birds apparently only a few days old, I did not succeed in finding the nest; and on inquiring of the natives, they could give me no information whatever respecting it or the period of incubation."

The sexes do not differ in colour or size.

All the upper surface brownish olive; wings and tail brown, margined with paler brown; throat white; all the under surface yellow; irides blackish brown; feet blackish grey.

Total length 3⅜ inches; bill ⅜; wing 2⅞; tail 2¼; tarsi ¼.

Genus MONARCHA, *Vigors* and *Horsfield*.

Several species of this genus occur in the Indian Islands, and two in Australia. They are insectivorous birds, and procure their food by quietly hopping about among the branches of the trees.

The members of the present form and those of the genus *Arses* are very nearly allied.

Sp. 152.   MONARCHA CARINATA.

CARINATED FLYCATCHER.

Muscipeta carinata, Swains. Zool. Ill., 1st ser. pl. 147.
Drymophila carinata, Temm. Pl. Col. 418. f. 2.
Monarcha carinata, Vig. and Horsf. in Linn. Trans., vol. xv. p. 255.

Monarcha carinata, Gould, Birds of Australia, fol., vol. ii. pl. 95.

This is a migratory bird in New South Wales, arriving in spring and departing again in March and April, the Australian autumn. It gives a decided preference to thick brushy forests, such as those at Illawarra and other similar districts extending from the Hunter to Moreton Bay. It is also equally abundant in the thick brushes which clothe the sloping mountains of the interior. During the spring or pairing-time it becomes very animated, and is continually flying about and beneath the branches of the trees; it does not capture insects, like the true Flycatchers, on the wing, but obtains them while hopping about from branch to branch, after the manner of the *Pachycephalæ*. It has a rather loud whistling note, which being often repeated tends considerably to enliven the woods in which it dwells.

The *Monarcha carinata* does not inhabit Tasmania or South Australia; its great nursery is evidently the south-eastern portion of the country.

Forehead, lores, and throat jet-black; all the upper surface grey; wings and tail brown; sides of the neck and the chest light grey; abdomen and under tail-coverts rufous; bill beautiful light blue-grey, the tip paler than the base; legs bluish lead-colour; irides black; inside of the mouth greyish blue.

In all probability, the females and the young males of the year are destitute of the black mark on the face.

Sp. 153.   MONARCHA TRIVIRGATA.

BLACK-FRONTED FLYCATCHER.

Drymophila trivirgata, Temm. Pl. Col. 418. fig. 1.
Monarcha trivirgata, Gould in Syn. Birds of Australia, part ii.

Monarcha trivirgata, Gould, Birds of Australia, fol, vol. ii. pl. 96.

Although the *Monarcha trivirgata* has been known to naturalists for many years, it is still a scarce bird, very few specimens occurring in any of the numerous collections sent home from Australia, which is doubtless occasioned by its true habitat not having been yet discovered. The specimens seen have been procured in the Moreton Bay district of the east coast.

All the examples that have come under my notice have been marked precisely alike, with the exception of one procured during the early part of Dr. Leichardt's expedition from

Moreton Bay to Port Essington, which differs in being destitute of the rufous tint on the flanks, and which may be a female.

I can perceive little or no difference between Australian examples and specimens brought by Mr. Wallace from the islands of Batchian and Timor.

In form and markings this species closely assimilates to the members of the genus *Arses*.

Forehead, throat, space round the eye, and the ears jet-black; upper surface dark grey; tail black, the three outer feathers on each side largely tipped with white; cheeks, chest, and flanks rufous; abdomen and tail-coverts white; bill lead-colour; feet black.

Sp. 154.   MONARCHA LEUCOTIS, *Gould*.

WHITE-EARED FLYCATCHER.

Monarcha leucotis, Gould in Proc. of Zool. Soc., part xviii. p. 201.

Monarcha leucotis, Gould, Birds of Australia, fol., Supplement, pl.

I have refrained from making the White-eared Flycatcher the type of a new genus until more information has reached us respecting it, and in the mean time have assigned it a situation with the other members of that form to which it seems to me to be most nearly allied. Like most of the other new birds figured in the Supplement to the Birds of Australia, it is a native of Cape York, and in all probability it ranges widely over the north coast. "Respecting this bird," says Mr. Macgillivray, "I regret to say I can afford you very little information. A specimen was obtained at Dunk Island, off the north-east coast of Australia, in lat. 17° 50′ S., where it was shot during its flight from one tree to another; a second individual was afterwards procured at Cape York, which renders it probable that its range extends between these two places."

Crown of the head, back of the neck, primaries and six middle tail-feathers black; three lateral tail-feathers on each side black, with white tips; lores, a broad mark over the eye, ear-coverts, sides of the neck, scapularies, and upper tail-coverts white; throat white, bounded below with black, the feathers lengthened and protuberant; chest and abdomen light grey; bill and feet lead-colour.

Total length 5¾ inches; bill ⅜; wing 2¾; tail 2¼; tarsi ⅝.

Family ——?

Genus GERYGONE, *Gould*.

The term *Psilopus* was originally proposed by me for this genus; but that name having been previously employed, *Gerygone* was substituted for it.

Several species inhabit Australia, and others, I believe, New Guinea and Polynesia. Their chief food consists of insects of the most diminutive size, such as aphides, gnats, and mosquitos. The more thickly-billed species may probably feed upon larger insects and their larvæ. They mostly frequent the thick umbrageous woods, where they flit about under the canopy of the dense foliage, or sally forth into the open glade like true Flycatchers. Their nests are of a domed form, with the entrance near the top, some species protecting the opening by constructing a projection above it like the peak of a cap; the eggs are generally four in number, and spotted with red like those of the *Maluri* and *Pari*.

All the members of the genus yet discovered are of small size, unobtrusive in colour, sprightly in their movements, and but little skilled in song. The sexes are similarly marked, and but slightly differ in outward appearance.

Sp. 155. GERYGONE ALBOGULARIS, *Gould*.

WHITE-THROATED GERYGONE.

Psilopus albogularis, Gould in Proc. of Zool. Soc., part v. p. 147.
—— *olivaceus*, Gould, ibid., p. 147, young?

Gerygone albogularis, Gould, Birds of Australia, fol., vol. ii. pl. 87.

This, so far as I know, is a stationary species, and is abundantly dispersed over all parts of New South Wales, but evinces a greater preference for the open forests of *Eucalypti* than for the brushes near the coast. I found it in considerable numbers in every part of the Upper Hunter district, nearly always among the gum-trees, and constantly uttering a peculiar and not very harmonious strain. It is very active among the small leafy branches of the trees, where it searches with the greatest avidity for insects, upon which it almost exclusively subsists, resorting for this purpose to trees of all heights, from the low sapling of two yards high to those of the loftiest growth.

I killed young birds in January, but was not so fortunate as to discover the nest.

The sexes are nearly alike in plumage; but the young of the year are distinguished from the adult by the throat being of the same colour as the breast, instead of white.

Crown of the head, ear-coverts, and all the upper surface olive-brown; throat white; chest and all the under surface bright citron-yellow; two centre tail-feathers brown, the remainder brown at the base, above which is a bar of white, succeeded by a broader one of deep blackish brown; the tips of all but the two middle ones buffy white on their inner web; bill blackish brown; irides scarlet; feet blackish brown in some specimens, and leaden brown in others.

Total length $4\frac{1}{4}$ inches; bill $\frac{1}{4}$; wing $2\frac{3}{8}$; tail $1\frac{3}{4}$; tarsi $\frac{3}{4}$.

Sp. 156.   GERYGONE FUSCA, *Gould.*

BROWN GERYGONE.

Psilopus fuscus, Gould in Proc. of Zool. Soc., part v. p. 147.
Gerygone fusca, Gould in De Strzelecki's Phys. Descr. of New South Wales and Van Diemen's Land, p. 821.

Gerygone fusca, Gould, Birds of Australia, fol., vol. ii. pl. 98.

The *Gerygone fusca* is an inhabitant of New South Wales, where it is to be found in all the brushes near the coast, as well as in those on the sides of the ranges in the interior. As its form would indicate, it has much of the habit of the Flycatcher, and lives almost exclusively upon insects, which are as frequently taken on the wing as they are from the under sides of leaves, &c. It particularly loves to dwell in the most retired and gloomy part of the forest, and is an active and lively little bird, flitting about from flower to flower, sometimes, like the true Flycatchers, sallying out into the open to capture an insect, and at others hanging to the under sides of the leaves, after the manner of the *Acanthizæ*.

Its feeble song is a pleasing, twittering sound, and is poured forth almost incessantly.

The breeding-season comprises the months of September, October, and November. The nest is a delicate and beautiful structure of a domed oblong form, the lower end terminating in a point, with the entrance at the side near the top covered with a well-formed spout, which completely excludes both sun and rain from the interior of the nest; it is about eight inches in height and ten in circumference, the spout projecting about two inches, and the entrance being scarcely an inch in diameter. The body of a nest found in the brushes of the Hunter was composed of green moss, mouse-eared lichen, soft wiry grasses, the inner bark of trees, and other materials, and was lined with extremely soft grasses. The eggs are three in number, and very similar, both in size and colour, to those of the *Malurus cyaneus*, being minutely speckled with red on a white ground; they are seven and a half lines long by five and a half lines broad.

The sexes are alike in colour.

Crown of the head, all the upper surface and wings dark fuscous brown, slightly tinged with olive; two centre tail-feathers brown; the remainder white at the base, succeeded by a broad band of deep blackish brown, round which is a broad stripe of white, which entirely crosses the outer feathers, but only the inner webs of the remainder, the tips pale brown; throat and chest grey; abdomen and under tail-coverts white; bill and feet deep blackish brown; irides bright brownish red.

Total length $3\frac{3}{4}$ inches; bill $\frac{1}{2}$; wing $2\frac{1}{4}$; tail $1\frac{3}{4}$; tarsi $\frac{3}{4}$.

Sp. 157. GERYGONE CULICIVORA, *Gould.*

WESTERN GERYGONE.

Psilopus culicivorus, Gould in Proc. of Zool. Soc., part viii. p. 174.
War-ryle-bur-dang, Aborigines of the lowlands of Western Australia.

Gerygone culicivorus, Gould, Birds of Australia, fol., vol. ii. pl. 69.

This species is plentifully dispersed over the colony of Swan River in Western Australia, where it inhabits forests, scrubs, and all situations where flowering trees abound, and where it is seen either in pairs or in small groups of four or five in number. Its food consists wholly of aphides and other small insects, which are captured on the wing or from off the flowers; it sometimes traverses the smaller branches, and even the upright boles of trees, prying about and searching for its prey with the most scrutinizing care. Its powers of flight are rarely exerted for any other purpose than to convey it from shrub to shrub, and for its little sallies in pursuit of insects, much after the manner of the true Flycatchers.

Its notes are very varied, being at one time a singing kind

of whistle, and at others a somewhat pleasing and plaintive melody; but it has a singular habit of uttering, when flitting from tree to tree, a succession of notes and half-notes, some of which are harmoniously blended, while others are equally discordant.

It is said by the natives to breed in September and October.

The nest is suspended by the top to the extremity of a branch, and is formed of threads of bark, small spiders' nests, green moss, &c., all felted together with cobwebs and vegetable fibres, and warmly lined with feathers; it is about eight inches in length, pointed at the top and at the bottom, and about nine inches in circumference in the middle; the entrance is a small round hole, about three inches from the top, with a slight projection immediately above it. I did not succeed in procuring the eggs.

The sexes are alike in plumage.

All the upper surface olive-brown; wings brown, margined with olive; two centre tail-feathers brown; the remainder white, crossed by an irregular band of black and tipped with brown, the band upon all but the external feathers so blending with the brown at the tip that the white between merely forms a spot on the inner web; lores blackish brown; line over the eye, throat, and chest light grey, passing into buff on the flanks, and into white on the centre of the abdomen and under tail-coverts; irides light reddish yellow; bill and feet black.

Total length $4\frac{1}{4}$ inches; bill $\frac{1}{2}$; wing $2\frac{1}{4}$; tail $1\frac{3}{4}$; tarsi $\frac{5}{8}$.

Sp. 158. GERYGONE MAGNIROSTRIS, *Gould*.

GREAT-BILLED GERYGONE.

Gerygone magnirostris, Gould in Proc. of Zool. Soc., part x. p. 183.

Gerygone magnirostris, Gould, Birds of Australia, fol., vol. ii. pl. 100.

Of this species I regret to say but little information has as yet been received; the two examples in my collection are all that have come under my notice; and these were shot by Gilbert on Greenhill Island near Port Essington, while hovering over the blossoms of the mangroves and engaged in capturing the smaller kinds of insects, during which occupation it gave utterance to an extremely weak twittering song: unfortunately he had no further opportunity of making himself acquainted with its habits and manners; but they doubtless resemble those of the other members of the genus.

All the upper surface brown; margins of the primaries slightly tinged with olive; tail-feathers crossed near the extremity by an indistinct broad band of brownish black; all the under surface white, tinged with brownish buff; irides light brown; bill olive-brown; the base of the lower mandible pearl-white; feet greenish grey.

Total length $3\frac{3}{4}$ inches; bill $\frac{7}{8}$; wing $2\frac{1}{4}$; tail $1\frac{3}{8}$; tarsi $\frac{3}{4}$.

Sp. 159. GERYGONE LÆVIGASTER, *Gould*.

BUFF-BREASTED GERYGONE.

Gerygone lævigaster, Gould in Proc. of Zool. Soc., part x. p. 183.

Gerygone lævigaster, Gould, Birds of Australia, fol., vol. ii. pl. 101.

Gilbert killed several specimens of this little bird on the Cobourg Peninsula, and on the islands in Van Diemen's Gulf, and sometimes observed a solitary individual among the mangroves near the settlement of Port Essington. He states that it has a very pleasing but weak piping note, and occasionally utters a number of notes in slow succession, but not so much lengthened as those of the *Gerygone culicivora* of Swan River; like that bird, it hovers before the smaller leafy branches of the trees and creeps about the thickets. It is very tame, and scarcely ever flies from the tree upon the approach of an intruder, but sits turning its little head about from side to side until the hand is almost upon it, when it merely hops upon another branch and again quietly looks about, apparently quite unconcerned.

The stomach is tolerably muscular, and the food consists of small insects, principally of the soft-winged kinds.

A narrow obscure line, commencing at the nostrils and passing over the eye, yellowish white; all the upper surface rusty brown; primaries brown, margined with lighter brown; tail whitish at the base, gradually deepening into nearly black, the lateral feather largely and the remainder, except the two middle ones, slightly tipped with white; all the under surface white, slightly washed with yellow; irides light reddish brown; bill olive-brown; base of lower mandible light ash-grey; feet dark greenish grey.

Total length $3\frac{3}{4}$ inches; bill $\frac{1}{2}$; wing 2; tail $1\frac{1}{4}$; tarsi $\frac{3}{4}$.

Sp. 160. GERYGONE CHLORONOTUS, *Gould*.

GREEN-BACKED GERYGONE.

Gerygone chloronotus, Gould in Proc. of Zool. Soc., part x. p. 133.

Gerygone chlorontus, Gould, Birds of Australia, fol., vol. ii. pl. 102.

This species is an inhabitant of the northern parts of Australia; it is tolerably abundant at Port Essington, where it dwells among the extensive beds of mangroves which stretch along the coast. It is of a very shy and retiring disposition; and as the colouring of its back assimilates very closely to that of the leaves of the mangroves, it is a very difficult bird to sight as it creeps about among the thick branches in search of insects, upon which it solely subsists. In form and in most of its habits and economy it offers some difference from the typical members of the genus *Gerygone*; and it would be no great stretch of propriety to assign to it a new generic appellation: the more lengthened form of its legs, the more rigid structure of its primaries, and the less development of the bristles at the gape are among the points in which it differs from the *Gerygone fusca* of the brushes of New South Wales.

The sexes are so precisely similar in plumage, and differ so little in size, that dissection must be resorted to to distinguish the one from the other.

Head and back of the neck brownish grey; back, wing-coverts, rump, upper tail-coverts, margins of the primaries, and the margins of the basal half of the tail-feathers bright olive-green; primaries and tail-feathers brown, the latter becoming much darker towards the extremity; under surface white; sides and vent olive-yellow; irides wood-brown; upper mandible greenish grey; lower mandible white; feet blackish grey.

Total length $3\frac{1}{2}$ inches; bill $\frac{9}{16}$; wing $2\frac{1}{4}$; tail $1\frac{3}{4}$; tarsi $\frac{3}{4}$.

Genus SMICRORNIS, *Gould*.

The members of this genus are the smallest birds of the Australian fauna. I have described two species, one inhabiting New South Wales, and the other Port Essington; and had I characterized the bird of this form found in Western Australia as distinct, I should probably not have been in error.

As it is impossible to convey a just conception of these diminutive birds by written descriptions, I must request those readers who have the opportunity to consult the Plates in the folio work, on which the species are represented.

Sp. 101. SMICRORNIS BREVIROSTRIS, *Gould*.

SHORT-BILLED SMICRORNIS.

Psilopus brevirostris, Gould in Proc. of Zool. Soc., part v. p. 147.
Geab-ter-but, Aborigines of the mountain districts of Western Australia.

Smicrornis brevirostris, Gould, Birds of Australia, fol., vol. ii. pl. 103.

This bird is a constant inhabitant of the leafy branches of the *Eucalypti*, and resorts alike to those of a dwarf stature and those of the loftiest growth. While searching for insects, in which it is incessantly engaged, it displays all the scrutinizing habits of the *Pari* or Tits, clinging about the finest twigs of the outermost branches, prying underneath and above the leaves and among the flowers, uttering all the while or very frequently a low simple song. I found it abundant in every part of South Australia I visited, particularly in the neighbourhood of Adelaide and in the gullies of the ranges skirting the belts of the Murray; in New South Wales it was frequently seen at Yarrundi, and other parts of the Upper Hunter district. Gilbert states that in Western Australia he only met with it in the York district, that it was always seen on the branches of trees, where it feeds on larvæ and small insects, that its flight was of very short duration, merely flitting from tree to tree, and that its note is a weak twitter, a good deal resembling that of the *Geobasileus chrysorrhous*.

It breeds in September and the two following months, and forms a nest of the downy buds of plants, mixed with green moss, the cocoons of spiders, &c., all matted and bound together very firmly and closely with spiders' webs, and the inside lined at the bottom with feathers; it is globular in form, and is attached by the back part to an upright branch, with the entrance in the side, the upper part over the entrance being carried out to a point, which shades the

T

opening like the eaves of a house. The eggs are three in number, of a dull buff, marked with extremely fine freckles at the larger end; they are six and a half lines long by four and a half lines broad.

A narrow stripe of yellowish white passes from the bill over each eye; crown of the head brownish grey, passing into olive at the back of the neck; back, rump, and upper tail-coverts olive, brightest on the latter; ear-coverts and sides of the face very pale reddish brown; throat and chest white, tinged with olive, with a faint longitudinal mark of brown down the centre of each feather, the remainder of the under surface pale citron-yellow; two centre tail-feathers brown; the remainder brown at the base, the middle being crossed by a broad band of blackish brown, which is succeeded by a spot of white on the inner webs, the tips pale brown; feet blackish brown; irides pale straw-yellow; bill varying from fleshy white to ashy grey.

Total length $3\frac{1}{4}$ inches; bill $\frac{3}{8}$; wing 2; tail $1\frac{1}{2}$; tarsi $\frac{1}{2}$.

Sp. 162. SMICRORNIS FLAVESCENS, *Gould*.

YELLOW-TINTED SMICRORNIS.

Smicrornis flavescens, Gould in Proc. of Zool. Soc., part x. p. 134.

Smicrornis flavescens, Gould, Birds of Australia, fol., vol. ii. pl. 104.

This is the least of the Australian birds I have yet seen, scarcely exceeding in size the smaller Humming-birds. It is tolerably abundant on many parts of the northern coasts of Australia, and particularly on the Cobourg Peninsula; it inhabits most of the high trees in the neighbourhood of Port Essington, keeping to their topmost branches, and there seeking its insect food among the leaves, over which it creeps and clings in every possible variety of position. From the circumstance of its confining itself exclusively to the topmost branches of the trees, it is not easily procured, its diminutive size preventing its being seen.

There is no outward difference in the sexes, either in size or plumage. Future research, and a longer sojourn in the country than has hitherto been afforded for the investigation of the natural productions of those distant parts, are requisite to determine whether it be migratory or not, and to procure correct information respecting its nidification.

All the upper surface bright yellowish olive; the feathers of the head with an indistinct line of brown down the centre; wings brown; tail brown, deepening into black near the extremity, and with a large oval spot of white on the inner web near the tip of all but the two central feathers; all the under surface bright yellow.

Total length $2\frac{3}{4}$ inches; bill $\frac{6}{8}$; wing $\frac{7}{8}$; tail $1\frac{1}{4}$; tarsi $\frac{9}{8}$.

Family SAXICOLIDÆ.

Genus ERYTHRODRYAS, *Gould*.

The birds of this form are much more delicate in structure than the members of the restricted genus *Petroica*, have their feeble bill strongly beset with bristles, and are more arboreal in their habits; their usual places of resort being the innermost recesses of the forest, where, in a state of quiet seclusion, they flit about in search of insects; the true *Petroicæ*, on the other hand, frequent open plains, are more bold and vigorous, and possess a structure which adapts them for the ground, over which they pass like the Wheatears.

The two species of this genus, all that are at present known, are confined to the south-eastern portions of Australia and Tasmania.

We may naturally conclude that, in their mode of nidification, the form of their nests, and in the number and colour of their eggs, they will very closely resemble the true *Petroicæ*.

Sp. 103. ERYTHRODRYAS RHODINOGASTER.

Pink-breasted Wood-Robin.

Saxicola rhodinogaster, Drap. Ann. Gén. des Sci. Phys. de Bruxelles.
Muscicapa lathami, Vig. in Zool. Journ., vol. i. p. 410, pl. 13.
Petroica rhodinogaster, Jard. and Selb. Ill. Orn. Add., vol. ii.
Erythrodryas rhodinogaster, Gould in Proc. of Zool. Soc., part i. p. 112.
Pink-breasted Robin, Colonists of New South Wales.

Erythrodryas rhodinogaster, Gould, Birds of Australia, fol., vol. iii. pl. 1.

The principal habitat of this species is Tasmania, where I shot several specimens in the gullies under Mount Wellington; it is also abundant on the Hampshire Hills of that island. In one instance only did I meet with it on the continent, in a deep ravine under Mount Lofty in South Australia; I shot the specimen, which on dissection proved to be a young male.

In habits and disposition this and the following species differ considerably from the Red-breasted Robins (*Petroicæ*), and are much less spirited in all their actions. They prefer the most remote parts of the forest, particularly the bottoms of deep gullies, the seclusion of which is seldom disturbed by the presence of man, and where animal life is almost confined to aphides and other minute insects. There are times, however, especially in winter, when they leave these quiet retreats, and even enter the gardens of the settlers; but this is of rare occurrence.

The food of the Pink-breasted Wood-Robin consists solely of insects, which it generally procures by pursuing them in the air.

The nest is formed of narrow strips of soft bark, soft fibres of decaying wood, and fine fibrous roots matted and woven together with vegetable fibres, and old black nests of spiders. The eggs are three in number, of a greenish white, thickly sprinkled with light chestnut and purplish brown; eight lines and a half long, by six lines and a half broad.

Like the true Petroicas, the sexes present considerable differences in their colouring.

The male has the head, neck, throat, and back sooty black; a small spot of white in the centre of the forehead; wings brownish black; a few of the primaries and secondaries with an oblong spot of reddish brown on the outer web near the base, and another near the tip, forming two small oblique bands when the wing is spread; breast and abdomen rose-pink, passing into white on the vent and under tail-coverts; irides and bill black; feet black, with the soles orange.

The female has an indication of the white spot on the forehead; all the upper surface brown; wings and tail brown, with the markings on the primaries and secondaries larger, and of a more buffy colour than in the male; throat brownish buff; chest and abdomen brownish grey; vent and under tail-coverts buff.

The young male during the first autumn closely resembles the female; for the first two months after they have left the nest, they have the centre of each feather striated with buff.

Sp. 164.   ERYTHRODRYAS ROSEA.
ROSE-BREASTED WOOD-ROBIN.

Petroica rosea, Gould in Proc. of Zool. Soc., part vii. p. 142.
Erythrodryas rosea, Gould, ibid. part x. p. 112.

Erythrodryas rosea, Gould, Birds of Australia, fol., vol. iii. pl. 2.

This pretty little Robin inhabits all the brushes skirting the south-eastern coast of New South Wales. I also observed it to be numerous in the cedar brushes of the Liverpool range. It is a solitary species, more than a pair being rarely seen at one time, is excessively quiet in its movements, and so tame that, in the course of my wanderings through the woods of Illawarra and in the neighbourhood of the Hunter, it frequently perched within two or three yards of me. What has been said respecting the habits and manners of the

Pink-breasted Robin is equally descriptive of those of the present bird.

Of its nidification and the number and colour of its eggs nothing is at present known.

Its cheerful song is very like that of the other Robins, but is much more feeble.

The male has the forehead crossed by a very narrow band of white; crown of the head, throat, and all the upper surface dark slate-grey; chest rich rose-red, inclining to scarlet; lower part of the abdomen and under tail-coverts white; wings and the six central tail-feathers blackish brown; the three outer ones on each side tipped with white, the white predominating over the inner webs, particularly on the two lateral feathers; bill and feet blackish brown; gape and soles of the feet yellow.

The female differs considerably from her mate, having the forehead crossed by a narrow band of buff; all the upper surface greyish brown; wings brown; secondaries crossed by two obscure bands of greyish buff; tail of a browner tint, but otherwise marked like that of the male.

Total length $4\frac{1}{4}$ inches; bill $\frac{1}{3}$; wing $2\frac{3}{4}$; tail $2\frac{1}{4}$; tarsi $\frac{3}{4}$.

Genus PETROICA, *Swains*.

Several species of this genus inhabit Australia, where they form a conspicuous feature in the landscape, their bright red colouring offering a strong contrast to the sombre tint of the ground, upon which they dwell. They are very sprightly in their actions, and, like the Wheatears, raise their breasts in such a manner as to show off their fine colours to the greatest advantage. They build round, cup-shaped nests, in the crevice of a tree or in a wall, and sometimes under a stone or at the base of a rock. The sexes are very dissimilar, the males being adorned with black, white, and scarlet, while the females are of a sombre brown.

Sp. 165.   PETROICA MULTICOLOR.

SCARLET-BREASTED ROBIN.

Muscicapa multicolor, Vig. and Horsf. in Linn. Trans., vol. xv. p. 243.
Red-breasted Warbler, Lewin, Birds of New Holl., pl. 17.
Petroica multicolor, Swains. Zool. Ill., 2nd ser. pl. 86.
Petroeca multicolor, Cab. Mus. Hein., Theil i. p. 11.
Goo-la, Aborigines of Western Australia.
Robin of the Colonists.

Petroica multicolor, Gould, Birds of Australia, fol., vol. iii. pl. 3.

This beautiful Robin enters the gardens of the settlers in New South Wales, and is a great favourite, its attractiveness being much enhanced by its gay attire—the strong contrasts of scarlet, jet-black, and white rendering it one of the most beautiful to behold of any of the birds of Australia. After a careful comparison of a large number of specimens, I feel fully satisfied that the scarlet breast of this species, like that of the Robin of Europe, is assumed during the first autumn, and that it is never again thrown off; but, as might be expected, it is much more brilliant and sparkling during the breeding-season than at any other period of the year. A slight difference exists in the depth of the colouring of specimens from the western and eastern coasts, those from the former, particularly the females, having the scarlet more brilliant and of greater extent than those from New South Wales and Tasmania; the difference, however, is too trivial to be regarded otherwise than as indicative of a mere variety.

Its song and call-note much resemble that of the European Robin, but are more feeble, and uttered with a more inward tone.

The nest is a very compact structure of dried grasses, narrow strips of bark, mosses, and lichens, all bound firmly together with cobwebs and vegetable fibres, and warmly lined with feathers and wool or hair; in some instances I have seen it lined entirely with opossum's hair; it is generally placed in the hollow part of the trunk of a tree, or in a slight cavity in the bark six or seven feet from the ground; but I have found it placed in a fork of a small upright tree more than thirty feet from the ground. The eggs, which are three or four in number, are greenish white, slightly tinged with bluish or flesh-colour, rather minutely freckled with olive-brown and purplish grey, the latter more obscure than the former; these freckles are very generally dispersed over the surface of the shell, but in some instances they also form a zone near the larger end: the medium length of the eggs is nine lines, and breadth seven lines.

It usually rears two or three broods in the year, the period of nidification commencing in August and ending in February.

The male has the head, throat, and upper surface black; forehead snowy white; a longitudinal and two oblique bands of white on the wings; breast and upper part of the belly scarlet; lower part of the belly dull white; irides very dark brown; bill and feet black.

The female has all the upper and under surface brown, with the breast strongly tinged with red.

Sp. 100.   PETROICA GOODENOVII.

Red-capped Robin.

Muscicapa goodenovii, Vig. and Horsf. in Linn. Trans., vol. xv. p. 245.
Petroica goodenovii, Jard. and Selb. Ill. Orn., Add. vol. ii.
Petræca goodenovii, Cab. Mus. Hein., Theil i. p. 11.
Mé-ne-gĕ-dang, Aborigines of the mountain districts of Western Australia.
Red-capped Robin of the Colonists.

Petroica goodenovii, Gould, Birds of Australia, fol., vol. iii. pl. 6.

Its red crown and much smaller size at once distinguish this Robin from every other species of the genus yet discovered. Although not plentiful in any part I have visited, it is very generally distributed over the whole of the southern portion of Australia; and was killed by Gilbert in Western Australia, where, however, it is very local, for he only met with it in two localities, one in the York district, and the other at Kojenup, about one hundred miles towards the interior from King George's Sound.

I generally observed it either singly or in pairs, and it appeared to give a decided preference to the beds of dry rivulets, and to thinly timbered plains, the dense brushes near the coast never being visited by it; it would seem therefore to be a species peculiar to the interior of the country.

The whole of the actions and economy of this bird closely assimilate to those of the *Petroica multicolor*; of its nidification but little information has yet been obtained; I possess an egg which may be described as of a bluish white, with numerous fine speckles, particularly at the larger end, of yellowish brown and purplish grey, the latter appearing as if beneath the shell; it is five-eighths of an inch long, by half an inch wide.

It possesses a peculiarly sweet and plaintive song, very much like that of the European Robin, but more weak and not so continuous.

The male has the upper surface, neck, upper part of the breast, and wings brownish black; wing-coverts and secondaries edged with white, forming a broad stripe along the wings; middle of the outer web of the quills with a narrow white margin; forehead, crown, and lower part of the breast bright scarlet, passing into white on the vent; irides, bill, and feet blackish brown; soles of the feet yellow.

The female, as is the case with the females of the other species, differs much from her mate in the colouring of the plumage.

Sp. 167.   PETROICA PHŒNICEA, *Gould.*

FLAME-BREASTED ROBIN.'

Petroica phœnicea, Gould in Proc. of Zool. Soc., part iv. p. 105.
Petroeca phœnicea, Cab. Mus. Hein., Theil i. p. 11.
Muscicapa erythrogaster, var., Lath.

Petroica phœnicea, Gould, Birds of Australia, fol., vol. iii. pl. 8.

Tasmania and the south-eastern portion of the Australian continent constitute the natural habitat of this species; in the former country it is very common, but in New South Wales and South Australia it is not so numerous, and is very local. It is far less arboreal than the *Petroica multicolor*, giving a decided preference to open wastes and cleared lands rather than to the woods: in many of its actions it much resembles the Wheatears and other true Saxicoline birds, often selecting a large stone, clod of earth or other substance, on which to perch and show off its flame-coloured breast to the greatest advantage. As the season of nidification approaches, it retires to the forests for the purpose of breeding, and builds its cup-shaped nest in the hole of a tree, in the cleft of a rock, or any similar situation. It is a very familiar species, seeking rather than shunning the presence of man, and readily taking up its abode in his gardens, orchards, and other cultivated grounds; I have even taken its nest from a shelving bank in the streets of Hobart Town.

It has a pretty, cheerful song, uttered somewhat low and inwardly; the male generally sings over or near the female while she is sitting upon her eggs.

The nest, which is thick and warm, is formed of narrow strips and thread-like fibres of soft bark, matted together with cobwebs and sometimes wool, and lined with hair and feathers, or occasionally with fine hair-like grasses. The general colour of the eggs is greenish white, spotted and freckled with purplish and chestnut-brown: much variety occurs in these markings, some assuming the form of large, bold irregular spots and blotches, while in others they are merely minute freckles; the eggs are three in number; their medium length nine lines, and breadth seven lines.

The male has the crown of the head and all the upper surface sooty grey, except a small white spot across the forehead, a patch of the same colour on the shoulders and the anterior edges of the tertials; primaries and tail-feathers greyish black, except the outer feathers of the latter, which are nearly all white; the second tail-feather on each side is also tinged with white; upper part of the throat sooty grey, the rest of the under surface rich scarlet; under tail-coverts white; irides, bill, and feet black.

The female is uniform brown above; wings dark brown; tertials and wing-coverts edged with reddish grey; tail brown; the outer tail-feathers on each side almost wholly white; all the under surface reddish grey; irides, bill, and feet black.

Genus MELANODRYAS, *Gould*.

For the Pied Robins, of which at least two species inhabit Australia, I propose the generic term of *Melanodryas*.

Sp. 168.   MELANODRYAS CUCULLATA.

Hooded Robin.

Muscicapa cucullata, Lath. Ind. Orn. Supp., p. 51?
Hooded Flycatcher, Lath. Gen. Syn. Supp., vol. ii. p. 223?
Petroica bicolor, Swains. Ill. Zool., 2nd ser. pl. 43.
Petrœca cucullata, Cab. Mus. Hein., Theil i. p. 11.
Grallina bicolor, Vig. and Horsf. in Linn. Trans., vol. xv. p. 223.
Jil-but, Aborigines of the mountain districts of Western Australia.
Goö-ba-môgin, Aborigines around Perth, Western Australia.
Black Robin of the Colonists.

Petroica bicolor, Gould, Birds of Australia, fol., vol. iii. pl. 7.

The *Melanodryas cucullata* inhabits New South Wales, Victoria, South Australia, and Swan River, but not Tasmania. It loves to dwell in the open parts of the country rather than in the thick brushes. I have always found it most numerous on such flats as were studded here and there with large trees, among the lower branches of which, as well as on the ground immediately beneath them, it might be observed darting about for insects in the most bold and active manner; the jet-black colouring of its upper surface, contrasted with the whiteness of the other parts, rendering it very conspicuous, particularly when its wings and tail are displayed to their full extent.

Its food consists solely of insects of various kinds, particularly coleoptera and their larvæ.

The breeding-season commences in September, and continues during the four following months; in this period two broods at least are reared. The nest, which is rather small and shallow, is formed of dried grasses, strips of bark, and fibrous roots, bound together and partly smoothed over with cobwebs, the inside being lined with fine wire-like fibres, and generally a little wool at the bottom; it is placed on the dried branch of a small tree, resting against the trunk, or in the fork of a fallen branch within two or three feet of the ground. The eggs, which are three in number and of a rather lengthened form, are light olive-green without any spots or markings, but occasionally washed with brown, particularly at the larger end; their medium length is ten lines and a half, and breadth seven lines and a half.

The male has the head, throat, neck, back, rump, upper tail-coverts, and the two centre tail-feathers deep velvety black; the next tail-feather on each side black on the inner web, white on the outer web, and largely tipped with black, the remainder of the tail-feathers white, largely tipped with black; feathers covering the insertion of the wing white; wings dull black, the secondaries edged with white; an oblique band of white across all but the two first primaries near their base; under surface of the shoulder, breast, abdomen, and under tail-coverts white; irides brownish black; bill black; feet blackish brown.

The female has the upper surface dark brownish grey; wings brown, with the oblique band less prominent than in the male; under surface light brownish grey, passing into white on the vent and under tail-coverts; tail brown, the lateral feathers white at the base, the white continuing to near the tip on the external web of the outer feather.

The young immediately after leaving the nest is dark brown, with a stripe of light brown down the centre of each feather, the markings of the wings and tail resembling those of the adult; under surface like the upper, but becoming white as it proceeds towards the vent.

Sp. 160. MELANODRYAS PICATA, *Gould*.

PIED ROBIN.

For many years I have had in my possession skins of two Pied Robins, one from the north-west, and the other, which is somewhat mutilated, and perhaps a female, from Port Essington. In all probability they are two distinct species, both differing from the *M. cucullata* of New South Wales; I shall here, however, only describe the one from the north-west coast. The specimen is that of a fully adult male. In its colour and general form it is very like the *M. cucullata*, but is much smaller than ornithologists admit to constitute a mere race or variety.

Head, throat, neck, back, and wings black; scapularies, bases of the innermost primaries and the secondaries and under surface white; tail black, the lateral feathers white for two-thirds of their length from the base; bill and legs black.

Total length 5¼ inches; bill ⅜; wing 3⅛; tail 2¼; tarsi ⅞.

Genus AMAURODRYAS, *Gould.*

The well-known Dusky Robin of Tasmania differs in several particulars from the true *Petroica*, not only in colour, but in the stouter and more robust or thicker form of the bill; its eggs are also very different from those of the *Petroica*. The sexes are alike in colouring.

Sp. 170.   AMAURODRYAS VITTATA.

DUSKY ROBIN.

Muscicapa vittata, Quoy et Gaim. Voy. de l'Astrolabe, pl. 8. fig. 2 ?

Petroica fusca, Gould, Birds of Australia, fol., vol. iii. pl. 8.

This plain-coloured species is very abundantly distributed over all those parts of Tasmania that are suitable to its habits; it gives preference to thinly-timbered hills, and all such plains and low grounds as are sterile and covered with thickets and stunted brushwood. In its manners and whole economy it assimilates to the Red-breasted Robins; I frequently observed it sitting on the stumps of dead and fallen trees, on the railings of inclosures, gardens, and other similar situations. Its food appeared to consist solely of insects, which it swallows entire, even coleoptera of a large size.

Its nest, which is rather large and of a cup-shape, is formed of coarse fibrous roots, small twigs, strings of bark and dried grasses intermixed with very fine hair-like fibrous roots, wool, and the soft seed-stalks of mosses. The size and form of the nest depend upon the nature of the situation chosen for a site; if a ledge or fissure of a rock, it is much spread out, but with the inside and top very neatly finished; the opening measures on an average about two inches and a half, and the nest is about one inch and a quarter in depth.

The eggs, which are three or four in number, differ in colour from those of every other member of the genus, but more nearly assimilate in tint and markings to those of

Petroica bicolor than of any other. They are of a light greenish blue, freckled and spotted with minute indistinct markings of brown; their medium length is ten lines, and breadth seven and a half lines.

Although I have paid considerable attention to the distribution of this species, I have never been able to meet with it on the continent of Australia, or in any other country than Tasmania. It is very numerous about Hobart Town, both in the gullies under Mount Wellington, and on the opposite side of the Derwent towards Clarence Plains.

Its note is low and monotonous, without any peculiar character.

The sexes are alike in colour.

Head, and all the upper surface reddish brown tinged with olive; wings and tail brown; primaries and secondaries crossed by a narrow line of white at the base; the outer tail-feather on each side margined externally and at the tip with white; under surface pale brown, passing into buffy white on the vent and under tail-coverts; irides, bill, and feet blackish brown.

The young is very dark brown above, striated with deep buff; beneath mottled brown and buffy white; the latter colour occupying the centre of the feathers.

Genus PŒCILODRYAS, *Gould*.

On reference to the figures of the birds I have called *Petroica? cerviniventris* and *P. superciliosa* in the third volume of the folio edition and in the Supplement, it will at once be seen that these two species cannot be associated with either of the preceding genera, and must be separated into a new one; this division I have accordingly made, and assigned to it the above appellation.

Sp. 171. PŒCILODRYAS CERVINIVENTRIS, *Gould*.

BUFF-SIDED ROBIN.

Petroica? cerviniventris, Gould in Proc. of Zool. Soc., part xxv. p. 221.

Petroica? cerviniventris, Gould, Birds of Australia, fol., Supplement, pl.

So far as regards Ornithological science, it was fortunate that Mr. Elsey remained for a long time encamped near the Victoria River, on the north-west coast of Australia, since it enabled him to pay much attention to the natural objects which surrounded him; and the discovery of the present bird, which is quite new to science, is one of the results of his long stay in that spot in charge of a portion of Mr. Gregory's Expedition. All who have read my work on the Birds of Australia, will have observed that a species of this form, collected by Gilbert in the neighbourhood of the Burdekin Lakes, towards the Gulf of Carpentaria, is figured in the third volume under the name of *P. superciliosa*; to this species the one here described is very nearly allied—so nearly, in fact, that, although I have treated them as distinct, a suspicion has arisen in my mind that they may be the sexes of one and the same species; they both differ in form from the typical or true *Petroicæ*, and are doubtless representatives of each other in the respective countries they inhabit, the *P. superciliosa* dwelling on the eastern parts of the continent, and the *P. cerviniventris* in the western.

The following is a correct description of the latter:—

All the upper surface, wings, and tail chocolate-brown; line over the eye, throat, tips of the greater wing-coverts, base of the primaries, base and tips of the secondaries, and tips of the tail white; breast grey; abdomen deep fawn-colour, becoming almost white in the centre; bill black; feet blackish brown; irides dark brown.

Total length $6\frac{1}{4}$ inches; bill $\frac{3}{4}$; wing $3\frac{1}{4}$; tail $3\frac{1}{4}$; tarsi $\frac{7}{8}$.

The original specimen from which the above description was taken is now in the British Museum.

Sp. 172. PŒCILODRYAS SUPERCILIOSA, *Gould.*

WHITE-EYEBROWED ROBIN.

Petroica superciliosa, Gould in Proc. of Zool. Soc., part xiv. p. 100.

Petroica superciliosa, Gould, Birds of Australia, fol., vol. iii. pl. 9.

For our knowledge of this species we are indebted to the researches of Gilbert, who, while in company with Dr. Leichardt, during his adventurous expedition from Moreton Bay to Port Essington, discovered it in the neighbourhood of the Burdekin Lakes towards the Gulf of Carpentaria. The following remarks in Gilbert's Journal comprise all that is at present known respecting it:—"May 14th. In a ramble with my gun I shot a new bird, the actions of which assimilate to those of the *Petroicæ* and the *Eopsaltriæ*: like the former it carries its tail very erect, but is more retiring in its habits than those birds; on the other hand, its notes resemble those of the latter. It inhabits the dense jungle-like vegetation growing beneath the shade of the fig-trees on the banks of the Burdekin. I succeeded in procuring two specimens."

Superciliary stripe, throat, abdomen, under surface of the shoulder, and the bases of the primaries and secondaries white; lores, ear-coverts, wing-coverts, and the primaries and secondaries for some distance beyond the white deep black; all the upper surface, wings, and tail sooty brown; all but the two central tail-feathers largely tipped with white; bill and feet black; irides reddish brown.

Total length 6 inches; bill ¾; wing 3; tail 2½; tarsi ⅞.

Genus DRYMODES, *Gould.*

Two species only of this genus have yet been discovered, and these, as their long legs would indicate, are denizens of the ground. One of them was figured in the folio edition, and the other in the Supplement; the latter was obtained near Cape York.

Sp. 173. DRYMODES BRUNNEOPYGIA, *Gould*.

SCRUB-ROBIN.

Drymodes brunneopygia, Gould in Proc. of Zool. Soc., part viii. p. 170.

Drymodes brunneopygia, Gould, Birds of Australia, fol., vol. iii. pl. 10.

I discovered this singular bird in the great Murray Scrub, where it was tolerably abundant; I have never seen it from any other part of the country, and it is doubtless confined to such portions of Australia as are clothed with a similar character of vegetation. It is a quiet and inactive species, resorting much to the ground, over which and among the underwood it passes with great ease; it appeared rarely to take wing, but to depend for security upon its dexterity in hopping away to the more scrubby parts. I have occasionally observed it mount to the most elevated part of a low bush, and there pour forth a sharp monotonous whistling note, not very unlike that of some of the *Pachycephalinæ*; indeed it was its note that first attracted my attention to it. When on the ground, and occasionally when perched on a twig, it elevates its tail considerably, but not to the extent of the *Maluri*.

The sexes are alike in colouring, but the female is much smaller than her mate; the young resemble the immature *Petroicæ* in the character of their plumage.

Head and all the upper surface brown, passing into rufous brown on the upper tail-coverts; wings dark brown, the coverts and primaries edged with dull white; primaries and secondaries crossed near the base on their inner webs with pure white; tail rich brown, all but the two middle feathers tipped with white; under surface greyish brown, passing into buff on the under tail-coverts; irides, bill, and feet blackish brown.

Total length 8 inches; bill $\frac{7}{8}$; wing $3\frac{7}{8}$; tail $4\frac{1}{4}$; tarsi $1\frac{1}{4}$.

Sp. 174. DRYMODES SUPERCILIARIS, *Gould.*

EASTERN SCRUB ROBIN.

Drymodes superciliaris, Gould in Proc. of Zool. Soc., part xviii. p. 200.
Trokāroo, Aborigines of Cape York.

Drymodes superciliaris, Gould, Birds of Australia, fol., Supplement, pl.

Perhaps one of the most interesting birds discovered by me in the brushes of South Australia was a species of this form, to which I gave the name of *Drymodes brunneopygia*; this second species of the genus is an inhabitant of the northeast coast; and it will be seen by the following notes by Mr. Macgillivray that the two birds, as might be supposed, accord as nearly in their habits as they are allied in structure.

" While traversing on the 17th of November, 1849, a thin open scrub of small saplings growing in a stony ground thickly covered with dead leaves, about five or six miles inland from Cape York, I observed a nest placed on the earth at the foot of a small tree; its internal diameter was four inches and a half; it was outwardly composed of small sticks, with finer ones inside, and lined with grass-like fibres, and was moreover surrounded with dead leaves heaped up to a level with its upper surface; it contained two eggs an inch long by seven-tenths of an inch broad, of a regular oval shape, and of a very light stone-grey thickly covered with small umber blotches, which increased in size and were more thickly placed at the larger end: they were placed side by side, with the large end of one opposite the small end of the other. After watching near the nest for some time, one of the owners appeared, and was procured; but putrefaction having commenced before my return to the ship, I could not ascertain the sex with certainty: it approached me within three or four yards, hopping with sudden jerks over the leaves, and moving by fits and starts like the Robin of Europe; it uttered no cry ornate during the time I was watching its motions. Two others were afterwards procured in the same kind of open scrub, and the birds, being probably in the immediate neighbourhood of their nest, hopped up quite close to the observer."

This is a much more gaily attired species than the last, its back and tail-feathers being rich reddish brown, which, with the black and white markings about its face and the two white bands across its wings, render it conspicuously different.

The sexes assimilate in colour, but the female is somewhat smaller than the male.

Lores white; immediately above and below the eye a black mark forming a conspicuous moustache; crown of the head and upper surface reddish brown, passing into chestnut-red on the rump and six middle tail-feathers; remainder of the tail-feathers black, tipped with white; wings black, with the base of the primaries and the tips of the coverts white, forming two bands across the wing; throat and centre of the abdomen fawn-white; chest and flanks washed with tawny; irides umber-brown; legs and feet flesh-colour.

Total length $6\frac{1}{4}$ inches; bill $\frac{7}{8}$; wing $3\frac{3}{4}$; tail 4; tarsi $1\frac{1}{8}$.

Genus EOPSALTRIA, *Gould*.

At least four species of this form are known; two of these are natives of Western Australia, and two inhabit the eastern portion of the country. Although generically distinct from, they are very nearly related to the *Petroicæ*. They are all more arboreal in their habits than those birds, and also differ from them in the silky character of their plumage and in the prevalence of yellow in their colouring. The females generally lay only two eggs, while those of the *Petroicæ* lay four.

Sp. 175.　　EOPSALTRIA AUSTRALIS.

YELLOW-BREASTED ROBIN.

Muscicapa australis, Lath. Ind. Orn. Supp., p. li.
Southern Motacilla, Motacilla australis, White's Journ., pl. in p. 239.
Southern Flycatcher, Lath. Gen. Syn. Supp., vol. ii. p. 219.
Sylvia flavigastra, Lath. Ind. Orn. Supp., p. liv ?
Todus flavigaster, Lath. Ind. Orn., vol. i. p. 168.
Pachycephala australis, Vig. and Horsf. in Linn. Trans., vol. xv. p. 242.
Muscipeta, sp. 15, *Muscicapa australis*, Less. Traité d'Orn., p. 385.
Eöpsaltria flavicollis, Swains. Class. of Birds, vol. ii. p. 250.
—— *australis*, G. R. Gray, List of Gen. of Birds, 2nd edit. p. 45.
—— *flavigastra*, G. R. Gray.
Yellow-breasted Thrush, Lewin, Birds of New Holl., pl. 23.
Eopsaltria parvula, Gould in Proc. of Zool. Soc., part v. 1837, p. 144, female.
Yellow Robin, Colonists of New South Wales.

Eopsaltria australis, Gould, Birds of Australia, fol., vol. iii. pl. 11.

This is a very common species in all the brushes of New South Wales; I also observed it in most of the gardens in the neighbourhood of Sydney, as well as in those of the settlers in the interior. It is very Robin-like in its actions, particularly in the habit of throwing up its tail, and in the sprightly air with which it moves about. It is by no means shy, and may often be seen crossing the garden walks, perching on some stump or railing, regardless of one's presence, at which time the fine yellow mark on its rump is very conspicuous. Its powers of flight are but feeble, and are seldom employed except to enable it to flit from bush to bush or from tree to tree. Its food consists entirely of insects, which are more frequently taken on the ground than on the trees.

It breeds in September and October. The nest is a beautiful, compact, round, cup-shaped structure, about three inches in diameter, and an inch and a half deep, composed of narrow strips of bark, wiry fibrous roots, and in some instances grasses; the outside held together with cobwebs, and sparingly speckled over with mouse-eared lichen and small pieces of bark hanging loosely about it; the inside of the nest is generally lined with leaves, but occasionally with portions of the broad blades of grasses. It is generally placed in the fork of some low tree in an open or exposed part of the brush, is a neat structure, and sometimes so nearly resembles the bark of the tree upon which it is constructed, as to be scarcely detectible. The eggs are usually two in number, of a bright apple-green, speckled and spotted all over with chestnut-brown and blackish brown, the latter tint being much less conspicuous than the former; they are nine lines long by seven and a half lines broad.

It is not migratory, and so far as is known, is confined to the southern and eastern portion of the country.

The sexes are very similar in colour; but the female is somewhat smaller in size, and has the rump olive instead of yellow: the young on leaving the nest has the plumage streaked and spotted very similar to that of young Robins, but obtains the plumage of the adult at an early period.

Head and all the upper surface, wings and tail, with the exception of the rump, very dark grey; chin white; all the under surface and rump wax-yellow; irides, bill, and feet black.

Sp. 176. EOPSALTRIA GRISEOGULARIS, *Gould*.

GREY-BREASTED ROBIN.

Eopsaltria griseogularis, Gould in Proc. of Zool. Soc., part v. p. 144.
Muscicapa georgiana, Quoy et Gaim. Voy. de l'Astrolabe, pl. 8. fig. 4.
Hum-boore, Aborigines of the lowland districts of Western Australia.

Eopsaltria griseogularis, Gould, Birds of Australia, fol, vol. iii. pl. 12.

The *Eopsaltria griseogularis* is abundant in every part of the colony of Swan River, inhabiting thickets and all spots clothed with vegetation of a brush-like character. "In its actions," says Gilbert, "this bird is very like the Robins, being much on the ground, and when feeding constantly flying up and perching on a small upright twig. It does not appear to be capable of great or continued exertion on the wing, as it is rarely seen to do more than flit from bush to bush. Its most common note much resembles the very lengthened and plaintive song of the *Estrelda bella*, but differs from it in being a double note often repeated; it also utters a great variety of single notes, and during the breeding-season pours forth a short but agreeable song.

"The nest is very difficult to detect, the situations chosen for it being the thickly-wooded gum-forests of the mountain districts and the mahogany-forests of the lowlands; from the forks of the younger of these trees a great portion of the bark generally hangs down in strips; and in the fork the bird generally makes its nest of narrow strips of the bark bound together with cobwebs, while around the outside a quantity of dangling pieces are suspended, giving it the exact appearance of other forks of the tree; the inside of the nest has no other lining than a few pieces of bark laid across each other, or a single dried leaf, large enough to cover the bottom. It breeds in September and October, and lays two eggs, which are more lengthened in form than those of *Eopsaltria australis*, and are of a wood-brown, obscurely freckled with yellowish red, ten lines long by seven lines and a half broad.

"Its stomach is muscular, and its food consists of insects of various kinds."

The sexes are precisely similar in outward appearance.

It is stationary in Western Australia, but the extent of its range over the continent is not yet known.

Crown of the head, ear-coverts, sides and back of the neck, and back grey; throat and chest greyish white; abdomen, rump, upper and under tail-coverts rich yellow; wings and tail greyish brown, the extreme tips of the latter edged with white; bill dark horn-colour; irides very dark reddish brown; legs and feet dark olive-brown.

Sp. 177. EOPSALTRIA LEUCOGASTER, *Gould.*
WHITE-BELLIED ROBIN.

Eopsaltria leucogaster, Gould in Proc. of Zool. Soc., part xiv. p. 19.
Muscicapa gularis, Quoy et Gaim. Voy. de l'Astrolabe, pl. 4. fig. 1.

Eopsaltria leucogaster, Gould, Birds of Australia, fol., vol. iii. pl. 13.

The White-bellied Robin is a native of Western Australia, but is only to be met with in the hilly portions of the country. Gilbert states that the first specimen he procured was killed on the Darling range, near the gorge of the River Murray, at an elevation of about seven or eight hundred feet, and that he afterwards met with it on the southern extremity of the same range, between Vasse and Augusta, but that he never observed it on the lower grounds between the mountain-range and the coast. Like the other species of the genus, it was constantly seen clinging to the bark of large upright trees, or straight and small stems, in search of its insect food. It is extremely quiet and secluded in its habits, is almost exclusively confined to the neighbourhood of small mountain-streams, where scarcely any other sound is heard than the rippling and gurgling of the water over the rocks, and on the slightest approach it immediately secretes itself among the thick scrub or brushwood. Its song very closely resembles that of the *Petroicæ.*

Immediately before the eye a small triangular-shaped spot of black; above the eye a faint line of greyish white; crown of the head, all the upper surface, wings, and tail dark slate-grey; the lateral tail-feathers largely tipped with white on their inner webs; all the under surface white; irides dark brown; bill and feet black.

Sp. 178. EOPSALTRIA CAPITO, *Gould*.

LARGE-HEADED ROBIN.

Eopsaltria capito, Gould in Proc. of Zool. Soc., part xix. p. 285.

Eopsaltria capito, Gould, Birds of Australia, fol., Supplement, pl.

The outer slopes of the high ranges which skirt the southern and eastern coasts of Australia, at a distance of from forty to sixty miles from the sea, have in the course of time changed into a soil so rich and deep as to be favourable, not only to the growth of the largest kinds of *Eucalypti*, but to magnificent cedars, fig-trees, and palms of two or three species. Favoured by an aspect which commands the rays of the sun, and by humidity from the sea, the vegetation here becomes of that dense and peculiar character technically known in New South Wales by the name of Brushes; these districts are tenanted by a bird-life equally peculiar; so that the fauna of the brushes is as distinct from that of the plains as if hundreds of miles of sea rolled between. The unobtrusively coloured bird here described is a native of the brushes of the south-east coast, and is tolerably plentiful in the neighbourhood of the Clarence, the Manning, and the Brisbane rivers. Its existence was not known to me when the 'Birds of Australia' was published; and its discovery is due to the late F. Strange, who sent me several specimens, two of which have been figured in the supplement to the folio edition. Its habits are doubtless very similar to those of the other *Eopsaltriæ*. Like them, the sexes do not differ in colour, but the female may generally be distinguished by her somewhat smaller size.

Upper surface olive-green, inclining to brown on the head; wings and tail slaty-brown, faintly margined with olive-green; ear-coverts grey; lores, a line below the eye, and the throat greyish white; under surface yellow; irides hazel; bill black; feet brownish flesh-colour.

Total length 5 inches; bill $\frac{1}{2}$; wing $3\frac{1}{4}$; tail $2\frac{1}{4}$; tarsi $\frac{3}{4}$.

Family MENURIDÆ.
Genus MENURA, *Davies.*

Two, if not three, species of this extraordinary form are known to inhabit the dense woods of the south-eastern portions of Australia; until very recently, however, the *M. superba* was the sole representative of the genus. Other species may yet be discovered when the country has been more thoroughly explored.

Sp. 170.   MENURA SUPERBA, *Davies.*
LYRE-BIRD.

Menura superba, Davies in Linn. Trans., vol. vii. p. 207, pl. 22.
Le Parkinson, Vieill. (Ois. Dor.) Ois. de l'arad., pls. 14, 15, 16.
Megapodius menura, Wagl. Syst. Av., sp. 1.
Menura lyra, Shaw, Nat. Misc., pl. 577.
—— *novæ-hollandiæ,* Lath. Ind. Orn. Supp., p. lxi.
Parkinsonius mirabilis, Bechst.
Menura vulgaris, Flem.
—— *paradisea,* Swains. Class. of Birds, vol. ii. p. 351.
Superb Menura, Lath. Gen. Syn. Supp., vol. ii. p. 271.
Pheasant of the Colonists.
Beleck-Beleck and *Balangara* of the Aborigines.

Menura superba, Gould, Birds of Australia, fol., vol. iii. pl. 14.

Were I requested to suggest an emblem for Australia among its avifauna, I should without the slightest hesitation select the Lyre-bird as the most appropriate, it being not only strictly peculiar to that country, but one which will always be regarded with the highest interest both by the people of Australia and by ornithologists in Europe, from whom it has received the specific appellations of *superba, paradisea,* and *mirabilis.*

In the structure of its feet, in its lengthened claws, and in its whole contour, the Lyre-bird presents the greatest similarity to the *Pteroptochus megapodius* of Kittlitz. The immense feet and claws of this bird admirably adapt it for the peculiar localities it is destined to inhabit.

The principal habitat of the *Menura superba* is New South Wales, and, from what I could learn, its range does not extend so far to the eastward as Moreton Bay, nor have I been able to trace it to the westward of Port Philip*; but further research can alone determine these points. It appears to inhabit alike the brushes on the coast and those that clothe the sides of the mountains in the interior; on the coast it was especially abundant at Western Port and Illawarra when I visited the colony in 1838. In the interior the cedar-brushes of the Liverpool range, and, according to Dr. Bennett, the mountains of the Tumut country, are among the places of its resort. Of all the birds I have ever met with, the *Menura* is by far the most shy and difficult to procure. While among the brushes I have been surrounded by these birds, pouring forth their loud and liquid calls, for days together, without being able to get a sight of them; and it was only by the most determined perseverance that I was enabled to effect this to me desirable object, which was rendered the more difficult by their often frequenting the almost inaccessible and precipitous sides of gullies and ravines, covered with tangled masses of creepers and umbrageous trees: the cracking of a stick, the rolling down of a small stone, or any other noise, however slight, is sufficient to alarm them; and none but those who have traversed the rugged, hot, and suffocating brushes can fully understand the excessive labour attendant on the pursuit of the *Menura*. Those who wish even to sight it must only advance when the bird's attention is occupied in singing, or in scratching up the leaves in search of food. To watch its actions, it is necessary to remain

---

* It will be seen that I consider the *Menura* from this part of the country to be different from the bird inhabiting New South Wales, and that, under this impression, I have named it *M. victoriæ* in honour of our gracious Sovereign.

perfectly motionless, or it vanishes from sight as if by magic. But the *Menura* is not always so alert; for in some of the more accessible brushes through which roads have been cut, it may frequently be seen, and on horseback even closely approached, the bird apparently evincing less fear of those animals than of man when thus unaccompanied. At Illawarra it is sometimes successfully pursued by dogs trained to rush suddenly upon it, when it immediately leaps upon the branch of a tree, and, its attention being attracted by the dog which stands barking below, it is more easily approached and shot.

The Lyre-bird is of a wandering disposition; and although it keeps to the same brush, it is constantly traversing it from one end to the other, from mountain-top to the bottom of the gullies, whose steep and rugged sides present no obstacle to its long legs and powerful muscular thighs; it is also capable of performing extraordinary leaps; and I have heard it stated that it will spring to the ledge of a rock or the branch of a tree ten feet perpendicularly from the ground. It appears to be of solitary habits, as I have never seen more than a pair together, and those only in a single instance; they were both males, and were chasing each other round and round with extreme rapidity, apparently in play, pausing every now and then to utter their loud shrill calls: while thus employed, they carried the tail horizontally, as they always do when running quickly through the brushes, that being the only position in which it could be conveniently borne. Among its many curious habits is that of forming small round hillocks, which are constantly visited during the day, and upon which the male is continually trampling, at the same time erecting and spreading out his tail in the most graceful manner, and uttering his various cries, sometimes pouring forth his natural notes, at others mocking those of other birds, and even the howling of the Dingo. The early morning and the evening are the periods when it is most animated and active.

It may truly be said that all the beauty of this bird lies in the plumage of his tail, the new feathers of which appear in February or March, but do not attain their full beauty and perfection until June; during this and the four succeeding months it is in its finest state; after this the feathers are gradually shed, to be resumed again at the period above stated. I am led to believe that they are all assumed simultaneously, by the fact of a native having brought to my camp a specimen with a tail not more than six inches long, the feathers of which were in embryo, and all of the same length. Upon reference to my journal I find the following notes upon the subject:—" Mar. 14, Liverpool range. Several *Menuras* killed to-day: their tails not so fine as they will be." " Oct. 25.—I find this bird is now losing its tail-feathers; and, judging from appearances, they will be all shed in a fortnight."

The food of the *Menura* consists of insects, particularly centipedes and coleoptera; I also found the remains of shelled snails in the gizzard, which is very strong and muscular.

I never found the nest but once, and this unfortunately was after the breeding-season was over; but all those of whom I made inquiries respecting it, agreed in assuring me that it is either placed on the ledge of a projecting rock, at the base of a tree, or on the top of a stump, but always near the ground; and a cedar-cutter whom I met in the brushes informed me that he had once found a nest, which, to use his own expression, was "built like that of a magpie," adding that it contained but one egg, and that upon his visiting the nest again some time afterwards he found in it a newly-hatched young, which was helpless and destitute of the power of vision. The nest seen by myself was placed on the prominent point of a rock, in a situation quite secluded from observation behind, but affording the bird a commanding view and easy retreat in front; it was deep and shaped like a basin, and had the appearance of having been roofed, was of a large size, formed outwardly of sticks, and lined with the inner bark of trees and fibrous roots.

General plumage brown; the secondary wing-feathers nearest the body, and the outer webs of the remainder, rich rufous brown; upper tail-coverts tinged with rufous; chin and front of the throat rufous, much richer during the breeding-season; all the under surface brownish ash-colour, becoming paler on the vent; upper surface of the tail blackish brown; under surface silvery grey, becoming very dark on the external web of the outer feather; the inner webs of these feathers fine rufous, crossed by numerous bands, which at first appear of a darker tint, but on close inspection prove to be perfectly transparent; the margin of the inner web and tips black; bill and nostrils black; irides blackish brown; bare space round the eye blackish lead-colour; legs and feet black, the scales mealy.

The female differs in wanting the singularly formed tail, and in having the bare space round the eye less extensive and less brilliantly coloured.

Sp. 160.   MENURA VICTORIÆ, *Gould*.

QUEEN VICTORIA'S LYRE-BIRD.

Menura victoriæ, Gould in Proc. of Zool. Soc., 1862, p. 23.

Those ornithologists who have examined specimens of the *Menura* from the neighbourhood of Melbourne must have noticed a great difference in the structure of their tails from this lyre-shaped organ in examples from New South Wales. Although on slender grounds, I admit, I have been induced to consider the Port Philip bird to be a distinct species; I say slender grounds, because I have not seen a sufficient number of specimens from that locality to enable me to say positively that it is really different. The specimens kindly sent to me by Professor M'Coy, the learned naturalist at the head of the zoological department of the public Museum at Melbourne, would, however, tend to warrant this view; and I would especially call the attention of Australians to the subject as one worthy of their attention. The chief difference of the bird I have named *M. victoriæ* is the diminished length of its outer tail-feathers, and their much stronger and broader markings.

Whether the bird be or be not distinct from *M. superba*, the following highly interesting notes kindly sent to me by the late Dr. Ludwig Becker have reference to it:—

"Bullan-Bullan is the name which the aborigines of the Yarra tribe give to this bird. The word has some similarity to the gurgling tone which the bird at times is heard to emit. The favourite place chosen by the Bullan-Bullon for building its nest is the dense scrub on the slopes of deep gullies, or in thickly grown small scrubs, lying between the bends of rivers, but still in the vicinity of mountains. Here the bird selects young trees standing close together; between the saplings, one or two feet from the ground, it makes fast its nest. Sometimes it may be found also upon the trunk of a tree, hollowed out by some bush-fire; or it selects a fern tree, of not too great a height, for the same purpose. The nest proper is ten inches in diameter, and is five inches high. It is closely woven together from fine but strong roots, and the inside is lined with the softer feathers of the bird. Round this nest the bird builds a rough covering, composed of sticks and pieces of wood, grass, moss, and leaves, in such a manner that it projects over the genuine nest, affording the sitting bird a shelter from above. An opening in the side serves as an entrance, through which the female enters backwards, with her tail laid over her back, and, with watchful eye and ear, keeps her head in the direction of the opening. She lays only one egg, of a purplish-grey tint, with numerous spots and blotches of purplish brown, especially at the larger end, as seen in the egg of the common Crow (*Corvus corone*); the colour resembles in fact so closely that of the feathers with which the nest is lined, that it is not easy to detect the egg. It is two and a half inches long by one inch and five-eighths broad.

"It is generally believed that the *Menura* makes use of the same nest for several years. A nest and egg, found on the 31st of August, arrived in Melbourne on the 4th of September, in a good state of preservation. This is somewhat astonishing, considering that the 'black fellow' carried them on his back day by day, wrapped up in his opossum-skin, while by night he had to protect them from the wild cats and other animals. In Melbourne, unfortunately, or rather fortunately, the egg was broken, and an almost fully developed young one dropped out, which would, in the course of two or three days, have broken through the shell.

"The young one is almost unfledged, having only here and there feathers, resembling black horsehair, of an inch in length. The middle of the head and spine are the parts most thickly covered, while the forearm and the legs are less so. A tuft is visible on its throat, and two rows of small and light-coloured feathers on its belly. The skin is of a yellowish-grey colour; feet dark; claws grey; beak black; eyelids closed.

"I believe that the period of incubation of the Lyre-bird begins in the first week of August, and that the young one breaks through the shell in the beginning of September."

Some further observations on this species were sent to me by Dr. Becker, on the 24th of September, 1859.

"In the month of October 1858, the nest of a Lyre-bird was found in the densely wooded ranges near the sources of the river Yarra-Yarra. It contained a young bird in a sickly state, and of a very large size compared with its helplessness. When taken out of the nest, it screamed loudly; the note was high, and sounded like 'tching-tching.' In a short time the mother, attracted by the call, arrived, and, notwithstanding the proverbial shyness of the species, she flew within a few feet of her young, trying in vain to deliver it from captivity by flapping her wings and making rapid motions in different directions towards the captor. A shot brought down the poor old bird, and, with its dead mother near it, the young *Menura* was soon silent and quiet. It was taken away, and kept at a 'Mia-Mia' erected in the midst of the surrounding forest.

"The following description will give you, as nearly as possible, a correct idea of this interesting bird:—

"Its height, from foot to crown of head, was sixteen inches. The body was covered with a brown down, but the wings and tail were already furnished with feathers of a dark brown colour. The head was thickly covered with a greyish-white down, from one to two inches in length. The eyes were hazel-brown; the beak blackish and soft; the legs nearly as large as those of a full-grown specimen; but it walked most awkwardly, with the legs bent inwards. When it rose, it did so with difficulty, the wings assisting; once on its legs, it ran sometimes, often falling down, however, in consequence of the want of strength to move properly the large and heavy bones of its legs. It constantly endeavoured to approach the camp-fire, attracted doubtless by the warmth, and it was a matter of some difficulty to keep it from that dangerous position. As I stated before, its cry was a high-sounding 'tching-tching,' often heard during daytime, as if recalling the parent bird. When this call was answered by its keeper, feigning the note 'bullan-bullan,' which is an imitation of the old bird's cry, it followed the voice at once, and was easily led away by it. It became quite tame very shortly after having been taken from the nest. It was always voracious, refusing no food when offered; it stood there with the bill gaping, awaiting the approaching hand which held the food, consisting principally of worms and the larvæ of ants, commonly called ants' eggs; but it did not refuse bits of meat, bread, &c. Sometimes it picked from the ground ants' eggs itself, but was never able to swallow them, as apparently the muscles of the neck had not attained sufficient power to produce the required jerk and throwing back of the head necessary for swallowing the grubs. It scarcely ever took water. It reposed in a nest made of moss, and lined with opossum-skin, where it appeared quite contented. While asleep, the head was covered by one of the wings. When called 'Bullan-Dullan' it awoke, looking for several seconds at the disturber, and soon put its head again under the wing, taking no notice whatever of other sounds or voices. A proof that the young of this bird often remain for a long time in their natural nest may be found in the manner in which they dispose of their droppings. The young captive always went backwards before discharging its dung, as if afraid of soiling the nest. It is probable that in its natural state and during daytime it leaves the nest, when the warmth of the weather invites it, but during the night, and if cold weather sets in, the mother will be with her young.

"Notwithstanding all the care bestowed upon this poor little bird, it died on the eighth day of its captivity, apparently in consequence of the excessively cold weather which set in, and which was even keenly felt by the possessor of the bird himself. At this time the young *Menura* had begun to change its plumage, feathers taking the place of the down with which it was previously covered; and the legs, enveloped in a sort of scaly scurf, which fell off as the bird grew older, already were of a blackish colour.

"There is no doubt that the Lyre-bird could be easily introduced into our menageries; they only require care while young, and, when full-grown and tamed, may be shipped to England with as little difficulty as any other Australian bird, none of which, however, offer such attractions as the *Menura*."

Sp. 181.   MENURA ALBERTI, *Gould.*
   PRINCE ALBERT'S LYRE-BIRD.
Menura alberti, Gould in Proc. of Linn. Soc., February 5, 1850.

Menura alberti, Gould, Birds of Australia, fol., Supplement, pl.

The dense, luxuriant, and almost impenetrable brushes which skirt the eastern coast of Australia from Sydney to Moreton Bay are, as might be supposed, tenanted by many forms both of mammalia and birds peculiarly their own; many of these districts are very partially known, and some of them may be said to be as yet untrodden; hence it is not surprising that an additional species of this extraordinary form should have been there discovered. I must fairly admit, however, that I was not prepared for the acquisition of so remarkable a bird within the limits of the colony of New South Wales.

The specific differences between the present bird and the *M. superba* are very apparent; they consist in the rufous colouring of the plumage, and in the total absence of the brown barrings of the outer tail-feathers, which, moreover, are much shorter than the others, while in *M. superba* they are the longest.

The first specimens that came under my notice were sent to me by the late F. Strange; my friend Dr. Bennett also forwarded to me almost simultaneously a fine example belonging to the Sydney Museum, which the Directors had at his request permitted to be sent to England for illustration. With reference to the latter, Dr. Stephenson, residing at York Station, Richmond River, wrote to Dr. Bennett, " You will perceive a very close affinity between it and the *M. superba*, except in the tail, which is very different. Since the idea of its being distinct occurred to me and to my friend Augustus A. Leycester, Esq., I have made every possible inquiry respecting the bird amongst the sawyers and others, all of whom agree that it is distinct; some of them had shot specimens of the *M. superba* at Camden Haven and other localities, but had never seen the present bird further to the south than the Nambucca River; they also state that the new bird is not so timid as the old one, and is consequently more easily shot. The locality it frequents consists of mountain-ridges not very densely covered with brush; it passes most of its time on the ground, feeding and strutting about with the tail reflected over the back to within an inch or two of the head, and with the wings dropping on the ground. Each bird forms for itself three or four '*corroborying places*,' as the sawyers call them; they consist of holes scratched in the sandy ground about two feet and a half in diameter by sixteen, eighteen, or twenty inches in depth, and about three or four hundred yards apart. Whenever you get sight of the bird, which can only be done with the greatest caution and by taking advantage of intervening objects to shelter yourself from its observation, you will find it in one or other of these holes, into which it frequently jumps and seems to be feeding, then ascends again and struts round and round the place, imitating with its powerful musical voice any bird it may chance to hear around it; the note of the *Dacelo gigas* it imitates to perfection; its own whistle is exceedingly beautiful and varied. No sooner does it perceive an intruder than it flies up into the nearest tree, first alighting on the lowermost branches and then ascending by a succession of jumps until it reaches the top, whence it instantly darts off to another of its play-grounds. The stomachs of those I dissected invariably contained insects, with scarcely a trace of any other material."

The late F. Strange informed me that he met with the bird "in the cedar-brushes which skirt Turunga Creek, Richmond River. Like the *M. superba*, it is of a shy disposition. I spent ten days in the midst of the cedar-brushes in the hope of learning something of its nidification, but did not succeed in finding any nest with eggs; I found, however, one large domed nest made of sticks and placed in the spur of a large fig-tree, which the natives assured me was that of the *Colwin*, their name for this bird; it resembled that of *Orthonyx*, except that the inside was not lined with moss, but with the litter from a large mass of parasitical plants that had fallen to the ground. The natives agree in asserting that the eggs are only laid in the cold weather, by which I apprehend they mean the spring, as I shot a young bird about four months old, on the 24th of November, which had the whole of the body still covered with a brown and greyish down. I have seen this species take some extraordinary jumps of not less than ten feet from the ground on to a convenient branch, whence it continues to ascend in successive leaps, until it has attained a sufficient elevation to enable it to take flight into the gully below."

The male has the crown of the head and back of a sooty black, with a tinge of chestnut on the forehead and some of the crest-feathers; all the upper surface, and particularly the upper tail-coverts, rich rusty chestnut; primaries blackish brown, tinged with rufous on their external edges; throat rusty red, passing into a paler tint of the same colour on the breast; abdomen grey, washed with sandy buff; thighs grey, slightly washed with buff; under tail-coverts bright rufous; upper surface of the tail-feathers slaty black, their under surface silvery grey; the large outer feather on each side much shorter than the corresponding feathers in *Menura superba*, and entirely destitute of the bare so conspicuous in that species; the two centre feathers narrow, prolonged, crossing each other at the base, curving outward at the tip, and webbed only on their external side.

The female is similar in colour to the male; but distinguishable by the feathers of the tail being much less filamentous in their structure, and by the two middle feathers being shorter, broader, and straighter than in the opposite sex, and broadly webbed on both sides of the shaft.

Since the above appeared in the folio edition, I have been favoured with several notes respecting this species which, in justice to the writers, I here insert.

The first is from A. A. Leycester, Esq., who says—

"These birds hitherto have been found only on the Richmond and Tweed rivers, in the dense brushes which clothe the mountains in those districts; and, what is most remarkable, though similar mountains and brushes exist on the rivers both north and south of those rivers, yet the *M. alberti* is never to be found in them, their boundary appearing to be limited to a patch of country not wider than eighty by sixty miles.

"The habits of *Menura alberti* are very similar to *M. superba*. Having seen and watched both on their play-grounds, I find the *M. alberti* is far superior in its powers of mocking and imitating the cries and songs of others of the feathered race to the *M. superba*; its own peculiar cry or song is also different, being of a much louder and fuller tone. I once listened to one of these birds that had taken up its quarters within two hundred yards of a sawyer's hut, and he had made himself perfect with all the noises of the sawyer's homestead—the crowing of the cocks, the cackling of the hens, and the barking and howling of the dogs, and even the painful screeching of the sharping or filing of the saw. I have never seen more than a pair together. Each bird appears to have its own walk or boundary, and never to infringe on the other's ground; for I have heard them day after day in the same place, and seldom nearer than a quarter of a mile to each other. Whilst singing, they spread their tails over their heads like a Peacock, and droop their wings to the ground, and at the same time scratch and peck up the earth. They sing mornings and evenings, and more so in winter than at any other time. The young cocks do not sing until they get their full tails, which, I fancy, is not until the fourth year, having shot them in four different stages; the two centre, curved feathers are the last to make their appearance. They live entirely upon small insects, principally beetles. Their flesh is not eatable, being dark, dry, and tough, and quite unlike other birds. They commence building their nests in May, lay in June, and have young in July. They generally place their nests on the side of some steep rock, where there is sufficient room to form a lodgment, so that no animals or vermin can approach.

"The nest is constructed of small sticks, interwoven with moss and fibres of roots, the inside being lined with the skeleton leaf of the parasitical tree fern, resembling horsehair, and covered in, with the entrance on the side. The single egg laid is of a very dark colour, appearing as if it had been blotched over with ink. The young bird for the first month is covered with down, and remains in the nest about six weeks before it takes its departure. Aboriginal name, 'Colwin.'"

Mr. Wilcox, in a letter dated Sydney, September 26, 1852, writes:—

"It gives me much pleasure to forward to you the nest and egg of *Menura alberti*, which I have just obtained from the Richmond River. It was placed on a rocky ledge, about one hundred feet above the stream, so difficult of access as to render its acquisition a task of no ordinary kind. Another nest was also found in the brush near the water; it would seem, therefore, that there is no rule as to the elevation of the locality in which it is placed. Only one egg was found in each nest; and, from all the information I could glean on the subject, the bird never lays but one.

"You will be as sorry to hear as I am to tell you, that by an accidental fire I have just lost four young birds which had been taken from nests *the moment they were ready to leave them*, and which had thriven well for four months on worms, insects, bread, and meat. Mr. Lonsdale, a gentleman who has paid much attention to the birds of Australia, tells me that while out shooting on Mount Kera he came upon a bird sitting on a nest at the base of a large tree on the side of a deep gully; on going to the spot, the bird got off and ran away; he pursued and captured it, when it proved to be a young *Menura superba*; on returning again to the spot, he found the nest to be a loose structure of large sticks, and lined with the fibres of the cabbage-tree leaf."

A nest and an egg sent to me by Mr. Turner were described in the 'Proceedings of the Zoological Society' for 1858:—

"The nest is oven-shaped in form, outwardly constructed of roots, tendrils, and leaves of palms, and lined with green mosses. It was about two feet in length by sixteen inches in breadth, and domed over except at one end. The eggs were barely two inches and a quarter in length by one inch and three quarters in breadth, and of a deep purplish chocolate, irregularly blotched and freckled with a darker colour. This nest and egg are now in the British Museum."

Genus PSOPHODES, *Vigors and Horsfield.*

This form is peculiar to Australia. Two species are known, one of which inhabits the eastern, and the other the western portion of the country.

Sp. 182. PSOPHODES CREPITANS, *Vig. and Horsf.*
COACH-WHIP-BIRD.

Muscicapa crepitans, Lath. Ind. Orn., Suppl. p. li.
Coach-whip Honey-eater, Lath. Gen. Hist., vol. iv. p. 187.
Psophodes crepitans, Vig. and Horsf. in Linn. Trans., vol. xv. p. 320.
Djou, Aborigines of New South Wales.
Corvus auritus et olivaceus, Lath. Ind. Orn., vol. i. p. 160, and Suppl. p. xxvi.
Pica olivacea, Vieill. Nouv. Dict. d'Hist. Nat., tom. xxix. p. 119.
—— *gularis*, Wagl. Syst. Av. *Pica*, sp. 13.
Dasyornis Abeillei, Less.

Psophodes crepitans, Gould, Birds of Australia, fol., vol. iii. pl. 15.

This bird, so renowned for the singularity of its note, is very abundant in many parts of New South Wales, to which portion of the Australian continent it appears to be confined. It is to be found only in dense brushes, such as those at Maitland, Manning, Illawarra, and the cedar-brushes of the Liverpool range; in fact, the localities that are suitable to the *Menura* and the Wattled Talegalla are congenial to the habits of the Coach-whip-bird. Its loud full note ending sharply like the cracking of a whip, with which the woods are constantly reverberating, appeared to me to be analogous to the peculiar call of the *Menura*; besides this peculiar whistle, it also gives utterance to a low inward song of considerable melody.

It is a shy and recluse species, rarely exposes itself to view, but generally keeps in the midst of the densest foliage and among the thickest climbing plants, frequenting alike those that have intertwined themselves with the branches of the tallest trees, and those that form almost impenetrable masses near the ground. It is extremely animated and sprightly in all its actions, raising its crest and spreading its tail in the most elegant manner. These actions become even more animated during the spring, when the males may often be seen chasing each other, frequently stopping to pour out their notes with great volubility.

The food consists of insects of various kinds, obtained almost entirely from the ground, and sought for by scratching up the leaves and turning over the small stones, precisely after the manner of the *Menura superba*.

The sexes are much alike in colour, but may be readily distinguished by the more obscure plumage and smaller size of the female. The young of the first year are of a much browner hue, a character of plumage that soon gives place to adult livery. On its nidification the late F. Strange sent me the following note:—" I found a nest on the 26th of November; it was placed in a small bush surrounded with a great number of weeds, at about two feet from the ground. It contained two young ones, which I looked at every day until they were half fledged, when they disappeared, having probably been taken out by the old birds, as I observed them in the neighbourhood for four days afterwards." More recently some eggs of this bird (which is said to lay two) have been sent to me from New South Wales, and may be thus described:—

They are lengthened and elegant in form, about an inch and an eighth in length by thirteen-sixteenths of an inch in breadth, and are greenish white, sparingly dotted with black and greyish black, the latter colour appearing as if beneath the surface of the shell, and the spots being most numerous at the larger end. In some specimens the markings assume the form of commas, small oblique dashes, and crooked Hebrew-like characters, reminding one somewhat of the markings of the eggs of the Buntings.

The male has the head, ear-coverts, chin, and breast black; a large patch of white on each side of the neck, all the upper surface, wings, flanks, and base of the tail-feathers olive-green; the remaining portion of the tail-feathers black, except that the three lateral feathers on each side are tipped with white; under surface olive-brown, some of the feathers on the centre of the abdomen tipped with white, and forming a conspicuous irregular patch; irides brownish red; bill, inside and out, and base of the tongue black; feet reddish brown.

Sp. 183. PSOPHODES NIGROGULARIS, *Gould.*

BLACK-THROATED PSOPHODES.

Psophodes nigrogularis, Gould in Proc. of Zool. Soc., part xii. p. 5.

Psophodes nigrogularis, Gould, Birds of Australia, fol., vol. iii. pl. 16.

The addition of a second species to the genus *Psophodes* will be hailed with pleasure by every one who makes the science of ornithology a matter of study; nor will its discovery be a subject of surprise, as it is only another illustration of that beautiful law of representation which is conspicuously carried out in Australia. The habitat of the present bird will doubtless be hereafter found to be as strictly confined to the western part of the continent as that of the *P. crepitans* is to the eastern. It is to Gilbert's perseverance that science is indebted for the knowledge of this new bird; and his notes respecting it I here transcribe:—"Inhabits thickets of a small species of *Leptospermum* growing among the sand-hills which run parallel with and adjacent to the beach. It utters a peculiar harsh and grating song which it is quite impossible to describe, and which is so different from that of every other bird I ever heard or am acquainted with, that I shall have no difficulty in recognizing it again wherever I may hear it. I heard it for the first time, together with the notes of many other birds equally strange to me, in the vicinity of the Wongan Hills a few weeks back, but could not then obtain a sight of the bird, although I knew that it was only a few yards from me."

Plumage of the upper surface olive; under surface ashy, passing into brown on the flanks and white on the centre of the abdomen; primaries brown; tail light olive-brown, the four lateral feathers crossed near the extremity with a band of black, and tipped with white; throat deep black, with a stripe of white from the angle of the lower mandible, just within the black; bill dark horn-colour; irides dark brown; feet dark horn-colour.

Total length $6\frac{1}{4}$ inches; bill $\frac{7}{8}$; wing $3\frac{1}{2}$; tail $4\frac{1}{2}$; tarsi $1\frac{1}{4}$.

Genus SPHENOSTOMA, *Gould*.

The only known species of this genus frequents the sterile parts of the interior of Australia generally, particularly those portions of the country clothed with low shrubs and bushes.

That this form and *Psophodes* are nearly allied there can, I think, be no doubt.

Sp. 184. SPHENOSTOMA CRISTATUM, *Gould*.

CRESTED WEDGE-BILL.

Sphenostoma cristatum, Gould in Proc. of Zool. Soc., part v. p. 150.
—— *cristata*, Gould, Birds of Australia, fol., vol. i. Introd. p. xliii.

Sphenostoma cristatum, Gould, Birds of Australia, fol., vol. iii. pl. 17.

The Crested Wedge-bill is an inhabitant of the low scrubby trees and *Polygonum* bushes which stud the hot plains of the interior of Australia, particularly those on the borders of the Lachlan and Darling: it has also been killed on the Lower Namoi. Whether it has any kind of loud sharp whistle analogous to that of the Coach-whip-bird (*Psophodes crepitans*), or if it has the same shy disposition, it would be interesting to ascertain; and to these points, as well as to all other details connected with its history, I would call the attention of those who may visit the interior, or may otherwise be favourably situated for observing them. The sombre tints of the bird are very like the colour of the earth of the plains it inhabits; and when the nature of its food shall have been ascertained, its wedge-shaped bill will doubtless be found admirably adapted for procuring it.

General plumage brown, lighter beneath; chin and centre of the abdomen greyish white; wings dark brown, edged with pale brown, the fourth and fifth primaries conspicuously margined with white; four centre tail-feathers dark brown, indistinctly barred with a still darker hue; the remainder brownish black, largely tipped with white; bill blackish brown; feet lead-colour.

A nest of this species now in the British Museum is rather large, round, cup-shaped, outwardly composed of fine twigs and lined with grasses. The eggs in my own collection are like those of *Psophodes crepitans*, lengthened and elegant in form, their ground-colour delicate greenish blue, thinly sprinkled with purplish-black specks, particularly at the larger end. In some instances these purple-black specks and markings assume forms similar to those described as occurring on the eggs of *Psophodes crepitans*.

Family ——— ?

Genus MALURUS, *Vieillot*.

The members of this genus are among the most beautiful of the Australian birds. Their gay attire, however, is only assumed during the pairing-season, and is retained for a very short period, after which the sexes are alike in colouring.

The genus is strictly an Australian one, and, with two or three exceptions, all the species are confined to the southern parts of the continent and Tasmania. They build dome-shaped nests, and are frequently selected to perform the office of foster-parents to the young of the Bronze Cuckoo, (*Chrysococcyx lucidus*).

Sp. 185.   MALURUS CYANEUS, *Vieillot*.

Superb Warbler.

Sylvia cyanea, Lath. Ind. Orn., vol. ii. p. 545.
Motacilla cyanea, Gmel. Syst. Nat., vol. i. p. 991.
——— *superba*, Shaw, Nat. Misc., pl. 10.
Superb Warbler, Shaw in White's Voy., pl. in p. 256, opp. fig.
Malurus cyaneus, Vieill. Gal. des Ois., p. 265, pl. 108.
Superb Warbler, Blue Wren, &c., of the Colonists.

Malurus cyaneus, Gould, Birds of Australia, fol., vol. iii. pl. 18.

Of the lovely group of birds forming the genus *Malurus*, the present species is the oldest known, being that described and figured in White's Voyage to New South Wales, under the name of Superb Warbler, a term by which the bird is still familiarly known in Australia. It is abundantly dispersed over the eastern portion of the country, and I observed it to be equally numerous on the plains of the interior; but how far its range may extend northwards, can only be determined when those parts of the continent shall have been fully explored. I killed many specimens in South Australia which I formerly believed to be identical with the present bird; but on a recent comparison, I find them to be more nearly allied to the *Malurus longicaudus*; a further knowledge of the South Australian bird is therefore necessary, before I can determine to which it is referable, or if it may not be distinct from both.

The *Malurus cyaneus* gives preference to those parts of the country which is thinly covered with low scrubby brushwood, and especially to localities of this description which are situated near the borders of rivers and ravines. During the months of winter it associates in small troops of from six to eight in number (probably the brood of a single pair), which continually traverse the district in which they were bred. At this period of the year the adult males throw off their fine livery, and the plumage of the sexes then becomes so nearly alike that a minute examination is requisite to distinguish them. The old males have the bill black at all seasons, whereas the young males during the first year, and the females, have this organ always brown; the tail-feathers also, which with the primaries are only moulted once a year, are of a deeper blue in the old male. As spring advances, the small troops separate into pairs, and the males undergo a total transformation, not only in their colour, but in the texture of their plumage; indeed a more astonishing change can scarcely be imagined. This change is not confined to the plumage alone, but extends also to the habits of the bird; for it now displays great vivacity, proudly shows off its gorgeous attire to the utmost advantage, and pours out its animated song unceasingly, until the female has completed her task of incubation, and the craving appetites of its newly-hatched young have called forth a new feeling, and given its energies a new direction.

During the winter months no bird can be more tame and familiar; for it frequents the gardens and shrubberies of the settlers, and hops about their houses as if desirous to court, rather than shun, the presence of man; but when adorned with his summer plumage, the male becomes more shy and retiring, appearing to have an instinctive consciousness of the danger to which his beauty subjects him; nevertheless they will frequently build their little nest and rear their young in the most populous places. Several broods are reared annually in the Botanic Garden at Sydney, and I saw a pair busily employed in constructing their nest in a tree close to the door of the Colonial Secretary's Office. The short and rounded wing incapacitates it for protracted flight, but the amazing facility with which it passes over the surface of the ground fully compensates for this deficiency; its mode of progression can scarcely be called running, it is rather a succession of bounding hops, performed with great rapidity: while thus employed its tail is carried perpendicularly, or thrown forward over the back; indeed the tail is rarely, if ever, carried horizontally.

The breeding-season continues from September to January, during which period two, if not three, broods are reared: the young of one being scarcely old enough to provide for themselves, before the female again commences laying: independently of rearing her own young, she is also the foster-parent of the Bronze Cuckoo (*Chrysococcyx lucidus*), a single egg of which species is frequently found deposited in her nest; but by what means, is (as in the case of its European prototype) unknown.

The nest, which is dome-shaped, with a small hole at the side for an entrance, is generally constructed of grasses, lined with feathers or hair: the site chosen for its erection is usually near the ground, in a secluded bush or tuft of grass. The eggs are generally four in number, of a delicate flesh-white, sprinkled with spots and blotches of reddish brown, which are more abundant and form an irregular zone at the larger extremity: they are eight lines long by five and a half broad.

The song is a hurried strain, somewhat resembling that of the Wren of Europe.

The stomach is muscular, and the food consists of insects of various kinds, collected on the ground, the trunks of fallen trees, &c.

The male in summer has the crown of the head, ear-coverts, and a lunar-shaped mark on the upper part of the back light metallic blue; lores, line over the eye, occiput, scapularies, back, rump, and upper tail-coverts velvety black; throat and chest bluish black, bounded below by a band of velvety black; tail deep blue, indistinctly barred with a darker hue, and finely tipped with white; wings brown; under surface buffy white, tinged with blue on the flanks; irides blackish brown; bill black; feet brown.

The female has the lores and a circle surrounding the eye reddish brown; upper surface, wings, and tail brown; under surface brownish white; bill reddish brown; feet fleshy brown.

Sp. 130. MALURUS LONGICAUDUS, *Gould.*
LONG-TAILED SUPERB WARBLER.

Malurus longicaudus, Gould in Proc. of Zool. Soc., part v. p. 148.

Malurus longicaudus, Gould, Birds of Australia, fol. vol. iii. pl. 19.

The Long-tailed Superb Warbler is so universally dispersed over Tasmania, as well as the islands in Bass's Straits, that to particularize any one part of the former island where it is found more than another would be vain, since it is present in every gully and every other place where low scrubby bushes and underwood are to be met with: I have also received from Kangaroo Island a single specimen in its winter dress which, I believe, is referable to this species. Active and cheerful, and possessing a sweet warbling song, the present bird is as much a favourite in Tasmania as the Superb Warbler is in New South Wales, and, like its congener, in the winter season it is equally tame and familiar. It is subject to the same changes of plumage, and its whole economy is so similar as to render a separate description unnecessary. Its nest is also similarly constructed, but is of a rather larger size, is usually composed of grasses and leaves, warmly lined with feathers, and in some instances with the fur of the Kangaroo and Opossum, and is placed either in a small bush near the ground, or artfully built in a tuft of grass. The season of reproduction commences in August and lasts until January, during which time two or three broods are reared. Like the *M. cyaneus*, it is also the foster-parent of the Bronze Cuckoo (*Chrysococcyx lucidus*). The eggs, which bear a similar character, but are proportionally larger than those of the *M. cyaneus*, are four or five in number, of a flesh-white, blotched and spotted with markings of reddish brown, particularly at the larger end, where these form an irregular zone: they are nearly nine lines long by six and a quarter broad.

The long legs of this species admirably adapt it for the ground, and for traversing the fallen trunks of trees, along which, with tail erect, it passes with the utmost activity: it is also frequently to be observed among the low trees and bushes, the male often selecting a small prominent bare twig, whereon to perch and warble forth his animated song.

The male in summer has the crown of the head, ear-coverts, and a broad lunar-shaped mark on the upper part of the back metallic blue; lores, line over the eye, occiput, scapularies, back, rump, and upper tail-coverts velvety black; throat and chest bluish black, bounded below by a band of velvety black; tail dark blue, indistinctly barred with a darker hue and finely tipped with white; wings brown; under surface buffy white, tinged with blue on the flanks; irides blackish brown; bill black; feet brown.

The female has the lores and a circle surrounding the eye reddish brown; upper surface, wings, and tail brown; under surface brownish white; bill reddish brown; feet fleshy brown.

Sp. 187. MALURUS MELANOTUS, *Gould*.
BLACK-BACKED SUPERB WARBLER.

Malurus melanotus, Gould in Proc. of Zool. Soc., part viii. p. 163.

Malurus melanotus, Gould, Birds of Australia, fol.; vol. iii. pl. 20.

The Belts of the Murray in South Australia were the only places in which I observed this species; but, although it was tolerably abundant there, it was so extremely shy and distrustful that specimens were obtained with the greatest difficulty. It was most frequently observed on the ground, particularly in the small open glades and little plains by which the outer belt of this vast scrub is diversified. The period of my visit was in winter; consequently the specimens I collected were all out of colour, or, more properly speaking, divested of the rich blue and black plumage, in which state a single specimen was afterwards forwarded to me by one of the party that accompanied His Excellency Colonel Gawler and Captain Sturt, when those gentlemen visited the Murray in 1839; and other examples have since been received. It is a most interesting species, inasmuch as it possesses characters intermediate between the *M. cyaneus* and *M. splendens*, having the blue belly and conspicuous pectoral band of the latter and the black back of the former; from both, however, it differs in the length of its toes, which are much shorter than those of its near allies: this difference in structure exerts a corresponding influence upon its habits and actions; for while the others run over the ground with great facility, the Black-backed Superb Warbler far exceeds them in this respect. Instead of exerting any power of flight, those I saw effected their escape by the extraordinary manner in which they tripped over the small openings and through the scrub, each troop appearing to have a leader, and keeping just beyond the range of the gun.

The male in summer has the crown of the head, chin, throat, abdomen, upper part of the back, upper and under tail-coverts beautiful metallic blue; ear-coverts verditer-blue; lores, back of the neck, band across the breast, and lower part of the back velvety black; external margins of all the wing-feathers green; tail bluish green, indistinctly barred with a darker tint, and slightly tipped with white; bill black; irides and legs blackish brown.

The female has the lores and circle surrounding the eye reddish brown; all the upper surface brown; under surface brownish white; wings brown; tail green, each feather slightly tipped with white; bill reddish brown; feet brown.

Total length $4\frac{3}{4}$ inches; bill $\frac{1}{2}$; wing 2; tail $2\frac{1}{2}$; tarsi $\frac{7}{8}$.

Sp. 188.   MALURUS SPLENDENS.
Banded Superb Warbler.

Saxicola splendens, Quoy et Gaim. Voy. de l'Astrol. Zool., tom. i. p. 197, pl. 10. fig. 1.
Malurus pectoralis, Gould in Proc. of Zool. Soc., part i. p. 100.
Djur-jeel-ya of the Aborigines of the lowland, and *Jeer-jal* of the Aborigines of the mountain districts of Western Australia.

Malurus splendens, Gould, Birds of Australia, fol., vol. iii. pl. 21.

The *Malurus splendens*, which may very justly be considered more gorgeous than any other of its race, its whole plumage sparkling with beautiful shining metallic lustre, is an inhabitant of the western coast of Australia, and is, I believe, very generally distributed over the Swan River settlement, where it inhabits scrubby places covered with underwood.

Its song very nearly resembles that of the Tasmanian species, *M. longicaudus*. It breeds in September and the three following months: the nest is constructed of dried, soft grasses, and lined either with hair, wool, or feathers, is of a dome-shape, the cover of the top resembling the peak of a cap, and is about six or eight inches in height: the eggs are generally four in number, of flesh-white, thickly blotched and freckled with reddish brown, especially at the larger end; eight and a quarter lines long by six and a quarter lines broad. The situation of the nest is much varied, being sometimes built among the hanging clusters of the stinkwood tree, at others among the upright reeds growing just above the water's edge on the borders of lakes and the banks of rivers.

The male in its summer dress has the crown of the head, back, scapularies, and upper tail-coverts deep metallic blue; ear-coverts verditer-blue; throat and all the under surface deep shining violet-blue; lores, crescent-shaped mark across the chest, and back of the neck deep velvet-black; external edges of the wing-feathers green; tail greenish blue, indistinctly barred with a darker tint; bill black; eyes and feet blackish brown.

The female has the bill, lores, and circle round the eyes reddish brown; crown of the head and all the upper surface brown; the external margins of the wing-feathers slightly tinged with green; tail as in the male, but paler, and slightly tipped with white.

Sp. 189.   MALURUS ELEGANS, *Gould*.

GRACEFUL SUPERB WARBLER.

Malurus elegans, Gould, Birds of Australia, part i. Aug. 1837.
Djur-jeel-ya, Aborigines of the lowland districts of Western Australia.

Malurus elegans, Gould, Birds of Australia, fol., vol. iii. pl. 22.

This is one of the largest species of the genus yet discovered, and is a most beautiful and elegant bird: the delicate verditer-blue of the centre of the back, and the larger size and more spatulate form of its tail-feathers, at once distinguish it from *Malurus Lamberti*, the species to which it is most nearly allied. It is an inhabitant of the western coast of Australia;

all the specimens I possess were collected at Swan River, where it is tolerably abundant.

The nest is dome-shaped, with a hole in the side for an entrance, and is generally formed of the thin paper-like bark of the Tea-tree (*Melaleuca*), and lined with feathers: it is also usually suspended to the foliage of this tree, and occasionally to that of other shrubs which grow in its favourite localities. The eggs are four in number, of a delicate flesh-white, freckled with spots of reddish brown, which are much thicker at the larger end; they are about eight lines long and six lines broad. The breeding-season commences in September, and continues during the three following months.

The males are subject to the same law relative to the seasonal change of plumage as the *Malurus cyaneus* and the other members of the genus. The gay nuptial costume of these birds renders them conspicuously different from the *Priniæ* of India, to which they have otherwise a seeming alliance.

The male has the forehead, ear-coverts, sides of the face, and occiput rich verditer-blue; centre of the back light verditer-blue; scapularies chestnut; throat, chest, back of the neck and rump deep velvety black, the throat in certain lights tinged with blue; wings brown; abdomen and under tail-coverts buffy white; tail dull bluish green, crossed by numerous indistinct bars, seen only in some positions, and very slightly tipped with white; bill black; eyes and feet blackish brown.

The female has all the upper surface and wings brown; throat and under surface buff-white; tail as in the male, but more dull, and devoid of the white at the extremity of the feathers; bill dull reddish brown, lighter beneath; space between the bill and eyes reddish brown; legs brown.

Total length $6\frac{1}{4}$ inches; bill $\frac{1}{2}$; wing $1\frac{7}{8}$; tail $3\frac{1}{2}$; tarsi 1.

Sp. 190. MALURUS PULCHERRIMUS, *Gould*.

BLUE-BREASTED SUPERB WARBLER.

Malurus pulcherrimus, Gould in Proc. of Zool. Soc., part xii. p. 106.

Malurus pulcherrimus, Gould, Birds of Australia, fol., vol. iii. pl. 23.

The Blue-breasted Superb Warbler is one of the variegated species of its genus, and is nearly allied to the *M. elegans* and *M. amabilis*, but is of larger size, and moreover differs from them in having the throat and breast of a rich deep blue instead of black.

For a knowledge of this species I am indebted to the researches of Gilbert, who informs me that "it appears to be exclusively confined to the thickets of the interior of Western Australia; in habits and manners it greatly resembles the other members of the genus, but its nest is somewhat smaller than that of either of them. A nest found on the 28th of October, in the vicinity of the Wongan Hills, was placed on the upper branches of a species of *Hakea*, about four feet from the ground; it contained two newly-laid eggs, which resembled those of the other species of the genus, but had the blotches very much larger."

Crown of the head and a broad band across the centre of the back rich glossy violet-blue; space surrounding the eye, and the ear-coverts, verditer-blue; throat intense indigo-blue; bounded below by an indistinct band of black; lores, collar surrounding the back of the neck, and the lower part of the back deep velvety-black; scapularies chestnut; wings brown; tail dull greenish blue, indistinctly barred with a darker tint, and slightly tipped with white; abdomen and under tail-coverts white; bill and feet black; irides dark brown.

Total length $5\frac{1}{4}$ inches; bill $\frac{7}{16}$; wing 2; tail $3\frac{1}{4}$; tarsi $\frac{7}{8}$.

Sp. 191. MALURUS LAMBERTI, *Vig. and Horsf.*

LAMBERT'S SUPERB WARBLER.

Malurus lamberti, Vig. and Horsf. in Linn. Trans., vol. xv. p. 221.
Superb Warbler, White's Journ., pl. in p. 250, low. fig.
Variegated Warbler, Lewin, Birds of New Holland, pl. xv.

Malurus lamberti, Gould, Birds of Australia, fol., vol. iii. pl. 24.

Although far less common and much more local than *M. cyaneus*, this species ranges over a greater extent of country, being an inhabitant of most parts of New South Wales, the districts near the coast, as well as those of the interior, but particularly those in the neighbourhood of the Namoi, where it is sometimes associated with its congener *M. cyaneus*.

The neighbourhood of Botany Bay is one of its most favourite resorts, and it is occasionally seen near Sydney, and even in the small gardens within the town. It does not inhabit Tasmania, nor did I observe it in South Australia, or hear of its ever having been seen there, neither have I received it from the colony of Swan River.

This is one of the few common birds of Australia of which I was not able to find the nest; but its changes of plumage, nidification, the number and colour of its eggs, are doubtless very similar to those of the other members of its family. Its food consists of insects of various kinds, which are sought for on the ground, over which it runs with great facility.

The male has the forehead, ear-coverts, sides of the head, occiput, and centre of the back beautiful violet-blue; throat, breast, crescent across the upper part of the back and rump black; scapularies chestnut; wings brown; abdomen white, tinged with brown on the flanks; tail dull greenish blue, indistinctly barred with a darker tint, and lightly tipped with white; bill black; eyes and feet dark brown.

The female has the body dull brown; the throat and under

surface much paler; tail-feathers as in the male, but less bright; bill and space round the eye reddish brown; feet brown.

Sp. 102. MALURUS AMABILIS, *Gould.*

LOVELY SUPERB WARBLER.

Malurus amabilis, Gould in Proc. of Zool. Soc., part xviii. p. 277.

Malurus amabilis, Gould, Birds of Australia, fol., Supplement, pl.

The officers of Her Majesty's Surveying Ship Rattlesnake so well employed their time in collecting the natural productions of the Cape York district, that they added very considerably to our knowledge of the fauna of that part of the continent. A single and somewhat imperfect specimen of this bird, bearing the words "Cape York, 1849," was transmitted by the late Captain Owen Stanley to the Zoological Society of London; and it is from this specimen that my description was taken. It is nearly allied to the *Malurus elegans,* but differs from that bird in its larger bill, in the deeper and more uniform blue of the cheeks and crown, in the darker colouring of the thighs, and in the much greater extent of the white on the tips and margins of the outer tail-feathers.

I feel assured the female of *M. amabilis* will be found to closely resemble that sex of *M. elegans* whenever it is our good fortune to have examples transmitted to us; and that this desideratum may soon be obtained, as well as additional skins of the male, is much to be wished. When the Cape York Peninsula is closely explored, not only this, but many other interesting species will reward the collector, and the fauna will probably be found to partake of that of the adjacent island of New Guinea, as well as of forms peculiar to New South Wales.

Head, ear-coverts, and centre of the back delicate violet-blue; lores, throat, breast, crescent across the upper part of the back, and the rump deep bluish-black; scapularies chestnut; wings brown, the secondaries slightly margined with white; abdomen white, very slightly tinged with buff on the flanks; tail dull greenish blue, the four lateral feathers margined externally and largely tipped with white; hinder part of the thighs black; bill black; irides and feet dark brown.

Sp. 103. . MALURUS CORONATUS, *Gould.*

CROWNED SUPERB WARBLER.

Malurus coronatus, Gould in Proc. of Zool. Soc., part xxv. p. 221.

Malurus coronatus, Gould, Birds of Australia, fol., Supplement, pl.

Charming as are many of the smaller Australian birds, I think the present species is entitled to the palm for elegance and beauty, not only among the members of its own genus, numerous and beautiful as they really are, but among all other groups of birds yet discovered; the charm, too, is considerably enhanced by the great novelty in the style of its colouring; for in how few birds do we find the lovely lilac tint which encircles and adorns the head of this bird! a similar tint, it is true, appears in the nape of the Bower-birds (*Chlamydoderæ*); but I scarcely know of a third instance

Having premised thus much respecting this new *Malurus*, I now come to the painful task of naming its collector; I say painful, because the gentleman who shot and brought it to this country has fallen, like many other Australian explorers, a victim to the climate of that country, congenial to Europeans as it generally is. It will be recollected by all those who take an interest in scientific explorations, that Mr. Elsey accompanied A. C. Gregory, Esq., as surgeon and naturalist on his great journey from the Victoria River to Moreton Bay. Soon after his return to England it became evident that he had contracted the disease called *hæmoptysis*, which speedily obliged him to remove to a warmer climate: he selected one of the West-Indian Islands, and, on arriving, commenced his investigations with his usual spirit; but he rapidly became worse, and science shortly had to deplore the loss of one of her most enthusiastic votaries. The little I saw of this gentleman impressed me with the belief that he had a true love for nature; and had he been spared, I feel assured he would have distinguished himself greatly in one or other branch of the natural sciences.

The *Malurus coronatus* is an inhabitant of the countries bordering the Victoria River. Both sexes were procured, and they now form part of the collection in the British Museum.

The male has the crown of the head rich lilac-purple, with a triangular spot of black in the centre, and bounded below by a band of velvety black, which, commencing at the nostrils, passes backwards through the eye, dilates upon the ear-coverts, and meets at the back of the neck; back and wings light brown; tail bluish green, becoming of a deeper hue towards the extremity; lateral feathers margined externally and tipped with white; under surface buffy white, becoming gradually deeper on the flanks and vent; irides brown; bill black; feet fleshy brown.

The female has all the upper surface light brown; lores and space behind the eye white; ear-coverts chestnut; in other respects she is similar to the male.

Total length $6\frac{1}{2}$ inches; bill $\frac{4}{8}$; wing $2\frac{1}{4}$; tail $3\frac{1}{2}$; tarsi $1\frac{1}{8}$.

Sp. 194. MALURUS LEUCOPTERUS, *Quoy et Gaim.?*

WHITE-WINGED SUPERB WARBLER.

Malurus leucopterus, Quoy et Gaim. Zool. de l'Uranie, p. 108, pl. 23. fig. 2?

Amytis leucopterus, Less. Traité d'Orn., p. 454.

Malurus leucopterus, Gould, Birds of Australia, fol., vol. iii. pl. 25.

I regret that I have not been able to clear up the doubt which exists in my mind, whether the present bird is or is not distinct from the one figured by Messrs. Quoy and Gaimard in the 'Voyage de l'Uranie,' since, on applying at the Museum of the Jardin des Plantes for the purpose of examining the original specimen, it could not be found: the figure above quoted, if intended for this bird, is by no means correct, and it is, moreover, said to be from Dirk Hatich's Island, on the western coast, a locality very distant from those in which my specimens were procured, New South Wales; which circumstance strengthens my belief that they may be distinct: besides which, the bird under consideration is supposed to be exclusively an inhabitant of the interior; for I never observed it between the mountain-ranges and the coast, and it is scarcely probable, therefore, that it should inhabit an island like that of Dirk Hatich. In case they should prove to be different, I propose the name of *Malurus cyanotus* for the bird from New South Wales.

The birds seen by me were either in pairs or in small troops, and evinced so much shyness as to render the acquisition of specimens a task of no little difficulty, particularly of the full-plumaged male, who appeared to be conscious that the display of his gorgeously coloured dress might lead to his detection. Its powers of flight are not great; but this is fully compensated for by the extraordinary manner in which it threads the bushes, and passes over the surface of the ground in a series of hopping bounds, whereby it readily eludes pursuit. The most successful mode of obtaining it is to ascertain the precise spot in which it is located, to approach it cautiously, and to remain silent for a short time, when the male will soon show himself by hopping out from the bush—the restless nature of his disposition not admitting of his remaining long concealed.

The nest is composed of grasses, rather large and dome-shaped, with a hole near the top for an entrance. The one sent me from South Australia contained two eggs, one of which was the Bronze Cuckoo's, thus showing that this little bird is also the foster-parent of those birds. The number of eggs laid by the *Malurus leucopterus* is in all probability four; the one I possess is flesh-white, finely freckled with reddish brown (forming a zone at the larger end), and is eight lines long by six lines broad.

The male has the whole of the head, body above and beneath, and the tail beautiful deep blue; scapularies, wing-coverts, and tertiaries snow-white; primaries brown, with their external edges silvery green; bill black; feet brown; eyes dark brown.

The female has the crown of the head and all the upper surface and flanks brown; throat and abdomen white, faintly washed with brown; external edges of the primaries and tail pale greenish blue; bill reddish brown.

Sp. 106.   MALURUS LEUCONOTUS, *Gould.*
White-backed Superb Warbler.

Malurus leuconotus, Gould in Proc. of Zool. Soc., 1865, p. 198.

In size this fine new species is very similar to the last, from which, however, it may be at once distinguished by its white back, which has suggested the specific name I have assigned to it.

It inhabits the interior of Australia, but the precise locality is unknown to me; it accompanied fine examples of *Geophaps plumifera*. My attention was called to it by Mr. Ward, of Vere Street. The example described is in the possession of Mrs. Elizabeth F. M. Craufuird, of Budleigh Salterton, Devon.

The entire head, neck, under surface, rump, and tail deep blue; back, shoulders, greater and lesser wing-coverts, and secondaries silky white; primaries brown; bill black; feet brownish black.

Total length 5½ inches; bill ½; wing 2; tail 3⅞; tarsi ⅞.

Sp.190. MALURUS MELANOCEPHALUS, *Vig. and Horsf.*

BLACK-HEADED SUPERB WARBLER.

Scarlet-backed Warbler, Lewin, Birds of New Holl, pl. xiv.
Malurus melanocephalus, Vig. and Horsf. in Linn. Trans., vol. xv. p. 222.
—— *brownii*, Vig. and Horsf. in Linn. Trans., vol. xv. p. 223.

Malurus melanocephalus, Gould, Birds of Australia, fol., vol. iii. pl. 26.

The Black-headed Superb Warbler, which probably inhabits all the south-eastern portion of Australia, is a local species, not being generally diffused over the face of the country, like several other members of the group, but confined to grassy ravines and gullies, particularly those that lead down from the mountain-ranges. I obtained several pairs of adult birds in very fine plumage in the valleys under the Liverpool range, all of which I discovered among the high grasses which there abound; but as the period of my visit was their breeding-season, I never observed more than a pair together, each pair being always stationed at some distance from the other, and in such parts of the gullies as were studded with small clumps of scrubby trees.

This Superb Warbler has many actions in common with the *M. cyaneus*, and like that species carries its tail erect: it also frequently perches on a stem of the most prominent grasses, where it displays its richly-coloured back, and pours forth its simple song. I did not succeed in finding the nest, although I knew they were breeding around me: it was probably placed among the grasses, but was so artfully concealed as to completely baffle my research.

One might suppose the greater development of feather on the back of this species to have been given it as a defence against the damp and dense grasses of the ravines, among which it usually resides; but from the circumstance of the female not possessing this character of plumage, and the rich garb being only seasonal in the male, this supposition falls to the ground. In their winter dress the sexes very nearly resemble each other; but the males may always be distinguished by the black colouring of the bill and tail-feathers. The young male of the year has the tail-feathers brown, like the females; and it is a curious fact, that at this age these feathers are much longer than in the adult.

The male has the head, all the under surface, wing-coverts, upper tail-coverts, and tail deep velvety black; back of the neck, scapularies, and remainder of the upper surface rich orange-scarlet; bill black; eyes blackish brown; feet fleshy brown.

Female brown above, paler beneath; bill brown; base of the under mandible reddish brown; feet flesh-brown.

Sp. 197. MALURUS CRUENTATUS, *Gould*.

Brown's Superb Warbler.

Malurus cruentatus, Gould in Proc. of Zool. Soc., part vii. p. 143.

Malurus Brownii, Gould, Birds of Australia, fol., vol. iii. pl. 27.

Among the species of which I sent home characters from Australia, for publication in the Proceedings of the Zoological Society, was the present pretty bird, to which I gave the specific name of *cruentatus*. It is a native of the north-western portion of the country, and formed part of the collection placed at my disposal by the officers of H.M.S. 'Beagle.' It differs from *Malurus melanocephalus* in the more intense blood-red colour of the back, and in its much smaller size.

We now know that this bird is common at Port Essington; and, as I have above stated, that it is also an inhabitant of the north-western coasts, and in all probability enjoys an extensive range over the north-western parts of the Australian continent, where grassy ravines occur.

The male in summer has the head, neck, wings, all the under surface, and tail black; primaries and secondaries brown; back and shoulders fine crimson; bill black; legs fleshy brown.

The female is uniform light brown, the abdomen inclining to white; bill and feet light brown.

Total length 4 inches; bill $\frac{1}{4}$; wing $1\frac{5}{8}$; tail $1\frac{3}{4}$; tarsi $\frac{3}{4}$.

Genus AMYTIS, *Lesson*.

A form nearly allied to *Malurus*, strictly Australian, and of which three species are known, inhabiting the southern half of the country, and not occurring in Tasmania.

Sp. 198.   AMYTIS TEXTILIS.

TEXTILE WREN.

Malurus textilis, Quoy et Gaim. Zool. de l'Uranie, p. 107, pl. 23. fig. 1.

Amytis textilis, Gould, Birds of Australia, fol., vol. iii. pl. 28.

The bird figured in the "Voyage de l'Uranie," doubtless represents the present species, while that figured by Lesson in the Atlas to his "Traité d'Ornithologie," and which seems to have been the subject from which he took his generic characters and description, as clearly belongs to *A. striatus*.

The only place in which I observed the Textile Wren was the plains bordering the Lower Namoi; and that its range extends far to the northward and westward is tolerably certain.

In the various positions it assumes, in the elevated carriage of its tail, and in its whole economy, it bears a close resemblance to the *Maluri*: like them also it wanders about in small troops of four or six in number, always keeping within a short distance, and returning towards the close of the day to its accustomed haunts. On the Lower Namoi, where it is very abundant, it is found in all those parts of the plains that are studded with scrubs and clumps of a low shrub-like tree, resembling the Barilla of the coast, through and among which it creeps with astonishing rapidity; indeed its mode of progression on the ground is such as no description can convey an accurate conception of, and must be seen to be understood: I cannot perhaps compare it with anything, unless with the motion of an Indian-rubber ball when thrown forcibly along the ground. While stealing from bush to bush, with this rapid movement, its head low and tail perfectly erect, it presents an exceedingly droll appearance. Like many others of its family, it seldom employs its power of flight.

On my arrival in Australia fresh from Europe, these birds and those of the preceding genus were regarded by me with the highest interest, as they must be by every person not born and bred in Australia, who sees them for the first time in a state of nature.

Of its nidification I have nothing to communicate: it doubtless builds a dome-shaped nest, and in all probability lays four spotted eggs; but to these points I would call the attention of those who are favourably situated for observing them, as also to confirm or refute the opinion I have elsewhere expressed of this and the following bird being distinct.

All the upper surface dark brown, each feather with a narrow stripe of white down the centre; under surface the same, but much paler; flanks and under surface of the shoulder rust-red; tail dark brown, indistinctly barred with a still darker hue and edged with pale brown; irides reddish hazel; base of lower mandible bluish horn-colour; remainder of the bill black; feet flesh-brown.

The male I dissected was destitute of the rusty red colouring on the flanks and the under surface of the shoulder.

Sp. 199. AMYTIS STRIATUS, *Gould.*
STRIATED WREN.

Amytis textilis, Less. Traité d'Orn., p. 454, pl. 67. fig. 2.
Dasyornis striatus, Gould in Proc. of Zool. Soc., part vii. p. 143.

Amytis striatus, Gould, Birds of Australia, fol., vol. iii. pl. 29.

The only specimen I procured of this little bird in a recent state was shot while I was traversing the Lower Namoi; it appeared to give preference to a loose sandy soil studded with high rank grass, which, growing in tufts, left the interspaces quite bare: through the natural labyrinth thus formed the Striated Wren ran with amazing rapidity; and it was only by forcing it to take wing that I succeeded in killing the one I obtained, which on dissection proved to be a male. All the specimens I have seen from New South Wales were in the red state of plumage, which goes far towards proving that this bird is really distinct from *Amytis textilis*.

Nothing has yet been ascertained respecting its nidification: its food, like that of the Textile Wren, consists of insects of various kinds. As might be conjectured from its form, its habits are terrestrial; and it rarely, if ever, mounts into the air, or flies except among the trees.

Upper surface fine rusty red, each feather with a line of buffy white bounded on each side by black down the centre; line beneath the eye black; ear-coverts black, striated with white; wings and tail brown, margined with light reddish brown; base of the primaries rust-red, forming a conspicuous patch; chin and throat white; feathers of the chest buffy white, with two lines of brown, one down each side the stem; under surface rust-red, some of the feathers with a stripe of white down the centre; tail dark brown, indistinctly barred with a still darker tint, margined with lighter brown; irides hazel; bill dark horn-colour; feet brownish lead-colour.

Total length $6\frac{1}{2}$ inches; bill $\frac{1}{2}$; wing $2\frac{3}{4}$; tail $3\frac{1}{2}$; tarsi 1.

Sp. 200.   AMYTIS MACROURUS, *Gould.*

LARGE-TAILED WREN.

Amytis macrourus, Gould in Proc. of Zool. Soc., part iv. p. 2.
Nyern-de and *Jee-ra*, Aborigines of the interior of Western Australia.

Amytis macrourus, Gould, Birds of Australia, fol., vol. iii. pl. 30.

The present is the only species of the genus that has been discovered in Western Australia; two examples were shot in the interior by Gilbert, who states that " it inhabits the thickets, and is almost always on the ground, moving about in families of from four to seven in number: it carries its tail more erect than any other bird I have seen, and certainly no bird runs or rather hops over the surface of the ground with greater rapidity."

It is evidently the representative of the *Amytis textilis* of the eastern coast, to which it is very nearly allied, but from which, as well as from the *A. striatus*, it may at once be distinguished by its more robust form, and by the much greater length and size of its tail.

All the upper surface brown, each feather with a narrow stripe of white down the centre; under surface the same, but much paler; under surface of the shoulder pale rusty red; tail brown, margined with pale brown; irides hazel; base of the lower mandible horn-colour, remainder of the bill black; feet flesh-brown.

Total length $5\frac{1}{4}$ inches; bill $\frac{1}{2}$; wing $2\frac{3}{4}$; tail $2\frac{1}{2}$; tarsi $\frac{7}{8}$.

Genus STIPITURUS, *Lesson.*

A form confined to Australia, where it frequents extensive grass-beds, particularly those which occur in humid situations. It runs quickly over the ground, and carries its tail erect like the *Maluri*. Some slight variation occurs in specimens from Tasmania, Southern and Western Australia; but I believe they are all referable to one species.

Sp. 201. STIPITURUS MALACHURUS, *Less.*

EMU WREN.

Muscicapa malachura, Lath. Ind. Orn., Supp. pl. 52.
Soft-tailed Flycatcher, Linn. Trans., vol. iv. p. 242, pl. 21.
Malurus malachurus, Vig. and Horsf. in Linn. Trans., vol. xv. p. 224.
—— *palustris,* Vieill., 2de édit. du Nouv. Dict. d'Hist. Nat., tom. xx. p. 214.
Stipiturus malachurus, Less. Traité d'Orn., p. 415.
Soft-tailed Warbler, Lath. Gen. Hist., vol. vii. p. 128.
Waw-gul-jelly, Aborigines of New South Wales.
Djur-jeel-ya, Aborigines of the lowlands of Western Australia.

Stipiturus malachurus, Gould, Birds of Australia, fol., vol. iii. pl. 31.

This curious little bird has a wide distribution, since it inhabits the whole of the southern portion of Australia, from Moreton Bay on the east to Swan River on the west, including Tasmania. Among the places where it is most numerous in the latter country, are the swampy grounds in the neighbourhood of Rocherche Bay in D'Entrecasteaux Channel, the meadows at New Norfolk, Circular Head, and Flinders Island in Bass's Straits; on the continent of Australia, Botany Bay and, indeed, all portions of the country having a similar character are favoured with its presence.

The Emu Wren is especially fond of low marshy districts covered with rank high grasses and rushes, where it conceals itself from view by keeping near the ground in the midst of the more dense parts of the grass-beds. Its extremely short round wings ill adapt it for flight, and this power is consequently seldom employed, the bird depending for progression upon its extraordinary capacity for running: in fact, when the grasses are wet from dew or rain, its wings are rendered perfectly unavailable. On the ground it is altogether as nimble and active, its creeping mouse-like motions, and the extreme facility with which it turns and bounds over the surface, enabling it easily to elude pursuit, and amply compensating for the paucity of its powers of flight. The tail is carried in an erect position, and is even occasionally retroverted over the back.

The nest, which is a small ball-shaped structure, with rather a large opening on one side, is composed of grasses lined with feathers, and artfully concealed in a tuft of grass or low shrub. One that I found in Recherche Bay contained three newly-hatched young: this being the only nest I ever met with, I am unable to give any description of its eggs from my own observation; but this want is supplied in the following account of this species from the pen of Mr. E. P. Ramsay, published in the 'Ibis' for 1865:—

"I had for many days visited the swamps upon Long Island, where these birds are very plentiful, in the hope of finding them breeding; but it was not until the 25th of September that I succeeded in discovering a nest, although I had watched them for hours together for several days. While walking along the edge of the swamp on that day, a female flew from my feet out of an overhanging tuft of grass growing only a few yards from the water's edge. Upon lifting up the leaves of the grass which had been bent down by the wind, I found its nest carefully concealed near the roots, and containing three eggs. They were quite warm, and within a few days of being hatched, which may account for the bird being unwilling to leave the spot; for upon my returning about five minutes afterwards, the female was perched upon the same tuft of grass, and within a few inches of whence I had taken the nest. The nest is of an oval form (but that part which might be termed the true nest is perfectly round), placed upon its side; the mouth very large, taking up the whole of the under part of the front. It is very shallow, so much so that, if tilted slightly, the eggs would roll out, being almost upon a level with the edge. It is outwardly composed of grass and the young dry shoots of the reeds which are so common in all the swamps near the Hunter River, lined with fine grass, roots, and, finally, a very fine green moss. It is very loosely put together, and requires to be moved very gently to prevent its falling to pieces.

"The eggs measure 0½ lines long by 4½ broad, sprinkled all over with minute dots of a light reddish brown, particularly at the larger end, where they are blotched with the same colour. One of the three had no blotches, but was minutely freckled all over. The ground-colour is a delicate white, with a blush of pink before the egg is blown.

"The only note of the bird, besides a slight chirp when flushed and separated, is a slight twitter, not unlike a faint attempt to imitate the *Malurus cyaneus*. While in the swamp, which at the time was nearly dry, I observed several separate flocks: of these some were hopping along the ground, picking up something here and there; others, whose appetites seemed appeased, were creeping along through the reeds about a foot from the ground, but as the reeds thickened I soon lost sight of them. They seldom took wing, except when disturbed, and not always then, seeming very averse to showing themselves. While watching them I observed one now and then hop to the top of a tall reed as if to get a glimpse at the world above. Upon coming suddenly upon a flock and following them, they keep to the reeds just in front of you, and never take wing unless hard driven, when they separate and do not collect again for some time."

The male is readily distinguished from the female by the blue colouring of the throat, and by a somewhat greater development of the tail-feathers. The decomposed or loose structure of these feathers, much resembling those of the Emu, has suggested the colonial name of Emu Wren for this species, an appellation singularly appropriate, inasmuch as it at once indicates the kind of plumage with which the bird is clothed, and the Wren-like nature of its habits.

Genus SPHENURA, *Lichtenstein*.

A group of birds adapted for situations covered with an impenetrable vegetation, reed-beds, &c. Two species are all that are at present known; of these one is from the eastern, and the other from the western part of Australia.

Sp. 202.   SPHENURA BRACHYPTERA.
BRISTLE-BIRD.

Turdus brachypterus, Lath. Ind. Orn., Supp. p. xliii.
Sphenura brachyptera, Licht. Verz. der Doubl., p. 40.
Malurus pectoralis, Steph. Cont. Shaw's Gen. Zool., vol. xiii. part i. p. 224.
Dasyornis australis, Vig. and Horsf. in Linn. Trans., vol. xv. p. 282.

Dasyornis australis, Gould, Birds of Australia, fol., vol. iii. pl. 32.

This bird inhabits reed-beds and thickets, particularly such as are overgrown with creepers and rank vegetation; I believe it to be found throughout New South Wales in all places suitable to its habits, although, from the recluse nature of its disposition, it is a species familiar to few, even of those who have been long resident in the colony. Its powers of flight are very limited, but it threads the thickets and runs over the ground with the greatest facility. It resembles the true *Maluri* in carrying the tail erect, and in many other of its actions. My own impression is that it is a stationary species, since its powers of flight are inadequate to enable it to pass over much extent of country, and the thick brushes near the coast afford it ample shelter in winter.

I did not succeed in finding its nest, but in its nidification it doubtless closely assimilates to the next species, the Long-billed Bristle-bird of the western coast.

The sexes present no difference in plumage, and but little in size; the female, however, is rather the smaller.

The food consists of insects of various orders.

All the upper surface brown; wings, tail-coverts, and tail rufous brown, the latter indistinctly barred with a darker tint; under surface grey, gradually passing into the brown of the upper surface; over the eye an indistinct buffy stripe; irides brown; bill brown, becoming much lighter on the lower mandible; legs greyish brown.

Sp. 203. SPHENURA LONGIROSTRIS, *Gould*.

LONG-BILLED BRISTLE-BIRD.

Dasyornis longirostris, Gould in Proc. of Zool. Soc., part viii. p. 170.
Djgr-dal-ya, Aborigines of the lowland districts of Western Australia.

Dasyornis longirostris, Gould, Birds of Australia, fol., vol. iii. pl. 33.

The present species assimilates very closely in the character and colouring of its plumage to its eastern analogue, the *Sphenura brachyptera*; but differs from that bird in being of a smaller size and in having a longer bill. It is a native of Western Australia, and is very generally distributed over the colony of Swan River, where it inhabits reed-beds and long grasses, and is occasionally seen in scrubby places. "It is so remarkably shy," says Gilbert, "that it is extremely difficult to get even a glimpse of it: it appeared to feed on the ground, where its actions are extremely quick, running over the surface with its tail erect. The only chance of procuring specimens is when it ascends to a small branch on the top of a scrub to sing. Its notes are loud, clear, and extremely varied.

"It flies very low; in fact the bird scarcely ever rises more than a few yards above the scrub or long grass it inhabits; it is consequently very rarely seen on a tree.

"The nest is formed of dry wiry grass, without any lining, more globular than those of the *Malwri*, but, like them, with an opening in the side; it is of rather a large size, and the only one I met with was built in a clump of coarse grass, sheltered by an overhanging dead bush. It contained two eggs, the ground-colour of which is dull brownish white, blotched and freckled with purplish brown, some of the blotches appearing as if beneath the surface, particularly at the larger end, where they are most numerous.

"Its food consists of seeds and insects."

The sexes so closely resemble each other, that a representation and description of one will suffice for both.

All the upper surface brown; wings, tail-coverts, and tail rufous brown, the latter indistinctly barred with a darker tint; under surface grey, gradually passing into the brown of the upper surface; irides bright reddish brown; upper mandible brown, lower mandible bluish green at the tip and greenish white at the base; legs bluish grey.

Total length 7½ inches; bill ⅞; wing 2⅝; tail 4; tarsi ⅞.

Genus ATRICHIA, *Gould*.

The only species of this genus yet discovered is as singular in its structure as it is shy and retiring in its habits; the total absence of vibrissæ in a bird apparently closely allied to *Sphenura*, in which they are so much developed, renders it one of the anomalies of the Australian fauna.

Sp. 204.   ATRICHIA CLAMOSA, *Gould*.

Noisy Scrub-bird.

Atrichia clamosa, Gould in Proc. of Zool. Soc., part xii. p. 2.

Atrichia clamosa, Gould, Birds of Australia, fol., vol. iii. pl. 34.

Few of the novelties received from Australia more interested me than the species to which I have given the generic name of *Atrichia*. Gilbert met with it among the dense scrubs of Western Australia, having had his attention attracted to it by its peculiar and noisy note long before he had an opportunity of observing it; and it was only after many days of patient and motionless watching among the scrubs, that he succeeded in obtaining specimens. Future research will doubtless furnish us with some interesting information respecting the habits of this curious form. It is a bird evidently destined to tenant the most dense thickets and tangled beds of dwarf trees.

The examples forwarded to me were killed between Perth and Augusta, and were all males. The females will doubtless, when discovered, prove to differ but little from their mates, except that the black mark on the breast will not be so large or conspicuous. I am led to offer this opinion from the circumstance of one of the specimens being a young male, which usually resembles the female during the first year.

All the upper surface, wings, and tail brown, each feather crossed by several obscure crescent-shaped bars of brown; the inner webs of the primaries very dark brown, without markings, and the tail freckled instead of barred; throat and chest reddish white, with a large irregular patch of black on the lower part of the throat; flanks brown; abdomen and under tail-coverts rufous; bill horn-colour; irides dark brown.

Total length $7\frac{3}{4}$ inches; bill $\frac{3}{8}$; wing 3; tail 4; tarsi 1.

Genus HYLACOLA, *Gould*.

A genus comprising two species peculiar to the southern parts of the country, one of which enjoys an extensive range from South Australia to Moreton Bay; the other has, as yet, only been found in the Great Murray Scrub. These birds carry their tail in an upright position, move quickly over the surface of the ground, and trip with agility along the horizontal branches of fallen trees. In size they are about equal to our well-known Hedge-Sparrow (*Accentor modularis*), to which they bear some resemblance when seen in their native country. The sexes are alike in plumage.

Sp. 205.   HYLACOLA PYRRHOPYGIA.

Red-rumped Hylacola.

Acanthiza pyrrhopygia, Vig. and Horsf. in Linn. Trans., vol. xv. p. 227.

Hylacola pyrrhopygia, Gould, Birds of Australia, fol., vol. III.
   pl. 39.

The situations most favourable to the habits of this bird are open sterile spots, here and there studded with clumps of brushes or dense herbage, the beds and sides of creeks, and the crowns of stony hills. I have generally observed it in small companies, probably the brood of a single pair.

Its agreeable song is poured forth while the bird is perched upon some conspicuous part of a bush, or some little spray among the branches of the large fallen trees, where it loves to dwell, as on the approach of an intruder it can readily and effectually secrete itself among the high grass and herbage which have grown up amidst the branches. The facility with which it creeps among or threads these little thickets is surprising. It rarely flies, but depends for progression more upon the rapidity with which it can pass over the ground, than upon the feeble powers of its small rounded wing.

I found it plentiful on the low hills to the north of the Liverpool Plains, as well as in most parts of South Australia, and believe it to be a stationary bird, for it appeared to be equally numerous in summer and winter.

Of its nidification I have nothing to communicate, its nest not having been discovered either by myself or by any of my party.

Its food consists of insects of various kinds.

The sexes present no visible difference in their plumage.

Crown of the head, all the upper surface, wings, and tail brown; lower part of the rump and upper tail-coverts chestnut-red; all but the two centre tail-feathers crossed near the tip with a broad band of black, beyond which the tips are greyish white; line over the eye and all the under surface greyish white, each feather of the latter with a line of black down the centre, except on the middle of the abdomen; bill dark brown; irides buffy white; legs flesh-brown.

Sp. 200.   HYLACOLA CAUTA, *Gould.*

CAUTIOUS HYLACOLA.

Hylacola cauta, Gould in Proc. of Zool. Soc., part x. p. 135.

Hylacola cauta, Gould, Birds of Australia, fol., vol. iii. pl. 40.

The only locality in which I have seen this species is the great scrub clothing the banks of the river Murray in South Australia, where it was not uncommon, but so excessively shy that I obtained but a single specimen during my stay in the district. Its timidity being so great, and its natural habitat the more dense parts of the scrub, it is a species which must for a long time be exceedingly scarce in our collections.

With the exception of its being even more shy, its whole habits and economy appeared to be very similar to those of the preceding species (*H. pyrrhopygia*). It carries its tail perfectly erect, and hops over the ground and threads the bushes with the greatest ease, generally keeping among the more dense parts of the low bushes, and only exposing itself on the outermost twigs when desirous of pouring forth its song, which is sweet and harmonious.

In size the *H. cauta* is rather less than the *H. pyrrhopygia,* has the markings of the under surface much bolder, and the chestnut-coloured mark on the rump of a much deeper tint.

Line from the base of the upper mandible along the side of the face and over the eye white; above this a narrow line of black; crown of the head and all the upper surface brown; upper and under tail-coverts bright chestnut; wing-coverts brown, edged with brownish white; primaries brown, with the outer web white at the base, forming a conspicuous spot in the centre of the wing; tail blackish brown, tipped with white; throat striated with black and white, produced by each feather being black down the centre and fringed with white; flanks mottled brown and white; abdomen white; bill dark brown; irides buffy white; feet flesh-brown.

When I characterized this species in the 'Proceedings of the Zoological Society of London,' I had only seen a single example; I have since received a second, proving the correctness of my view of its being quite distinct from the *H. pyrrhopygia*, a fact disputed by the late Mr. Strickland, who stated it to be his opinion that *H. pyrrhopygia* and *H. cauta* were one and the same species, but who, upon an examination of the specimens themselves, acknowledged he was in error.

Total length 5¾ inches; bill ₇⁄₈; wing 2¼; tail 2½; tarsi ⅞.

Genus PYCNOPTILUS, *Gould*.

Of this form only a single species is known, all the information respecting which will be found below. Although I do not doubt that it is really an inhabitant of Australia, I have no positive evidence on this point. The type specimen is in the British Museum.

Sp. 207. PYCNOPTILUS FLOCCOSUS, *Gould*.

DOWNY PYCNOPTILUS.

Pycnoptilus floccosus, Gould in Proc. of Zool. Soc., part xviii. pp. 95, 279.

Pycnoptilus floccosus, Gould, Birds of Australia, fol., Supplement, pl.

I know nothing of the habits and economy of this bird, nor what part of Australia it inhabits. I purchased it of Mr. Warwick, who had obtained it in a small collection of birds said to have been formed in the interior of New South Wales towards the river Morumbidgee. Judging from its very thick clothing and overhanging back-feathers, I conclude that, like the members of the genus *Dasyornis*, it is a frequenter of the ground in dense and scrubby places—a conjecture which I should be happy to have verified by residents in New South Wales who may be favourably situated for observing it.

General plumage brown, inclining to rufous on the lower part of the back, upper tail-coverts, and tail; forehead, lores, throat, and breast dark reddish buff, with a very narrow crescent of dark brown at the tip of each feather; centre of the abdomen greyish brown, crossed by crescentic bands of black; flanks and vent brown, passing into deep rufous on the under tail-coverts; bill brown; base of under mandible fleshy brown; legs and feet fleshy brown.

Total length $6\frac{3}{4}$ inches; bill $\frac{3}{4}$; wing $2\frac{3}{4}$; tail $2\frac{3}{4}$; tarsi $1\frac{1}{4}$.

Genus CISTICOLA, *Kaup*.

These little birds are most perplexing, and the due elucidation of the Australian members of this form can only be effected by resident ornithologists; to this subject I would therefore direct the special attention of Mr. Ramsay of New South Wales, Mr. White of South Australia, or any other person favourably located for investigating them. A knowledge of the changes of the plumage, if any, of a single species would be a key to the whole. By closely watching the birds while breeding, obtaining the mated pairs, ascertaining the sex of each by dissection, and by observing the young from youth to maturity, the matter might easily be determined.

Sp. 203.   CISTICOLA MAGNA, *Gould*.

GREAT GRASS-WARBLER.

Cysticola campestris, Gould in Proc. of Zool. Soc., part xiii. p. 20.

Cysticola magna, Gould, Birds of Australia, fol., vol. iii. pl. 41.

This is one of the largest species of the group, and hence

I have assigned to it the distinctive appellation of *magna*. Nothing whatever is known of its habits and manners; but we may reasonably infer that they are very similar to those of its congeners. The precise locality it inhabits is also unknown, the specimen from which my description was taken having been obtained from a general collection of Australian birds, without the situation in which it had been procured being attached to it.

Head rusty red; back and wing-coverts brownish grey; all the feathers of the upper surface with a broad stripe of dark brown down the centre; wings blackish brown, the primaries margined externally with rusty red, and the secondaries edged all round with brownish grey; tail reddish brown, all but the two centre feathers with a large spot of black near the tip; all the under surface pale buff.

Total length 5¾ inches; bill ⅜; wing 2⅜; tail 2¾; tarsi ¾.

Sp. 209.   CISTICOLA EXILIS.
   Exile Grass-Warbler.

Exile Warbler, Lath. Gen. Hist., vol. vii. p. 136.
Malurus exilis, Lath. MS., Vig. and Horsf. in Linn. Trans., vol. xv. p. 223.

Cysticola exilis, Gould, Birds of Australia, fol., vol. iii. pl. 42.

This species appears to have been first noticed by Latham in the seventh volume of his "General History of Birds" under the title of Exile Warbler, and to have been subsequently placed in the genus *Cisticola* by Vigors and Horsfield while engaged in naming the collection of Australian birds in the possession of the Linnean Society. Its natural habitat is New South Wales and South Australia, in both of which colonies I observed it to be abundantly dispersed among the thick beds of grasses which clothe the valleys and open plains. I have never received it from either of the other colonies, all of which, however, are inhabited by nearly allied species. It is very retiring in its habits, generally creeping about among the grasses, and will almost admit of being trodden upon before it will rise and take wing; during the months of spring the male becomes somewhat bolder, and early in the morning will frequently perch on the highest of the grasses and pour forth a pretty but feeble song, resembling that of the *Maluri*. As some confusion existed respecting the sexes of the various species of this genus, I was particular in dissecting all the individuals I shot, and I can therefore state with certainty that the plumage of both sexes of this species is perfectly similar, and that the only outward difference between them consists in the female being somewhat smaller than her mate.

Crown of the head, back, wing-coverts, scapularies, and tail-feathers brownish black, each feather narrowly margined with buff; sides and back of the neck and all the under surface sandy buff, fading into white on the throat and centre of the abdomen; bill and feet flesh-brown.

Sp. 210. CISTICOLA LINEOCAPILLA, *Gould.*

Lineated Grass-Warbler.

Cysticola lineocapilla, Gould in Proc. of Zool. Soc., part xv. p. 1.
Cisticola linricapilla, Bonap. Consp. Gen. Av., tom. i. p. 287, *Cisticola*, sp. 7.

Cysticola lineocapilla, Gould, Birds of Australia, fol., vol. iii. pl. 43.

The *Cisticola lineocapilla* is a much smaller and more delicately formed species than the *C. exilis*, and may, moreover, be distinguished from that and every other member of the genus with which I am acquainted by the lineated form of the markings of the head. It is a native of the north coast of Australia, and all the specimens I have seen were from the neighbourhood of Port Essington. Gilbert states that it " is very rarely seen, in consequence of its generally inhabiting the long grass of the swamps, where it creeps about more like a mouse than a bird, and if once alarmed it is no easy task to get a sight of it again; its note is a short and feeble but very pleasing song.

"The stomach is muscular, and the food consists of insects of various kinds."

General plumage pale rufous, with broad and conspicuous striæ of blackish brown forming lines down the centre of the feathers of the head and back; the under surface fading into white on the throat and centre of the chest; tail-feathers with a conspicuous blackish spot on the under surface near the tip; irides light reddish brown; bill and feet flesh-brown.

Total length $3\frac{3}{4}$ inches; bill $\frac{1}{2}$; wing $1\frac{3}{8}$; tail $1\frac{1}{4}$; tarsi $\frac{3}{4}$.

Sp. 211.   CISTICOLA ISURA, *Gould.*

SQUARE-TAILED GRASS-WARBLER.

Cysticola isura, Gould in Proc. of Zool. Soc., part xv. p. 32.

Cysticola isura, Gould, Birds of Australia, fol. vol. iii. pl. 44.

I am uncertain whether this bird may not prove to be a female, or an example in some peculiar state of plumage of the *Cisticola ruficeps*. Without a further knowledge of the subject, I can only view it as distinct, and I have therefore assigned to it the specific appellation of *isura*, as indicative of the shorter and more truncated form of its tail, the principal character by which it may be distinguished. Like the other species of the group, it appears to enjoy an extensive range over the grassy districts of the country, the specimens in my possession having been killed on the Liverpool Plains and at Port Phillip.

Sides and back of the neck and rump pale rufous; crown of the head, back, and secondaries deep brownish black, each feather margined with buff; tail dark brown margined with buff, and crossed on the under side near the tip with a broad conspicuous band of black; under surface deep buff, becoming paler on the chin and centre of the abdomen; bill brown; feet yellowish brown.

Total length 4 inches; bill ¼; wing 1¾; tail 1½; tarsi ¾.

Sp. 212.   CISTICOLA RUFICEPS, *Gould*.

RUFOUS-HEADED GRASS-WARBLER.

Cysticola ruficeps, Gould in Proc. of Zool. Soc., part v. p. 150.

Cysticola ruficeps, Gould, Birds of Australia, fol., vol. iii. pl. 45.

It would give me great pleasure could I communicate any particulars respecting this pretty little bird, but unfortunately I am unable so to do, no information of any kind having as yet reached me; I can only say therefore that I possess three examples, one from the Liverpool Plains, another from the district of Port Philip, and a third from the north coast, which proves that it enjoys a widely extended range of habitat. The uniform rufous colouring of the head and occiput at once distinguishes it from all the other Australian members of the genus. In its habits, manners, and general economy it doubtless closely assimilates to its congeners, and like them inhabits the open grassy glades between the forests, the grassy crowns of thinly-timbered hills, and all similar situations.

Crown of the head, and back of the neck, rump, chest, flanks, and thighs delicate fawn-colour, becoming deeper and redder on the crown and the rump; upper part of the back, secondaries, and tail deep brownish black, each feather margined all round with buff; throat and centre of the abdomen white; bill brown; feet yellowish brown.

Total length 4 inches; bill ½; wing 1⅞; tail 1¾; tarsi ¾.

Genus SERICORNIS, *Gould*.

A group of small birds peculiar to Australia, and confined almost exclusively to the southern portion of the country. Their habits lead them to frequent the most retired parts of the forests, damp and secluded places, and scrubby gullies where the herbage is thick and dense; but some species are found on the flat islands near the coast, covered with *Salsolæ* and other shrub-like trees; they usually frequent the ground, over which they pass with celerity, and when their haunts are intruded upon conceal themselves under the fallen or dried herbage. They all build domed nests like that of the common Wren (*Troglodytes Europæus*); and their plumage is of a soft and silky character, impervious to wet.

The members of this genus, like the *Cisticolæ*, require to be more closely investigated than I had opportunities of doing during my brief sojourn in the localities they frequent. It is just possible that one or two of them must be united; but, after having had numerous examples before me for nearly thirty years, I can come to no other conclusion than that the species hereafter described are really distinct.

Sp. 213. SERICORNIS CITREOGULARIS, *Gould.*

YELLOW-THROATED SERICORNIS.

Sericornis citreogularis, Gould in Proc. of Zool. Soc., part v. p. 133.
Muscicapa barbata, Lath. Ind. Orn., Supp. p. li?

Sericornis citreogularis, Gould, Birds of Australia, fol., vol. iii. pl. 46.

This is the largest and most attractive species of the genus yet discovered, and, so far as I am aware, its habitat is restricted to the south-eastern portions of Australia, where it dwells exclusively in the districts known by the name of "brushes." I personally observed it in those of Illawarra and the Hunter, and in the cedar-brushes of the Liverpool range. It frequents the most retired parts of the forest, living in gullies and under the canopy of lofty trees, hopping about among the stems of the tree fern, fallen trunks of patriarchal gums, and moss-covered stones. It rarely flies, and, when disturbed, seeks seclusion and safety by hopping away among the underwood. Its food, which consists of insects of various kinds, is obtained on the ground or among the trunks of the prostrate trees, over which and the large stones it passes with much ease and agility.

The sexes are very similar in colour, but the female may at all times be distinguished by her smaller size and the less strongly contrasted tints of her plumage, particularly in the hue of the streak running through the eye and extending over the ear-coverts, which is neither so dark nor so broad as in the male.

One of the most interesting points connected with the history of this species is the situations chosen for its nest. All those who have rambled in the Australian forests must have observed that in their more dense and humid parts there is a redundant growth of mosses of various kinds, and that these mosses not only grow upon the trunks of decayed trees, but are often accumulated at the extremities of the drooping branches, in masses of sufficient size to admit of the bird constructing a nest in the centre of them with so much art that it is impossible to distinguish those selected for this purpose from any of the other pendulous masses in the vicinity. These bunches are frequently a yard in length, and in some instances hang so near the ground as to strike the head of the explorer during his rambles; while in others they are placed high up upon the trees, but only in those parts of the forest where there is an open space entirely shaded by overhanging foliage. As will be readily conceived, in whatever situations they are met with, they at all times form a remarkable and conspicuous feature in the forest scenery. Although the nest is constantly disturbed by the wind and liable to be shaken when the tree is disturbed, so secure does the inmate consider itself from danger or intrusion of any kind, that I have frequently captured the female while sitting on her eggs.

The nest is formed of the inner bark of trees, intermingled

with green moss, dried grasses, and fibrous roots, and is warmly lined with feathers. The eggs, which are three in number and much elongated in form, vary considerably in colour, the most constant tint being a clove-brown, freckled over the larger end with dark umber-brown, frequently assuming the form of a complete band or zone: their medium length is one inch, and their breadth eight lines.

Lores, circle around the eye, and the ear-coverts deep black; a conspicuous line of yellowish white above and for some distance beyond the eye; crown of the head, and all the upper surface, secondaries, wing-coverts, and tail reddish brown, becoming more rufous on the upper tail-coverts and tail; outer edges of the primaries olive; spurious wing blackish brown; throat yellow; chest and flanks olive-brown; centre of the abdomen white; bill brownish black; irides reddish brown; legs purplish flesh-colour, in some specimens flesh-white.

Total length 5½ inches; bill ½; wing 2¾; tail 2⅜; tarsi 1⅛.

Sp. 214. SERICORNIS HUMILIS, *Gould*.

SOMBRE-COLOURED SERICORNIS.

Sericornis humilis, Gould in Proc. of Zool. Soc., part v. p. 139.

Sericornis humilis, Gould, Birds of Australia, fol., vol. iii. pl. 47.

This species is very generally dispersed over Tasmania; and as I have found it on some of the islands in Bass's Straits, it is not improbable that it may also extend its range to the southern coast of the continent of Australia. Ravines, deep glens, water-courses covered with dense herbage, and thickly-wooded copses are the situations congenial to its habits. Although abundant and generally distributed, it is a bird that is less seen, and one whose habits are less known than most others inhabiting the island. In many of its actions it closely resembles the *Troglodytes Europæus*, particularly in its manner of hopping about on the ground, and from stone to stone, with its tail erect, in search of insects, upon which it solely subsists. It rarely flies more than a few yards at a time, but secretes itself in the midst of the little thicket in which it has taken up its abode. The male constantly cheers his mate with a pretty lively song, which, although neither loud nor voluminous, serves to give life to its secluded abode, which in many instances is in the depths of the forests, where few sounds are heard except the monotonous note of the Honeysucker, and the perpetual rippling of the rivulet as it steals over the stony bed of the gully.

The sexes presenting no difference in the colouring of the plumage, by dissection alone can they be distinguished.

There is but little difficulty in finding the nest; for although it is in general very artfully concealed among the herbage at the base of a tree, on the edge of a shelving bank, or among the thick tangle of the scrub, the actions of the old birds soon indicate its site. It is of rather a large size and of a domed form, outwardly composed of any coarse materials at hand, such as leaves, tufts of grass, roots, &c., the interior being formed of similar substances, but of a finer kind, and the whole carefully lined with feathers. The eggs, which are large for the size of the bird, are three in number, of a reddish white, curiously freckled and marked all over with reddish brown, particularly at the larger end, where the markings assume the form of a zone; they are ten and a half lines long by eight lines broad.

Lores blackish brown, above which an obscure stripe of white; crown of the head and all the upper surface, wings, and tail dark olive-brown with a tinge of red, which becomes more conspicuous on the rump and tail-feathers; spurious wing blackish brown, each feather margined with white; throat greyish white, spotted with blackish brown; chest and centre of the abdomen brownish yellow, the former singularly but more obscurely spotted than the throat; flanks chestnut-brown; bill blackish brown; legs dark brown; irides straw-yellow.

Total length 5 inches; bill ½; wing 2¼; tail 2½; tarsi 1.

Sp. 215.   SERICORNIS OSCULANS, *Gould.*

ALLIED SERICORNIS.

Sericornis osculans, Gould in Proc. of Zool. Soc., part xv. p. 2.

Sericornis osculans, Gould, Birds of Australia, fol., vol. iii. pl. 49.

The *Sericornis osculans* inhabits South Australia, where it frequents underwoods, scrubby places, and the bottoms of dry water-courses; it is naturally shy and retiring in its habits, and evades pursuit by creeping beneath the herbage and making its exit on the other side. It is nearly allied to the *S. frontalis* and *S. humilis*; but differs from the former in having at all times numerous longitudinal blotches of black on the throat, and from the latter in these spots being much more distinct. I have seen specimens in which the yellow tint which pervades the centre of the abdomen has given place to grey or greyish white.

The sexes present the usual characteristic of the genus, in the absence of any black mark on the lores of the female, which are similar to the other parts of the body.

All the upper surface, wings, and tail dark brown, all but the two centre feathers of the latter crossed by an obscure band of black near the extremity; spurious wing-feathers black, margined with white; lores black, above which on each a patch of white, continued in a fine line over the eye; throat and centre of the abdomen greyish white in some and yellowish white in others, marked with a few oblong black spots on the throat.

The female is somewhat smaller in size, and has the lores brown instead of black.

Total length 4½ inches; bill ½; wing 2¼; tail 2; tarsi ⅞.

Sp. 210. SERICORNIS FRONTALIS.

WHITE-FRONTED SERICORNIS.

Acanthiza frontalis, Vig. and Horsf. in Linn. Trans., vol. xv. p. 226.
Sericornis parvulus, Gould in Proc. of Zool. Soc., part v. p. 134.

Sericornis frontalis, Gould, Birds of Australia, fol., vol. iii. pl. 49.

This little bird inhabits the brushes, and those humid situations which are clothed with thick underwood, such as the sides of creeks, gullies, &c. The locality in which it is most abundant is the south-eastern part of Australia, where it is very numerous in all the dense forests which stretch along the coast between Sydney and Moreton Bay; and I believe I may safely state that its range does not extend westward of the 134th degree of East longitude, beyond which a nearly-allied species is found; the species, therefore, inosculate about Spencer's and St. Vincent's Gulfs in South Australia. Like the other members of the genus, this bird generally hops about the bottoms of the brushes, selecting in preference the most humid parts, where rotten wood and moss-covered stones afford some peculiar species of insect food, upon which it is destined to live. The present is one of the smallest species of its form yet discovered, and was always a favourite with me; for in the inmost recesses of the forest the presence of this little bird, hopping about from stone to stone in search of its insect food, now and then broke the monotony of the scene with its inward warbling strain.

The sexes present so little difference in colour that they cannot be distinguished with certainty; the female, however, is somewhat smaller than the male. The young birds differ from the adult in having a few faint spots on the throat, which are entirely lost as they advance in age.

The nest of this species is made of leaves, moss, and fibrous roots, and lined with feathers; it is sometimes placed under the shelving of a bank, and at others at the foot of a tuft of grass or herbage, beneath a stone, &c.; it is spherical in form, with a small neatly-made hole for an entrance. The breeding-season includes August and the three or four following months, during which period two or three broods are usually reared. The eggs, which are generally three in number, are of a dull flesh-white, freckled and streaked with purplish brown, particularly at the larger end; their medium length is ten lines, and breadth seven and a half lines.

Centre of the forehead, lores, and a line beneath the eye black; over the eye a line of greyish white; crown of the head, all the upper surface, wings, and tail olive-brown; wing-coverts tipped with white; spurious wing blackish brown; throat white, striated with black; centre of the chest and abdomen citron-yellow; flanks olive-brown; bill blackish brown; feet yellowish white.

Sp. 217. SERICORNIS LÆVIGASTER, *Gould*.
BUFF-BREASTED SERICORNIS.

Sericornis lævigaster, Gould in Proc. of Zool. Soc., part xv. p. 3.

Sericornis lævigaster, Gould, Birds of Australia, fol., vol. iii. pl. 50.

This species, although nearly allied to the *S. maculatus*, is distinguished by the entire absence of spots on the throat and chest, and by having the tail-feathers largely tipped with white.

The acquisition of a male and a female is part of the results of Dr. Leichardt's overland expedition from Moreton Bay to Port Essington, an example of each sex having been killed by Gilbert on the 30th of November 1844; but there is no information whatever respecting them in his Journal.

All the upper surface brown; tail deepening into black near the extremity and tipped with white; spurious wing-feathers dark brown, margined with white on their inner webs; lores and mark under the eye brownish black; above the eye an indistinct line of white; all the under surface washed with yellowish buff; irides greenish white.

The female presents the usual differences, being somewhat smaller in size and wanting the black mark on the lores.

Total length 4¼ inches; bill ⅝; wing 2¼; tail 2; tarsi ⅞.

Sp. 219. SERICORNIS MACULATUS, *Gould.*
Spotted Sericornis.

Sericornis maculatus, Gould in Proc. of Zool. Soc., part xv. p. 2.
Goor-gal, Aborigines of the mountain districts of Western Australia.

Sericornis maculatus, Gould, Birds of Australia, fol., vol. iii. pl. 51.

The present bird, to which I have assigned the specific term of *maculatus*, has always been a source of perplexity to me, from the circumstance of its varying considerably in its markings; after mature consideration, however, I am induced to regard the specimens from Southern and Western Australia and the north coast as referable to one and the same species, each however possessing trivial differences by which it may be known from whence it was received. Specimens from the Houtman's Abrolhos are of a rather smaller size, of a much greyer tint on the back, and have much darker-coloured legs. I believe that the bright yellow wash on the under surface of some individuals is characteristic of newly moulted birds: in this species, not only is the throat spotted with black, but the spotting extends over the chest and some distance down the flanks; it has at all times the tail tipped with white, a character which serves at once to distinguish it from *S. osculans* and *S. frontalis*. Scrubby places and ravines covered with dense herbage, whether in sterile or humid situations, are its favourite resort. It has the same shy disposition and retiring habits as the other members of the genus, depending for safety rather upon its creeping, mouse-like habits than upon its powers of flight, which are indeed seldom resorted to.

Its note is a harsh, grating kind of twitter, often repeated.

The nest is a warm, dome-shaped structure, formed of leaves and grasses, and lined with feathers; the eggs, which are reddish white, minutely freckled and streaked with reddish brown, particularly at the larger end, are three in number, and nine lines long by seven lines broad.

All the upper surface, wings, and tail brown; the latter crossed near the tip with a broad band of blackish brown, and the outer feathers slightly tipped with white; forehead and lores deep black; stripe above and a small patch below the eye white; spurious wing-feathers black, margined on their inner web with white; under surface in some specimens greyish white, in others washed with yellow; the feathers of the throat and chest spotted with black on a light ground; irides greenish white.

The female is somewhat smaller than her mate, and has the lores brown instead of black; in other respects her plumage is very similar to that of the male.

Total length 4½ inches; bill ⅝; wing 2¼; tail 2; tarsi ⅞.

Sp. 219. SERICORNIS MAGNIROSTRIS, *Gould.*

LARGE-BILLED SERICORNIS.

Acanthiza magnirostra, Gould in Proc. of Zool. Soc., part v. p. 146.

Sericornis magnirostris, Gould, Birds of Australia, fol., vol. iii. pl. 52.

The *Sericornis magnirostris* is an inhabitant of the brushes of New South Wales, both those which clothe the gullies and sides of the mountain-ranges of the interior, and those near the coast, such as occur at Illawarra and on the banks of the Hunter, the Clarence, the Mackay, and other rivers. Although it has nothing either in its form or colouring to recommend it to notice, it must always be an object of interest, from the very singular nest it constructs, and which, like that of *Sericornis citreogularis*, forms a remarkable object in the scenery of the portion of the country it inhabits. It is formed of a large loose mass of moss, and, being attached to the extreme tips of the pendant branches, waves about with every wind that blows; it is very frequently constructed within reach of the hand, but is more often suspended at about ten and sometimes as high as thirty feet from the ground; occasionally two or three are constructed together under a dense canopy of foliage, overhanging water or a deep and gloomy gully, and then present a very singular appearance. I procured several examples by shooting the branch asunder just above the nest, which so perfectly resembles the tufts of living moss attached to many of the extremities of the branches of the trees, that it is impossible to distinguish the one from the other; and it is a question whether the bird purposely builds its nest in imitation of these hanging masses, or whether, by a little architectural skill, it converts one of them into a receptacle for its eggs. The breeding-season commences in August and continues until February, during which period many broods are reared. I procured a nest in September, out of which flew three young birds, and others during the same month which contained eggs so recently laid that they could scarcely have been sat upon. The eggs are generally two or three in number; their ground-colour varies from bluish white to dull reddish white, with the larger end sparingly washed, freckled, and streaked with dark brown; they are large for the size of the bird, being nine and a half lines long by seven lines broad.

The large-billed Sericornis is a very active but shy bird, keeping much among the branches of the high trees, where it gains a plentiful supply of insect food; it may, however, be easily enticed into view by imitating the squeak of its young.

The sexes do not differ in external appearance, nor do the young when fully fledged offer any variation in colour from the adult.

Crown of the head, all the upper surface, wings, and tail

olive-brown, the forehead and tail becoming rufous brown; throat and chest brownish white; abdomen greyish white, passing into bright olive-green on the lower part of the flanks; bill black; feet light brown; irides brown.

Total length 4¾ inches; bill ¾; wing 2¼; tail 1⅞; tarsi ⅞.

Genus ACANTHIZA, *Vigors and Horsfield.*

With the exception of the north coast, the *Acanthizæ* are dispersed over all the wooded districts of the continent of Australia and Tasmania: some species frequent the brushes, while others tenant the shrubs and belts of trees on the plains; others again are only found in such districts as the belts of the Murray.

Like some other groups, the *Acanthizæ* admit of division into two or more sections, some being feeble in structure, and strictly arboreal in their habits, while others resort to the ground; for two of the latter, *A. chrysorrhœa* and *A. reguloides,* M. Cabanis has proposed the generic appellation of *Geobasileus*, which I shall accordingly adopt. The nests of all the species that I have seen are of a domed form.

The members of this genus are frequently the fosterparents of the Shining Cuckoo (*Chrysococcyx lucidus*).

Sp. 220.   ACANTHIZA PUSILLA.

LITTLE BROWN ACANTHIZA.

Sylvia pusilla, Lath. Ind. Orn., Supp. p. lvi.
Motacilla pusilla, White's Journ., pl. in p. 257.
Dwarf Warbler, Lath. Gen. Syn., Supp. vol. ii. p. 251.
Acanthiza pusilla, Vig. and Horsf. in Linn. Trans., vol. xv. p. 227, note.

Acanthiza pusilla, Gould, Birds of Australia, fol., vol. iii. pl. 53.

The present bird is very generally dispersed over New South Wales, where it inhabits the brushes, thickets, and gardens. It is most nearly allied to the *A. diemenensis*, but may be distinguished from that species by its more diminutive size, by its much shorter bill, and smaller tail. It is an active prying little bird, and spends much of its time amid the smaller leafy branches of the trees, from among which it collects its insect food: the tail is generally carried above the line of the body. The nest is of a dome-shaped form, and is constructed of fine dried grasses and hairy fibres of bark, intermingled and bound together with the hairy cocoons of a species of Lepidopterous insect, and lined with feathers. The eggs are four or five in number, of a beautiful pearly white, sprinkled and spotted with fine specks of reddish brown, forming in some instances a zone near the larger end; their medium length is eight lines and a half, by six lines in breadth.

The sexes are so precisely similar in outward appearance that dissection must be resorted to to distinguish the one from the other.

Forehead buff, each feather edged with brown; all the upper surface and wings brown, tinged with olive; tail reddish olive, crossed near the tip by a narrow band of black; throat and chest greyish white, each feather margined with black, giving that part a mottled appearance; flanks, abdomen, and under tail-coverts buff; irides brownish red; bill dark brown; feet brown.

Sp. 221. ACANTHIZA DIEMENENSIS, *Gould*.

TASMANIAN ACANTHIZA.

Acanthiza diemenensis, Gould in Proc. of Zool. Soc., part v. p. 140.
—— *ewingii*, Gould in Proc. of Zool. Soc., part xv. p. 32; Id. Birds of Aust. fol., vol. iii. pl. 55?
Brown-tail, Colonists of Tasmania.

Acanthiza diemenensis, Gould, Birds of Australia, fol., vol. iii. pl. 54.

I believe this species to be peculiar to Tasmania, over the

whole of which country it is rather numerously dispersed, and where it inhabits forests and open woodlands, but evinces a preference to low and shrub-like trees rather than to those of a higher growth. It also frequents the gardens and shrubberies of the colonists; it is consequently one of the commonest and one of the best-known birds of the island. Active and sprightly in its actions, it pries about the foliage with the most scrutinizing care in search of insects and their larvæ, which constitute its sole food. It frequently utters a rather loud harsh note, which is sometimes changed for a more full and clear strain; still its vocal powers are by no means conspicuous. It has a much more lengthened bill, and is altogether a larger bird than the *Acanthiza pusilla*, whose habitat seems restricted to the south-eastern portion of the Australian continent. The nest of this little bird, which is usually built in a low shrub, is rather a dense structure, being formed of grasses, fibrous roots, and the inner bark of trees, warmly lined with feathers; it is of a globular form, with a small hole in the side near the top for an entrance. The eggs are four or five in number, of a beautiful pearly bluish white, sprinkled and spotted with reddish brown. In some instances the spots form a zone round the larger end. The medium length of the eggs is eight lines and a half, and breadth six lines.

Independently of the task of incubating its own offspring, this species very frequently has to perform the additional labour of hatching and rearing the young of the Bronze Cuckoo (*Chrysococcyx lucidus*), whose single egg or young is often found in the nest. It is a very early breeder, commencing in August and continuing until January, during which period two or three broods are generally reared by each pair.

The plumage of the sexes is alike, and their size and general appearance so similar, that without the aid of dissection it was impossible to distinguish them.

Forehead rufous brown, each feather with a crescent-shaped mark of bright buff near its extremity and tipped with blackish brown; all the upper surface and wings deep olive-brown; upper tail-coverts reddish brown; tail olive-brown, crossed by a band of blackish brown; cheeks, throat, and chest greyish white, each feather margined with a broken line of deep brown; abdomen and under tail-coverts greyish white, tinged with rufous, which is deepest on the flanks and under tail-coverts; bill dark brown; irides lake-red; feet brown.

Total length 4 inches; bill $\frac{3}{16}$; wing $2\frac{1}{4}$; tail 2; tarsi $\frac{3}{4}$.

Sp. 222. ACANTHIZA UROPYGIALIS, *Gould*.

Chestnut-rumped Acanthiza.

Acanthiza uropygialis, Gould in Proc. of Zool. Soc., part v. p. 146.

Acanthiza uropygialis, Gould, Birds of Australia, fol., vol. iii. pl. 56.

I received examples of this new and well-marked species from New South Wales, and believe that they had been collected either on the Liverpool Plains or the country immediately to the northward of them; but as there is some degree of uncertainty as to the locality in which they were procured, a knowledge of the true habitat of the species is very desirable, and I should be happy if this could be ascertained.

The chestnut colour pervading the basal half of the tail and the tail-coverts forms a very conspicuous mark, and presents a strong contrast to the remainder of the plumage. That its habits, actions, and economy are very similar to those of the other members of the genus, there can be no doubt; but on these points also I am compelled to silence, no notes of any kind having been sent with the specimens.

Head, upper surface, and wings brown, slightly tinged with olive; the feathers on the forehead tipped with a lighter colour; rump and upper tail-coverts rich reddish chestnut; tail-feathers brownish black, largely tipped with white, which on the two centre feathers is tinged with brown; throat, chest, and centre of the abdomen greyish white; flanks and under tail-coverts buffy white; bill and feet black.

Total length 3¾ inches; bill ½; wing 2; tail 1¾; tarsi ¾.

Sp. 223. ACANTHIZA APICALIS, *Gould.*

WESTERN ACANTHIZA.

Acanthiza apicalis, Gould in Proc. of Zool. Soc., part xv. p. 31.
Djool-be-djööl-bung, Aborigines of the lowland districts of Western Australia.
Wren, Colonists of Swan River.

Acanthiza apicalis, Gould, Birds of Australia, fol., vol. iii. pl. 57.

This species, which is a native of Western Australia, is distinguished from those immediately allied to it, *A. diemenensis* and *A. pusilla*, by its large size, by its larger and rounder tail, by the broad and distinct band of black which crosses the tail-feathers near their extremities, and by their being largely tipped with white.

It occurs in great abundance in the colony of Western Australia, both at Swan River and King George's Sound, and is to be met with in all wooded situations. Like the other members of the genus, it is active and sprightly in its actions, leaping about from branch to branch with its tail erect, and often repeating a note which very much resembles the syllables *Gee-wo-wut.* Its stomach is somewhat muscular, and the food consists of small insects of various kinds.

It breeds in September and October. The nest, which is usually placed in a thickly-foliaged bush, or in a clump of the Tea-tree, is of a domed form, with an entrance in the side, and is composed of dried grasses and strips of Tea-tree bark, and lined with feathers. The eggs are from three to five in number, of a flesh-white, thickly freckled with reddish chestnut, the freckles becoming so numerous at the larger end as to form a complete zone; their medium length is eight lines, and breadth six lines.

The sexes are alike in plumage, but the female is somewhat smaller than the male.

Feathers of the forehead deep buff, edged with dark brown; all the upper surface, wings, and tail light olive-brown; tail crossed with a broad and distinct band of brownish black near the extremity, and largely tipped with white; upper tail-coverts tinged with rufous; throat and chest greyish; tail-coverts pale buff white, each feather margined with black, giving that part a mottled appearance; flanks, abdomen, and under irides light red; bill, legs, and feet dark brown.

Total length 4 inches; bill $\frac{1}{4}$; wing 2; tail 2; tarsi $\frac{7}{8}$.

Sp. 224. ACANTHIZA PYRRHOPYGIA, *Gould.*

RED-RUMPED ACANTHIZA.

Acanthiza pyrrhopygia, Gould, Birds of Australia, fol., vol. iii. pl. 58.

This species differs from the *Acanthisæ diemenensis, pusilla,* and *apicalis,* in having a shorter and more robust bill, and in the greater depth of the red colouring on the rump and upper tail-coverts; it also differs from the two former in having the tail tipped with white, in which respect it assimilates to the *A. apicalis* and *A. uropygialis,* to the former of which it is most nearly allied.

I discovered this species in the Belts of the Murray, where it inhabits the small shrubby trees; upon first seeing it, I at once perceived that it was a distinct species by the red colouring of the rump, which showed very conspicuously at the distance of several yards, and also by the peculiarity of its note. In its actions it very closely assimilates to the other members of the genus, being an alert and quick little bird, carrying its tail above the level of the back, and showing the red colouring of the coverts to the greatest advantage. I succeeded in killing both sexes, and found that they exhibit no outward difference, and are only to be distinguished with certainty by dissection.

All the upper surface and wings olive-brown, the feathers of the forehead margined with buff; wings brown with pale edges; throat white, each feather margined with black; abdomen whitish; flanks pale buff; upper tail-coverts rufous; tail olive, crossed by a broad band of black, and tipped on the outer webs with pale olive, on the inner webs with white; bill blackish brown, under mandible somewhat lighter; feet brown; irides reddish brown.

Total length 4 inches; bill $\frac{1}{4}$; wing 2; tail $1\frac{3}{4}$; tarsi $\frac{3}{4}$.

Sp. 225.   ACANTHIZA INORNATA, *Gould*.

PLAIN-COLOURED ACANTHIZA.

Acanthiza inornata, Gould in Proc. of Zool. Soc., part viii. p. 171.

Djo-bal-djo-bal, Aborigines of the lowland districts of Western Australia.

Acanthiza inornata, Gould, Birds of Australia, fol., vol. iii. pl. 59.

Although neither elegant in form nor characterized by any beauty of plumage, the present little bird demands as much of our attention as any other species of the group. Its true habitat seems to be the south-western parts of Australia, for it is numerously dispersed over the colony of Swan River; it is equally abundant at King George's Sound; and as I killed specimens on the small low islands at the mouths of Spencer's and St. Vincent's Gulfs, it is most probable that its range extends all along the coast between those localities. Independently of its plainer colouring, the truncated form of its tail serves at once to distinguish it from the *Acanthiza apicalis*, with which it is often seen in company; unlike the latter bird however it does not erect its tail, but carries it in a line with the body.

Its note is a little feeble song somewhat resembling that of the *Maluri*. It feeds solely on minute insects of various kinds, in searching for which it assumes the usual clinging and prying positions of other insectivorous birds which seek their food among the leaves and branches of shrubs and trees.

It breeds in November; the nest, which is of a domed form, being placed in some low shrub, often in that of the jam-wood, and composed of grasses lined with a few feathers.

The eggs are five in number, and of a white colour, slightly tinged with greenish grey; they measure seven and a half lines long by five and a half lines broad.

No visible difference is observable in the sexes.

All the upper surface, wings, and tail olive-brown; primaries dark brown; tail crossed by a broad band of brownish black; all the under surface light buff; irides greenish white; bill and feet black.

Total length $3\frac{1}{2}$ inches; bill $\frac{1}{4}$; wing $1\frac{7}{8}$; tail $1\frac{1}{2}$; tarsi $\frac{1}{16}$.

Sp. 226. ACANTHIZA NANA, *Vig. and Horsf.*
LITTLE ACANTHIZA.

Dwarf Warbler, var. A, Lath. Gen. Hist., vol. viii. p. 184?

Acanthiza nana, Vig. and Horsf. in Linn. Trans., vol. xv. p. 220.

Acanthiza nana, Gould, Birds of Australia, fol., vol. iii. pl. 60.

This little bird, which is very generally distributed over the colonies of New South Wales and South Australia, frequents the extremities of the branches of the various trees, without, so far as I could observe, evincing a partiality for any particular kind; the *Casuarinæ* on the banks of creeks, the *Eucalypti* of the plains, and the belts of *Banksia* being equally resorted to by it. Minute insects constitute its sole food, and in the capture of these it exhibited many lively and varied actions, which strongly reminded me of those of the *Regulus cristatus* of England.

The nest is a neat domed structure with a small entrance

near the top, and is composed of fine grasses; its site varies according to circumstances, but is generally among the smaller branches of the trees. The number and colour of its eggs are unknown to me.

As its name implies, the *Acanthiza nana* is one of the more diminutive, although not the least of the Australian birds.

There is no outward difference by which the sexes can be distinguished, neither do they undergo any seasonal change, nor is there any great variation in the colouring of the young and the adult.

All the upper surface bright olive; tail greyish brown tinged with olive, and crossed by a broad band of blackish brown; throat and under surface yellow; irides brown with a very narrow rim of yellowish white; bill and feet blackish brown.

Sp. 227. ACANTHIZA LINEATA, *Gould*.

STRIATED ACANTHIZA.

Acanthiza lineata, Gould in Proc. of Zool. Soc., part v. p. 140.

Acanthiza lineata, Gould, Birds of Australia, fol., vol. iii. pl. 61.

This pretty little species inhabits most of the wooded districts of South Australia, particularly the gullies among the mountain ranges; it is also tolerably abundant among the brushes and trees near the brooks and rivulets of the Liverpool range in New South Wales. It is very active and animated in its actions, clinging and prying about among the branches in search of insects in every variety of position. It is a permanent resident in the countries above-mentioned, but is not found in Tasmania or Western Australia. Unfortunately I did not succeed in procuring its nest, but judging from those of the other members of the genus, it is doubtless of a domed form.

Its food consists entirely of insects, which are procured from the leaves and flowers of the various trees.

The sexes can only be distinguished by dissection, for no perceptible difference whatever is observable either in their size or the colouring of their plumage.

This species, the least of the genus to which it belongs, may be thus described:—

Crown of the head brownish olive, with a fine line of white down the centre of each feather; back and wings greenish olive; tail the same, crossed by a broad band of brownish black near the tip, beyond which the extremities are brownish grey; throat and chest grey, tinged with olive, the margins of the feathers spotted with dark brown, giving these parts an irregular spotted appearance; bill and feet black; irides brown.

Total length $3\frac{3}{4}$ inches; bill $\frac{3}{8}$; wing 2; tail $1\frac{3}{4}$; tarsi $\frac{5}{8}$.

Sp. 228.   ACANTHIZA MAGNA, *Gould.*

GREAT ACANTHIZA.

Acanthiza magna, Gould, Birds of Australia, Supplement, pl.

For the knowledge of this new and very distinct species of *Acanthiza* I am indebted to Ronald C. Gunn, Esq., a gentleman who has long resided in Tasmania, and whose name will be for ever perpetuated in the annals of science for the numerous botanical discoveries made by him in the island he has adopted as his home. I have carefully compared the specimen sent to me by Mr. Gunn, and which had been collected in the northern part of Tasmania, with other members of the genus, and have no hesitation in pronouncing it to be previously unknown. In size it approaches the smaller species of *Sericornis*; but in its structure and the character of its plumage, it is closely allied to the members of the genus in which I have placed it.

Head, all the upper surface, sides of the neck, and flanks olive-brown, becoming of a more rufous hue on the rump and upper tail-coverts; wings blackish brown, washed with olive on the external webs; coverts, particularly the greater ones, tipped with white; primaries narrowly edged with grey, innermost secondaries margined all round the tip with white; tail olive, crossed near the tip by a broad band of dusky brown, beyond which the external feathers are margined on both webs with greyish white; lores black; ear-coverts slaty brown; throat and under surface straw-yellow; bill blackish brown; feet fleshy brown.

Total length 4¾ inches; bill ⅜; wing 2⅜; tail 2; tarsi ¾.

Genus GEOBASILEUS, *Cabanis*

In my remarks on the members of the genus *Acanthiza*, I stated that those birds might be divided into two or three sections, and this view has been taken by M. Cabanis, who has proposed the above generic title for the birds figured in the folio edition as *Acanthiza chrysorrhœa* and *A. reguloides*.

Sp. 229. GEOBASILEUS CHRYSORRHOUS.

Yellow-rumped Geobasileus.

Saxicola chrysorrhœa, Quoy and Gaim. Voy. de l'Astrolabe, p. 198, pl. 10. fig. 2.
Geobasileus chrysorrhous, Cab. Mus. Hein., Theil i. p. 32.
Jee-da, Aborigines of the lowland districts of Western Australia.

Acanthiza chrysorrhœa, Gould, Birds of Australia, fol., vol. iii. pl. 63.

This well-known species inhabits Tasmania, Western and Southern Australia, and New South Wales, in all of which countries it is a permanent resident. It is generally met with in small companies of from six to ten in number, and is so tame that it may be very closely approached before it will rise, and then it merely flits off to a short distance and alights again; during these short flights the yellow of the rump shows very conspicuously.

It commences breeding very early, and rears at least three broods a year. The nest is somewhat carelessly constructed of leaves, grasses, wool, &c., and is of a domed form, with a small hole for an entrance. But the most curious feature connected with it is, that a small cup-shaped depression or second nest, as it were, is frequently formed on the top or side of the other, and which is said to be either the roosting-place of the male, or where he may sit in order to be in company with the female during the task of nidification. I have myself found many of these double nests, but have not had opportunities for satisfactorily ascertaining the use of the upper one. The bird very readily resorts to the gardens of the settler, and constructs its curious nest in any low shrub. In Tasmania one of the trees most frequently selected for the purpose is the prickly *Mimosa*: in Western Australia it is suspended from the overhanging branches of the *Xanthorrhœa*, and in the district of the Upper Hunter upon the apple-trees (*Angophoræ*). It varies very much in size. The eggs are generally of a beautiful uniform flesh-colour, but occasionally they are found sprinkled over with very minute specks of reddish yellow, which in some instances form a zone at the larger end; they are four or five in number, their medium length being nine lines, and breadth six lines.

This is one of the species to which the Bronze Cuckoo (*Chrysococcyx lucidus*) delegates the task of rearing its young. I have several times taken the egg of the Cuckoo from the nest of this bird and also the young, in which latter case the parasitical bird was the sole occupant.

The song of the *Geobasileus chrysorrhous* is extremely pretty, many of its notes closely resembling those of the Goldfinch of Europe (*Carduelis elegans*). Its food consists of small coleopterous and other kinds of insects.

The sexes are alike in plumage, and may be thus described:—

Forehead black, with a spot of white at the tip of each

feather; cheeks, throat, and a line from the nostrils over each eye greyish white; chest and under surface yellowish white, passing into light olive-brown on the flanks; upper surface and wings olive-brown; rump and upper tail-coverts bright citron-yellow; base of the tail-feathers white, tinged with yellow; the external margin of the outer feathers and the tips of all brownish grey, the central portion blackish brown; bill and feet blackish brown; irides very light grey.

Sp. 230.   GEOBASILEUS REGULOIDES.

Buff-rumped Geobasileus.

Acanthiza reguloides, Vig. and Horsf. in Linn. Trans., vol. xv. p. 226.
Geobasileus reguloides, Cab. Mus. Hein., Theil i. p. 82, note.
Dwarf Warbler, var. β?, Lath. Gen. Hist., vol. vii. p. 185.

Acanthiza reguloides, Gould, Birds of Australia, fol., vol. iii. pl. 62.

Many of the actions of this little bird offer a close resemblance to those of the *Geobasileus chrysorrhous*; like that species, it moves about in small flocks of from eight to fifteen in number; when flushed shows the yellow or buff of the rump very conspicuously; always spreads its tail while flying; flits along with a jerking motion, and is very tame. It is extremely common in South Australia, where I observed it in every part of the country I visited; in New South Wales I found it in the interior beyond the ranges, and also on the bare ridges between Patrick's Plains and the Liverpool range. I did not meet with it in Tasmania. It evinces a decided preference for the open country or hills slightly covered with brush, where it can feed on the ground and fly to the low shrub-like trees when disturbed; I have also seen it busily engaged among the branches, apparently in search of insects, in the pursuit of which, like the other members of the genus, it displays unusual alertness and address.

Its domed nest is placed among the foliage of the gum, swamp-oak, and other trees, and is composed of fine grasses, interwoven with cobwebs, and slightly lined with feathers. The breeding-season comprises the months of September, October, and November, and the eggs are four in number.

Crown, back of the neck, upper surface, and wings olive-brown, the feathers of the forehead tipped with a lighter colour; rump, upper and under tail-coverts pale ochre; throat and chest white, each feather with a very slight broken margin of brown; base of all the tail-feathers pale buff, the external margin of the outer feathers and the tips of all brownish buff, the central portion blackish brown; bill brown, the under mandible paler than the upper; feet olive-brown; irides beautiful straw-yellow.

Genus EPHTHIANURA, *Gould.*

The three species of this form at present known inhabit the southern part of Australia, where they frequent open districts studded with bushes and low trees.

Sp. 231.   EPHTHIANURA ALBIFRONS.

WHITE-FRONTED EPHTHIANURA.

Acanthiza albifrons, Jard. and Selb. Ill. Orn., vol. ii. pl. 50. figs. 1 and 2.
Ephthianura albifrons, Gould, Birds of Australia, vol. i. Introd. p. xlvii.
Cysura torquata, Brehm (Cabanis).

Ephthianura albifrons, Gould, Birds of Australia, fol., vol. iii. pl. 64.

This species appears to range over the whole of the southern portion of the Australian continent, for I have specimens in my collection which were killed at Swan River, in South Australia, and in New South Wales. It does not inhabit Tasmania; but is very common, and breeds, on some of the islands in Bass's Straits.

It is a most sprightly and active little bird, particularly the male, whose white throat and banded chest render him much more conspicuous than the sombre-coloured female. As the structure of its toes and lengthened tertiaries would lead us to expect, its natural province is the ground, to which it habitually resorts, and decidedly evinces a preference to spots of a sterile and barren character. The male frequently perches either on the summit of a stone, or on the extremity of a dead and leafless branch. It is rather shy in its disposition, and when disturbed flies off with considerable rapidity to the distance of two or three hundred yards before it alights again.

From some interesting notes on Australian birds by E. P. Ramsay, Esq., published in the 'Ibis' for 1863, I learn that "These birds arrive in New South Wales about the beginning of September and October. In the latter month they commence building on open lands studded with low bushes. The stunted *Bassariæ*, the prickly twigs of which are often used to form the framework of their nests, seem their favourite building-places. The nests are usually situated a few inches from the ground, are cup-shaped, placed upon a strong framework of twigs, and neatly lined with grass, hair, &c. I have frequently found them among the dead leafy tops of a fallen *Eucalyptus* which had been left by the wood-cutters when clearing a piece of new ground.

"The eggs are usually three but sometimes four in number, from 0½ to 7 lines long by 5 broad, of a beautiful white, some spotted and others irregularly marked with bright deep reddish brown at the larger end, where occasionally they form an indistinct zone. In some specimens the spots are crowded at the top, and very sparingly sprinkled on the other parts of the egg.

"These birds readily betray the position of their nests or young by their anxiety and attempts to draw one from the spot by feigning broken wings, and by lying struggling on the ground as if in a fit. They have two broods (and perhaps more) in the year, after which the young accompany the parent birds to feed generally on the salt marshy grounds near the water's edge. About Botany and the Paramatta River, upon the borders of the Hexham swamps, &c., they are plentiful. They evince a decided preference for open half-cleared patches of land. I never found more than four or six together, doubtless the offspring of one pair; still it is not unusual to find them in pairs only. As far as I am aware, they have but one very plaintive note, which is emitted chiefly while flying or when the nest is approached."

Mr. White, of the Reed-beds near Adelaide, informs me that a nest taken by him on the 10th of July was placed in a tuft of rushes near the ground, but he has met with it in a small bush at a height of two or three feet; the ground-colour of the eggs, like many others with thin shells, are of a pinkish tint before being blown.

The male has the forehead, face, throat, and all the under surface pure white; occiput black; chest crossed by a broad crescent of deep black, the points of which run up the sides of the neck and join the black of the occiput; upper surface dark grey, with a patch of dark brown in the centre of each feather; wings dark brown; upper tail-coverts black; two centre tail-feathers dark brown; the remainder dark brown, with a large oblong patch of white on the inner web at the tip; irides, in some, beautiful reddish buff, in others yellow with a slight tinge of red on the outer edge of the pupil; bill and feet black.

The female has the crown of the head, all the upper surface, wings, and tail greyish brown, with a slight indication of the oblong white spot on the inner webs of the latter; throat and under surface buffy white; and a slight crescent of black on the chest.

Sp. 232. EPHTHIANURA AURIFRONS, *Gould.*
ORANGE-FRONTED EPHTHIANURA.

Ephthianura aurifrons, Gould in Proc. of Zool. Soc., part v. p. 148.
Ephthianura aurifrons, Gould, Birds of Aust., vol. i. Introd. p. xlvii.

Ephthianura aurifrons, Gould, Birds of Australia, fol., vol. iii. pl. 65.

Skins of this species, which were very rare when I first described it, are now common, being sent in abundance from Victoria; that it also inhabits South Australia we know from the circumstance of Mr. White informing me that, in an ornithological trip made by him to the north of Adelaide, he saw this bird in great numbers from Port Augusta to the 27th degree of latitude; he states that it lives chiefly on caterpillars, builds in low shrubs, and that the eggs, which are four or five in number, are white or pinkish white spotted with rust red.

Head, upper tail-coverts, sides of the neck, breast, and all the under surface fine golden orange, which is richest on the forehead and centre of the abdomen; back olive; wings brown, margined with olive; tail brownish black, each feather except the two middle ones having an oval spot of white on the inner web at the tip; chin and centre of the throat black; bill black; feet brown.

Total length 4 inches; bill $\frac{3}{8}$; wing $2\frac{1}{2}$; tail $1\frac{1}{2}$; tarsi $\frac{3}{4}$.

Sp. 233. EPHTHIANURA TRICOLOR, *Gould.*
TRI-COLOURED EPHTHIANURA.

Ephthianura tricolor, Gould in Proc. of Zool. Soc., part viii. p. 159.

Ephthianura tricolor, Gould, Birds of Australia, fol., vol. iii. pl. 66.

While traversing, soon after sunrise on the 11th of December 1839, the forest lands near Peel's River to the eastward of Liverpool Plains, a fine male specimen of this bird attracted my notice by the beauty of its colouring, the sprightliness and activity of its actions, and the busy manner in which it was engaged in capturing small insects. As may be supposed, the sight of a bird of such beauty, which, moreover, was entirely new to me, excited so strong a desire to possess it that scarcely a moment elapsed before it was dead and in my hand. In a small collection procured in South Australia by the late F. Strange, two other specimens occurred which I supposed to be male and female; unfortunately they were unaccompanied by any information respecting the habits or economy of this rare bird; more recently, however, both G. French Angas, Esq., and Mr. White have favoured me with brief notes on the subject.

"A nest and eggs of the Tricoloured Ephthianura," says Mr. Angus, "were taken on the 27th of October 1862, in a low bush at Evandale, about three miles from Collingrove, Angaston, South Australia. It had never been seen in that locality before, the farthest south being the head of Spencer's Gulf, where I obtained it in the scrub in September 1860."

"This," says Mr. White, "is a very rare species; and from all I can learn, I imagine its true haunt to be in the far north and west of South Australia, and that it occasionally comes southward to breed. On a journey made in October last to the head of Spencer's Gulf and Mount Brown, I first observed it in a flock of ten or twelve males, females, and young birds, and at once saw that it was a species I had not previously noticed; and on inquiring in various northern localities found that it had not been seen by the residents until this spring. Those I saw were in the scrubby country on the western slopes of Flinder's Ranges, close to the head of Spencer's Gulf; they were hopping about on the low bushes and on the ground."

In another note by this gentleman he states that he saw this species and *E. aurifrons* in considerable numbers about latitude 27° and 28°.

The male has the crown of the head, upper tail-coverts, breast, and abdomen bright scarlet; lores, line above and beneath the eye, ear-coverts, occiput, and back dark brown; wings brown, each feather margined with brownish white; tail dark brown, each feather having a large spot of white on the inner web at the tip; chin, throat, and under tail-coverts white; irides straw-white; bill and feet blackish brown.

The female is similar in colour, but has only a slight wash of the scarlet colouring, except on the upper tail-coverts, where it is as brilliant as in the male.

Total length $3\frac{1}{4}$ inches; bill $\frac{9}{16}$; wing $2\frac{3}{4}$; tail $1\frac{1}{2}$; tarsi $\frac{3}{4}$.

Genus XEROPHILA, *Gould.*

A curious form, of which only one species is known, and the situation of which in the natural system is quite undetermined. The single species known has many of the actions and manners of the *Acanthizæ*, but its robust and gibbose bill precludes its being placed with that genus. It is mainly terrestrial in its habits and builds a domed nest.

Sp. 234. XEROPHILA LEUCOPSIS, *Gould.*

WHITE-FACED XEROPHILA.

Xerophila leucopsis, Gould in Proc. of Zool. Soc., part viii. p. 175.

Xerophila leucopsis, Gould, Birds of Australia, fol. vol. iii. pl. 67.

I found this species tolerably abundant in all parts of the colony of South Australia that I visited, both in the interior and in the neighbourhood of the coast. It was generally met with in small flocks of from six to sixteen in number, and more frequently on the ground than among the trees. It hops over the surface very quickly and appears a busy little bird, prying among the herbage for its food, which principally consists of the seeds of the grasses and small annuals which abound on the plains and low hills of South Australia. In disposition it is so remarkably tame that it will allow of a very near approach before it will rise, and then it merely flies to the nearest bush or low tree.

The male offers no external difference by which it can be distinguished from the female, neither do the young exhibit any contrast to the adults in their plumage; it has in fact little to recommend it to the notice of the general observer either in its colouring or in the quality of its song.

The nest forwarded to me by Strange was of a rather large size, of a domed form, with a hole for an entrance near the top, and composed of dried grasses, moss, spiders' webs, wool, the soft blossoms of plants, and dead leaves matted together and warmly lined with feathers; it was about seven inches in height and four inches in diameter. The eggs received with the nest were three in number, of a fleshy white, eight and a half lines long and six lines broad.

Forehead and lores white; upper surface olive-brown; wings and tail brown, the latter passing into black near the extremity, and tipped with white; all the under surface pale buff; bill and feet black; irides light straw-colour.

Total length 4 inches; bill $\frac{3}{8}$; wing 2$\frac{1}{4}$; tail 1$\frac{3}{4}$; tarsi $\frac{3}{4}$.

Genus PYRRHOLÆMUS, *Gould.*

A singular form, the structure of which does not approximate very nearly to that of any other genus, but is perhaps most nearly allied to *Acanthiza.* The only species known frequents scrubby places and thick underwood; is much on the ground, but occasionally mounts on a small branch to sing.

Sp. 235. PYRRHOLÆMUS BRUNNEUS, *Gould.*
Red-Throat.

Pyrrholæmus brunneus, Gould in Proc. of Zool. Soc., part viii. p. 173.
Acanthiza brunnea, Gray & Mitch. Gen. of Birds, vol. i. p. 189, *Acanthiza*, sp. 20.
Ber-rit-ber-rit, Aborigines of the mountain districts of Western Australia.

Pyrrholæmus brunneus, Gould, Birds of Australia, fol., vol. iii. pl. 68.

I found this bird tolerably abundant in the Delta of the Murray, about forty miles to the northward of Lake Alexandrina, where it gave a decided preference to low stunted bushes and fallen trunks of trees overgrown with herbage, under which it secreted itself; it sometimes rose to the top of a bush to sing, pouring forth a melody equal to any of the smaller birds of Australia, which must render it a general favourite when that portion of the country becomes colonized. It passes much of its time on the ground, hopping about with great celerity, and with its tail elevated considerably above the level of its back.

Specimens were also obtained by Gilbert in Western Australia, from whose notes I learn that it is there an inhabitant of the underwood and the thickest scrub; and that "it possesses a very sweet and melodious song, which it generally utters while perched on the extreme topmost branch of a small scrubby tree, and having repeated it two or three times, dives down into the impenetrable bush. While feeding it utters a weak, piping, call-like note. I never saw it fairly on the wing, for it seems averse to flying, but generally prefers creeping from bush to bush, and even if closely hunted merely flits a few yards. It makes its nest on the ground, precisely like the members of the genus *Calamanthus*. I found a pair building in the month of September; upon visiting the spot again after an interval of a week, the nest appeared finished, being lined with feathers; but there were no eggs; unfortunately from this time the birds deserted the nest."

Lores greyish white; all the upper surface and wings brown; tail brownish black, the three lateral feathers on each side largely tipped with white; centre of the throat rufous; the remainder of the under surface brownish grey, passing into sandy buff on the flanks and under tail-coverts; irides reddish brown, with an outer ring of yellowish white; upper mandible reddish brown; lower mandible greenish white; legs and feet dark greenish grey.

Genus ORIGMA, *Gould*.

The only species of this form yet discovered, the structure, habits, and manners of which are all equally singular, inhabits New South Wales, where it frequents stony gullies and rocky situations in the neighbourhood of caverns, to the roofs of which it attaches its pendant nest.

Sp. 236.   ORIGMA RUBRICATA.
Rock-Warbler.

Sylvia rubricata, Lath. Ind. Orn. Supp., p. li.
Ruddy Warbler, Lath. Gen. Syn. Supp., vol. ii. p. 249.
Motacilla solitaria, Lewin, Birds of New Holl., pl. 16.
Solitary Flycatcher, Lath. Gen. Hist., vol. vi. p. 220.
Saxicola solitaria, Vig. and Horsf. in Linn. Trans., vol. xv. p. 236.
Origma solitaria, G. R. Gray, List of Gen. of Birds, 2nd edit. p. 30.

Origma rubricata, Gould, Birds of Australia, fol., vol. iii. pl. 69.

The true habitat of this species is New South Wales, over which part of the country it is very generally distributed wherever situations occur suitable to its habits; water-courses and the rocky beds of gullies, both near the coast and among the mountains of the interior, being equally frequented by it; and so exclusively in fact is it confined to such situations, that it never visits the forests, nor have I ever seen it perching on the branches of the trees. It does not even resort to them as a resting-place for its nest, but suspends it to the ceilings of caverns and the under surface of overhanging rocks in a manner that is most surprising; the nest, which is of an oblong globular form, and composed of moss and other similar substances, is suspended by a narrow neck, and presents one of the most singular instances of bird-architecture that has yet come under my notice.

It was one of the birds which excited the notice and interest of Mr. Caley, who, in his 'Notes,' says, "*Cataract Bird*; an inhabitant of rocky ground. While at the waterfall of Carrung-gurring, about thirty miles to the southward of Prospect Hill, I saw several of them. I have also seen them in the North Rocks, about a couple of miles from Paramatta."

Mr. Ramsay, in his "Notes on Australian Birds," published in the 'Ibis' for 1863, states "that the Rock Warbler is a very pleasing and lively little bird, and seems to love solitude. I have never seen it perch on a tree, although I have spent several evenings in watching it. It runs with rapidity over the ground, and over heaps of rubbish left by floods, where it seems to get a good deal of its food. Sometimes it will remain for a minute on the point of a rock, then, as if falling over the edge, repeat its shrill cry, and dash off into some hole in the cliffs.

"The nest is of an oblong form, very large for the size of the bird, with an entrance in the side about two inches wide. It is generally suspended under some overhanging rock, and is composed of fibrous roots interwoven with spiders' webs; the bird evincing a preference for those webs which contain the spiders' eggs, and that are of a greenish colour. The moss does not assume the shape of a nest until a few days before it is completed, when a hole for entrance is made, and the inside is warmly lined with feathers; but when finished, it is a very ragged structure, and easily shaken to pieces. The birds take a long time in building their nests: one found on the 6th of August was not finished until the 25th of that month; on the 30th three eggs were taken from it. This nest was suspended from the roof of a small cave in the gully of George's River, near Macquarie Fields, and was composed of rootlets and spiders' webs, warmly lined with feathers and opossum-fur; it contained three eggs of a pure and glossy white, each of which was 8½ lines in length by 6¼ in breadth.

"The breeding-season lasts from August to December, during which two broods are reared.

"I have never found more than *one* nest or one pair of birds near the same part of the gully; and I do not think they will make their nests near each other, much less under the same rock."

Its food consists of insects of various kinds.

Its note is a low squeaking sound, which it utters while hopping about the rocks with its tail raised above the level of the body, after the manner of some of the *Acanthizæ*.

The sexes are precisely similar in their plumage, which may be thus described:—

All the upper surface and wings dull brown; tail brownish black; throat grey; under surface dark rusty red; forehead slightly washed with ferruginous red; irides dark reddish brown; bill and feet brownish black, the former rather lighter than the latter.

Genus CALAMANTHUS, *Gould*.

This genus comprises two species, one inhabiting Tasmania, the other Southern and Western Australia; they are terrestrial in their habits, but occasionally perch on the smaller branches of the trees.

Sp. 237. CALAMANTHUS FULIGINOSUS.

STRIATED CALAMANTHUS.

Anthus fuliginosus, Vig. and Horsf. in Linn. Trans., vol. xv. p. 230.
Praticola fuliginosa, G. R. Gray, List of Gen. of Birds, 2nd edit. p. 27.

Calamanthus fuliginosus, Gould, Birds of Australia, fol., vol. iii. pl. 70.

This species is very generally dispersed over Tasmania, where it frequents open forests and sandy land covered with scrub and dwarf shrub-like trees. It carries its tail erect, like the *Maluri*, but differs from the members of that group in moving that organ in a lateral direction whenever it perches, and at the termination of a succession of hops on the ground, over which it passes with great celerity, depending at all times for safety more on this power than on that of flight. It eludes pursuit by running through a bush to the opposite side, and hopping off to another beyond, which it does quite unseen unless closely watched. It builds a dome-shaped nest, which is placed on the ground, and frequently so hidden by the surrounding grass as to be with great difficulty discovered; a small narrow avenue of a yard in length, like the run of a mouse, being frequently resorted to by the bird, expressly, as one would suppose, to avoid detection. The eggs are three or four in number, rather large and somewhat round in form, of a reddish wood-brown, obscurely clouded with markings of reddish brown, the larger end of the eggs being the darkest; their medium length is ten lines and a half, and breadth eight lines and a half.

The nest is formed of dried grasses and leaves, and is warmly lined with feathers. The breeding-season commences in September and lasts until January.

This species emits so strong an odour, that pointers and other game-dogs stand to it as they do to a quail, and that too at a considerable distance. It possesses a clear and pretty song, which it frequently pours forth while sitting on a bare twig, or the summit of a low bush or shrub among the thickets, to a part of which it dives on the least alarm.

The sexes are precisely similar in colour, and nearly so in size.

All the upper surface olive, with a broad mark of sooty black down the centre of each feather; wings sooty black, narrowly margined with olive; tail olive, all but the two centre feathers crossed near the tip by a broad band of sooty black; line over the eye white; throat greyish white; breast, abdomen, and flanks deep buff, each feather of the throat breast, and flanks with a narrow line of sooty black down the centre; irides light sandy buff; bill and feet brownish flesh-colour.

Sp. 239. CALAMANTHUS CAMPESTRIS.

FIELD CALAMANTHUS.

Praticola campestris, Gould in Proc. of Zool. Soc., part viii. p. 171.

Calamanthus campestris, Gould, Birds of Australia, fol., vol. iii. pl. 71.

The *Calamanthus campestris* is a native of Southern and Western Australia, where it inhabits open plains and scrubby lands, particularly such as are interspersed with tufts of coarse grasses. It has never yet been discovered within the colony of New South Wales. Like its near ally of Tasmania it is a rather shy and recluse species, running mouse-like over the ground among the herbage with its tail perfectly erect, and is not easily forced to fly, or even to quit the bush in which it has secreted itself.

Its song is an agreeable and pretty warble, which is poured forth while the bird is perched upon the topmost twig of a small bush.

This species also emits so very powerful an odour, that my dog frequently pointed at it from a very considerable distance.

The nest, which is placed on the ground, is a globular structure, composed of grasses and feathers. The eggs are three or four in number, of a light chestnut-colour, thickly blotched with deep chestnut-brown, particularly at the larger end.

Forehead rufous, passing into the reddish brown of the crown and upper surface, with a stripe of blackish brown down the centre of each feather; wings sandy brown; internal webs of the primaries dark brown; two centre tail-feathers reddish brown, the remainder reddish brown at the base, crossed towards the extremity with a broad band of brownish black and broadly tipped with white; over the eye a line of white; ear-coverts mingled rufous and white; throat white, gradually passing into the buff of the under surface; all the feathers of the under surface with a stripe of brownish black down their centre; bill blackish brown, lighter at the base of the under mandible; irides rufous brown; feet blackish brown.

Total length 4½ inches; bill ⅜; wing 2¼; tail 2; tarsi ⅞.

Genus CHTHONICOLA, *Gould*.

The single species known of this genus combines in a remarkable manner the outward appearance, habits, and manners of the *Acanthizæ* and *Anthi*, but is, I believe, more nearly allied to the former than to the latter.

Sp. 239. CHTHONICOLA SAGITTATA.

LITTLE CHTHONICOLA.

Sylvia sagittata, Lath. Ind. Orn., Supp. p. liv.
Anthus minimus, Vig. and Horsf. in Linn. Trans., vol. xv. p. 230.

Chthonicola minima, Gould, Birds of Australia, fol., vol. iii. pl. 72.

This pretty little bird is usually seen on the ground in small companies of five or six in number, and is so very tame in disposition as to admit of the nearest approach, and when flushed merely flits off to the distance of a few yards. Its distribution, so far as we yet know, is confined to New South Wales and South Australia, in both of which countries it is a stationary and abundant species. It is very active in its actions, passing with great celerity over the gravelly ridges of the ground beneath the shade of the apple- and gum-trees.

The nest is of a domed form, and is placed among withered grass in a depression of the ground, so as to be on a level with the surface, and being formed of the same material as that with which it is surrounded, it is all but impossible to discover it; the entrance is an extremely small hole close to the ground. The eggs, which are four in number, are of a light cochineal-red, with a zone of blackish-brown spots at the larger end; their medium length is nine lines by seven lines in breadth.

The sexes are very similar; some individuals however are distinguished by the superciliary stripe being brown instead of white; whether this be characteristic of youth or maturity, I have not satisfactorily ascertained; I can scarcely conceive that so trivial a difference should indicate a difference of species.

General plumage olive-brown, the feathers of the back with darker centres, and of the head with a longitudinal stripe of buff down the middle of each; primaries narrowly edged with whitish; tail slightly tipped with white; under surface white, washed with yellow, each feather with a broad stripe of blackish brown down the centre, except on the middle of the abdomen, which is nearly pure white and without stripes; irides straw-yellow; bill brown; feet fleshy brown.

Family MOTACILLIDÆ.

Genus ANTHUS, *Bechstein.*

Whether this Old World form is represented in Australia by more than a single species is a point I have not satisfactorily determined; every part of its extra-tropical regions, including Tasmania, is inhabited by Pipits which differ somewhat in size in almost every colony; still their difference is so slight that I have hitherto regarded and still consider them to be mere varieties or local races of one and the same species.

Sp. 240. ANTHUS AUSTRALIS, *Vig. and Horsf.*

AUSTRALIAN PIPIT.

Anthus australis, Vig. and Horsf. in Linn. Trans., vol. xv. p. 229.
—— *pallescens,* Vig. and Horsf., id., p. 229.
Wăr-ra-joo-lon, Aborigines of the lowlands of Western Australia.
Common Lark of the Colonists.

Anthus Australis, Gould, Birds of Australia, fol., vol. iii. pl. 73.

The *Anthus australis* has all the habits and actions of its European prototypes; its note is also very similar; when flushed from the ground it rarely flies to any great distance before it descends again rather abruptly, to the earth, to the branch of a tree, or a small bush.

The nest is a rather deep and compactly formed structure of dried grasses; it is placed in a hole in the ground, sometimes beneath the shelter of a tuft of grass, but more frequently in a clear, open and exposed situation, the top of the nest being level with the surface. The eggs, which are three and sometimes four in number, are of a lengthened form, being eleven lines long by seven and a half lines broad, and are of a greyish white, blotched and freckled with light chestnut-brown and purplish grey, the latter colour appearing as if beneath the surface of the shell.

The breeding-season commences in the early part of September and continues until January, during which season two or three broods are reared.

The stomach is very muscular, and the food consists of insects of various kinds and small seeds.

The sexes are alike in plumage and may be described as follows:—

All the upper surface dark brown, each feather broadly margined with reddish brown; wings and two centre tail-feathers brown, margined with whitish brown; two lateral tail-feathers white, margined on the inner webs with blackish brown and with blackish-brown shafts, the remaining tail-feathers blackish brown; stripe over the eye light buff; ear-coverts brown; under surface dull white, washed with buff on the under surface of the shoulder and on the under tail-coverts; the feathers of the breast, flanks, and sides of the neck with a streak of dark brown down the centre, these marks being most conspicuous on the sides of the neck and across the upper part of the breast, where they are arranged in the form of a gorget, the points of which proceed upward to the angle of the lower mandible; irides very dark brown; bill and feet fleshy brown.

Freshly moulted individuals differ in having a rich tint of rufous pervading the whole of the upper surface, the breast, and flanks.

Genus CINCLORAMPHUS, *Gould*.

The members of this genus are closely allied to the Indian genus *Megalurus*, and present even a greater disparity in the size of the sexes; they are all confined to Australia, where they frequent the grassy plains and open districts. The song of the males is more animated than that of any other bird inhabiting the country.

Sp. 241. CINCLORAMPHUS CRURALIS.

Brown Cincloramphus.

Megalurus cruralis, Vig. and Horsf. in Linn. Trans. vol. xv. p. 228.
Cincloramphus cruralis, Gould in Proc. of Zool. Soc., part v. p. 150.

Cincloramphus cruralis, Gould, Birds of Australia, fol., vol. iii. pl. 74.

As there are two, if not three, species of this very singular genus inhabiting the southern portion of Australia, which bear a great resemblance to each other, it becomes necessary to state that this is the one commonly seen during the months of spring and summer in all the open districts of New South Wales, where it arrives in August, and after performing the task of incubation, departs again in January or February. Open downs, grassy flats, and fields of corn are its favourite places of resort. It is certainly one of the most animated of the Australian birds. Had I not visited Australia and personally studied its habits, my credulity would have been severely taxed upon being informed that the two birds, differing so greatly in size, were the male and female of the same species, many genera having been instituted upon much slighter grounds of difference; I had abundant proofs, however, that such is really the case, having seen many of the nests and eggs with the parent bird in the act of incubation, during the two seasons I spent in the country. In most of its habits and in its economy this bird closely assimilates to the Skylark of Europe. During the early months of spring it trips over the ground in the most sprightly manner with its tail nearly erect; mounts on the dead limbs of trees and the fences of enclosures, and runs along them with the greatest dexterity; at this season of the year also the male may be frequently seen running beside its diminutive partner, and so busily engaged in pouring forth his song for her amusement, as to be apparently unconscious of the presence of any other object. After the female has chosen the place for her nest, which is always on the ground, the male, like the Skylark, frequently mounts in the air with a tremulous motion of the wings, and after cheering her with his animated song, descends again to the ground or skims off to a neighbouring tree, and incessantly pours forth his voluble and not unpleasing notes.

I found it very abundant in all the Upper Hunter districts, as well as in all the surrounding country, both to the north and south: I killed numerous examples of both sexes, but not one male with the throat and under surface black, like specimens I have seen from Port Philip and South Australia, and which I consider to be specifically distinct.

The male has the entire plumage brown, each feather margined with brownish white; a large patch of dark brown on the centre of the abdomen; bill, inside of the mouth and tongue black; irides hazel; feet flesh-brown.

The female, which is less than half the size of the male, is similar in colour, but the feathers being more broadly margined with brownish white gives her a paler hue than her mate; the under surface is also much lighter, and the patch in the centre of the abdomen is much smaller.

Sp. 242. CINCLORAMPHUS CANTILLANS, *Gould*.

BLACK-BREASTED CINCLORAMPHUS.

Cincloramphus cantatoris, Gould in Proc. of Zool. Soc., part x. p. 135.
Ye-jul-lap, Aborigines of the mountain districts of Western Australia.
Sky-Lark of the Colonists.

Cincloramphus cantillans, Gould, Birds of Australia, fol. vol. iii. pl. 75.

Specimens killed at Port Philip in South Australia and others procured at Port Essington are precisely similar; but they differ from *C. cruralis* in their smaller size and in their darker colouring, a character which is confined to the male sex, and which is, I believe, strictly a summer livery. At Swan

River the individuals are still smaller, and like the *C. cruralis* are never so black on the breast.

I possess no information respecting the habits of the Port Philip bird. The following notes are the result of Gilbert's observations of the bird in Western Australia:—

"The *Cincloramphus cantillans* is a summer visitor, a remarkably shy and wary species, and a most difficult bird to procure, from its generally perching on a part of a tree whence it can command an uninterrupted view all round, rarely admitting any one to approach it within gun-shot. On being flushed from the ground it immediately takes to a tree, where, with its tail erect, and its head stretched out to the full extent of its neck, it presents a most grotesque appearance. It often ascends perpendicularly to a considerable height in the air, and then floats horizontally without any apparent motion of the wings to the distance of three hundred yards. While flying it utters a most disagreeably harsh and grating note, which is exchanged for an inward, rather plaintive tone when perched among the branches. The nest, which is deposited in a slight depression of the ground, is formed of dried grasses, and is so loosely put together that it is extremely difficult to preserve it entire; the eggs are four in number, and are similar to, but larger and of a lighter colour than those of the *C. rufescens*."

All the upper surface sandy brown, the centres of the feathers darker; primaries and tail greyish brown, slightly margined with reddish brown; immediately before the eye a triangular spot of brownish black; throat and chest dull white, the latter with a stripe of brown down each feather; under surface light brown; in the centre of the abdomen a patch of dark brown, each feather margined with pale brown; bill and feet fleshy brown.

Total length 8 inches; bill $1\frac{1}{8}$; wing $4\frac{1}{8}$; tail $4\frac{1}{8}$; tarsi $1\frac{1}{2}$.

Genus PTENOEDUS, *Cabanis*.

M. Cabanis has instituted the above genus for the bird named by Messrs. Vigors and Horsfield *Anthus rufescens*, and placed by me with the *Cincloramphi*; I admit the justice of the separation, and therefore adopt his generic appellation. A single species only has as yet been discovered in Australia, and so far as I am aware the form does not exist in other countries.

Sp. 243.   PTENOEDUS RUFESCENS.

RUFOUS-TINTED CINCLORAMPHUS.

Anthus rufescens, Vig. and Horsf. in Linn. Trans., vol. xv. p. 230.
Megalurus rufescens, Gray and Mitch. Gen. of Birds, vol. i. p. 109,
   Megalurus, sp. 3.
Ptenoedus rufescens, Cab. Mus. Hein., Theil i. p. 89.
E-role-del, Aborigines of the mountain districts of Western Australia.
Singing Lark of the Colonists.

Cincloramphus rufescens, Gould, Birds of Australia, fol., vol. iii. pl. 76.

If Australia be not celebrated for its singing-birds, it has still some few whose voices serve to enliven the monotony of its scenery; and of these no one deserves greater attention than the bird here described, which is a very sweet songster, and whose note somewhat resembles, but is much inferior to that of our own Skylark. With the exception of Tasmania, where I believe it is never seen, it appears to be distributed over all parts of Australia, specimens having been obtained in every locality yet visited by Europeans. In New South Wales and Western Australia it is strictly migratory, and only a summer visitor, arriving in August and departing in February; on the other hand, I met with it on the sandhills at Holdfast Bay, in South Australia, in the month of July, the period of winter: although not exclusively a terrestrial bird, it evinces a great partiality to open grassy plains here and there studded with trees, and spends much of its time on the ground, from which it makes perpendicular ascents to a great height in the air, and then descending to the tops of the highest trees, flies horizontally from one tree to another, singing all the time with the greatest volubility; the female, which is not more than half the size of the male, remaining all the while on the ground, from which she is not easily aroused, and consequently not so often seen. It breeds in October, November, and December, and generally rears two broods during the season. The nest is placed in a depression of the earth, most frequently at the foot of a slightly raised tuft of grass, and is externally composed of strong grasses and lined with very fine grasses, and sometimes with hairs. The eggs are four in number, ten lines long by seven and a half lines broad, and are of a purplish white, very broadly marked with freckles and small blotches of deep chestnut-brown, so much so as frequently to render the blotches more conspicuous than the ground-colour.

The male has all the upper surface dark brown, each feather margined with olive brown; upper tail-coverts rufous; lores black; stripe above the eye and throat whitish; all the under surface pale brownish grey, deepening into buff on the under tail-coverts, and with a series of minute spots of brown on the breast; irides hazel; bill dark lead-colour in summer, fleshy brown in winter; tarsi yellowish grey; feet bluish ashy grey.

The female is smaller, and is destitute of the black lores; in other respects she is so like the male that a separate description is unnecessary. She is said to frequently utter a sharp shriek during the night.

Family ?

Genus SPHENŒACUS, *Strickland*.

A group of reed- and grass-frequenting birds, which are found not only in every part of Australia, but also in the Indian Islands and India.

Sp. 244.   SPHENŒACUS GALACTOTES.

TAWNY GRASS-BIRD.

Malurus galactotes, Temm. Pl. Col., 65.
Megalurus galactotes, Vig. and Horsf. in Linn. Trans., vol. xv. p. 228.

Sphenoeacus galactotes, Gould, Birds of Australia, fol., vol. iii. pl. 35.

This is a scarce species in New South Wales, the few individuals I have seen being from the grassy districts of the Liverpool Plains; in all probability, however, it ranges along the eastern and over the whole of the northern portion of Australia. Gilbert's notes inform me that he found it "tolerably abundant on the islands at the head of Van Diemen's Gulf, where it inhabits the long grass or rushes growing in or adjacent to the swamps; it is so shy that it is very rarely seen; when closely hunted it takes wing, but flying appears to be a difficult action at all times; at least I have never seen it sustain a flight of more than a hundred yards at the utmost, and even in that short distance it seemed ready to sink into the grass with fatigue. The only note I have heard it emit is a harsh and rapidly repeated *chuick*. The stomachs of those I dissected were extremely muscular, and contained the remains of insects of various kinds and what appeared to be vegetable fibres."

General plumage pale brown, deepening into rufous on the crown of the head and fading into dull white on the throat and centre of the abdomen; all the feathers of the upper surface with blackish brown centres; secondaries blackish brown, broadly margined with pale brown; tail pale brown, crossed with indistinct bars of a darker tint; irides light brown; upper mandible olive-brown, the cutting edges light yellowish white; lower mandible bluish white; tarsi and feet light reddish flesh-colour.

Sp. 245. SPHENŒACUS GRAMINEUS, *Gould*.

LITTLE GRASS-BIRD.

Sphenæacus gramineus, Gould in Proc. of Zool. Soc., part xiii. p. 19.
Megalurus gramineus, Gray and Mitch. Gen. of Birds, vol. i. p. 169,
 Megalurus, sp. 5.
Poodytes gramineus, Cab. Mus. Hein., Theil i. p. 42.

Sphenœacus gramineus, Gould, Birds of Australia, fol., vol. iii. pl. 36.

Although the present species is very generally dispersed over the whole of the southern and western portions of Australia and Tasmania, in all situations suitable to its habits, it is as little known to the colonists as if it were not in existence, which is readily accounted for by its recluse nature and the localities it frequents, the thick beds of grasses, rushes, and other kinds of herbage growing in low, damp, and wet places on the mainland, and on such islands as those of Green and Actæon, in D'Entrecasteaux's Channel, being its favourite places of resort. It is a very shy species, and will almost allow itself to be trodden upon before it will quit the place of its concealment; in the open grassy beds of the flats it is more easily driven from its retreat, but even then it merely flies a few yards, and pitches again among the herbage.

Its song consists of four or five plaintively-uttered notes, repeated five or six times in succession.

The nest is generally a very compact structure; in Western Australia it is formed of the soft tops of the flowering part of the reeds, and the thin skin-like coating of the reed-stalks, but occasionally of fine swamp grasses, and is always lined with feathers; in some instances two large feathers are made to meet over the opening, near the top of the nest, and thus protects the inside from cold or rain: it is attached to two or three upright reeds about two feet from the surface of the water. The eggs, which are laid during the months of August and September, are four in number, nearly eight lines long and six lines broad; they are of a fleshy-white, freckled and streaked all over, particularly at the larger end, with purplish red: in some instances large obscure blotches of reddish grey appear as beneath the surface of the shell.

The sexes present no difference in size or colour, and there is scarcely any variation in specimens from Tasmania, Swan River, and New South Wales.

Stripe over the eye white: all the upper surface brown, the centre of the feathers being dark brown; secondaries brownish black, margined with buff; tail pale reddish brown, with dark brown shafts; under surface grey, passing into black on the flanks and vent; each feather of the breast with a very minute line of dark brown down the centre; bill and tarsi fleshy brown.

Total length 5¼ inches; bill ⅝; wing 2¼; tail 2⅜; tarsi ¾.

Family SYLVIADÆ.

Genus CALAMOHERPE, *Meyer.*

Of this European and Indian form two species inhabit Australia, where they frequent the reed-beds and the dense herbage of marshy situations.

Those who are acquainted with the habits of the Great Sedge-Warbler of Europe (*Calamoherpe turdoides*) will have a just idea of what the present and the following species are like.

Sp. 246. CALAMOHERPE AUSTRALIS, *Gould.*

REED-WARBLER.

Reed-Warbler, Lewin, Birds of New Holland, pl. 18.

Acrocephalus australis, Gould, Birds of Australia, fol., vol. iii. pl. 37.

This bird does not inhabit Tasmania, but is universally dispersed among the sedgy sides of rivers and lagoons, both in South Australia and New South Wales; I also observed it in great abundance on the banks of the rivers to the northward of Liverpool Plains; in all these localities it is strictly migratory, arriving in September, and departing again before the commencement of winter. In its general economy it closely resembles its European congeners, but possesses a still louder and more melodious song than any of them, except the *Calamoherpe turdoides*. It is rather a late breeder, scarcely ever beginning this natural duty before the month of November. The nest, like that of the Reed-Warbler of Europe, is suspended from two or three reeds at about two feet above the surface of the water, and is composed of the soft skins of reeds and dried rushes. The eggs, which are four in number, ten lines long by seven lines broad, are of a greyish white, thickly marked all over with irregular blotches and markings of yellowish brown, umber brown, and bluish grey, intermingled together without any appearance of order or arrangement.

The food consists of insects of various kinds.

The sexes are so precisely alike that dissection must be resorted to to distinguish them.

All the upper surface olive-brown; wings and tail brown, margined with olive-brown; all the under surface tawny or deep buff, fading into white on the throat; under mandible fleshy white, remainder of the bill and the legs olive horn-colour; irides brown.

Sp. 247. CALAMOHERPE LONGIROSTRIS, *Gould.*

LONG-BILLED REED-WARBLER.

Calamoherpe longirostris, Gould in Proc. of Zool. Soc., part xiii. p. 20.
Goor-jee-goor-jee, Aborigines of the lowland districts of Western Australia.

Acrocephalus longirostris, Gould, Birds of Australia, fol., vol. iii. pl. 38.

The present bird, which I have designated *longirostris*, is the largest of the two species of *Calamoherpe* known to inhabit Australia.

It is a native of the western portion of the country, where I learn from Gilbert's notes that "it is to be found in all the dense reed-beds bordering the river and lakes around Perth, but is so shy that it scarcely ever shows itself above the reeds. I have remarked also that it never wanders many yards from the nest, which is placed on four or five upright reeds growing in the water at about two feet from the surface. It is of a deep cup-shaped form, and is composed of the soft skins of reeds and dried rushes. The breeding-season comprises the months of August and September. The eggs are four in number, of a dull greenish white, blotched all over, but particularly at the larger end, with large and small irregularly shaped patches of olive, some being darker than the others, the lighter-coloured ones appearing as if beneath the surface of the shell; they are three-quarters of an inch in length by five-eighths of an inch in breadth.

"It sings both night and day, and its strain is more beautiful and melodious than that of any other Australian bird with which I am acquainted; being in many parts very like to that of the far-famed Nightingale of Europe.

"The stomach is tolerably muscular, and the food consists of coleopterous and other kinds of insects."

Faint line over the eye fawn-colour; all the upper surface reddish brown, becoming more rufous on the upper tail-coverts; primaries and tail deep brown, fringed with rufous; chin whitish; all the under surface deep fawn-colour; irides yellowish brown.

Total length 6½ inches; bill ⅝; wing 3; tail 3; tarsi 1.

Genus MIRAFRA, *Horsfield*.

One, if not two, species of this well-defined genus inhabit Australia. At present one only has been characterized; but the bird of this form, frequenting the intertropical portions of the country, may prove to be a distinct species.

Sp. 248.   MIRAFRA HORSFIELDII, *Gould*.

HORSFIELD'S BUSH-LARK.

Mirafra Horsfieldii, Gould in Proc. of Zool. Soc., part iv. p. 1.

Mirafra Horsfieldii, Gould, Birds of Australia, fol., vol. iii. pl. 77.

This species, which I have named *horsfieldii* after the founder of the genus, is sparingly dispersed over all the plains and open districts of New South Wales, but is more abundant on the inner side of the mountain ranges towards the interior than between the ranges and the sea; I have also a specimen procured during Dr. Leichardt's overland expedition from Moreton Bay, and one from the neighbourhood of Port Essington: both of these, although possessing characters common to each other, differ from specimens obtained in New South Wales in being larger, redder in colour, and in having a stouter bill—features which will probably hereafter prove them to be distinct, and which exhibit a near alliance to the *Mirafra javanica*.

The bird here described is from New South Wales, where I found it more abundant on the Liverpool Plains than elsewhere; I also met with solitary individuals in the district of the Upper Hunter.

In its habits it is more terrestrial than arboreal, and will frequently allow itself to be almost trodden upon before it will rise, and then it merely flies to a short distance and descends again; it may often be seen perched upon the strong blades of grass and occasionally on the trees; it frequently mounts high in the air after the manner of the Skylark of Europe, singing all the time very melodiously, but with a weaker strain than that favourite bird; it also occasionally utters its pleasing song while perched on the branches of the trees.

The sexes are alike in colour and size.

General plumage ashy brown, the centre of the feathers dark brown, the latter colour predominating on the head, lower part of the back and tertiaries; wings brown margined with rufous; over the eye a stripe of buff; chin white; under surface pale buff; throat crossed by a series of dark brown spots arranged in a crescentic form; under surface of the wing rufous; bill flesh-brown at the base and dark brown at the tip; feet fleshy brown.

Family FRINGILLIDÆ.

The Finches of Australia comprise about twenty well-marked species, pertaining to several genera or subgenera, each of which exhibits a slight difference in structure, accompanied as is always the case, by a difference in habit, and in the districts inhabited; thus the *Stictopterœ* frequent grassy patches in the glades of the forests, the open parts of gullies, &c.; the *Steganopleuræ*, the stony hills and flats; the *Poëphilæ*, the grass beds of the open plains; and the *Donacolæ*, the marshy districts and reed-beds: of the habits of *Emblema* nothing is known; its pointed bill indicates some peculiarity in its economy differing from those of the other genera.

Most if not all the species build large grassy nests, some with a spout-like opening.

Genus ZONÆGINTHUS, *Cabanis*.

Of this genus, established by M. Cabanis, two species inhabit Australia, both of which are very similarly marked, and differ in this respect from the other species of this extensive family.

Sp. 249.   ZONÆGINTHUS BELLUS.

FIRE-TAILED FINCH.

Loxia bella et *nitida*, Lath. Ind. Orn., Supp. pp. xlvi, xlvii.
Black-lined Grosbeak, Lath. Gen. Syn. Supp., vol. ii. p. 108.
Fringilla bella, Vig. and Horsf. in Linn. Trans., vol. xv. p. 257.
Amadina nitida, Gray and Mitch. Gen. of Birds, vol. ii. p. 370;
· *Amadina*, sp. 15.
Zonæginthus nitidus, Cab. Mus. Hein., Theil i. p. 171.
Wee-bong, Aborigines of New South Wales.
Fire-tail, Colonists of Tasmania.

Estrelda bella, Gould, Birds of Australia, fol., vol. iii. pl. 78.

Tasmania may be considered the principal habitat of this species, for it is universally and numerously dispersed over all parts of that island suited to its habits and economy. It also inhabits New South Wales, but is there far less abundant. I generally observed it in small communities varying from six to a dozen in number, searching on the ground for the seeds of grasses and other small plants which grow on the plains and open parts of the forest. It also frequents the gardens and pleasure-grounds of the settlers, with whom it is a favourite, few birds being more tame or more beautifully coloured than this little Finch; the brilliant scarlet of the rump and the base of the tail-feathers strongly contrasting with the more sombre hue of the body. Its flight is extremely rapid and arrow-like, particularly when crossing a plain or passing down a gulley. It is a stationary species in Tasmania, and probably also in New South Wales. In the former country I constantly found it breeding in communities, my attention being usually attracted by the enormous nests which they build among the branches of shrubby trees without the slightest attempt at concealment. They are constructed entirely of grasses and stalks of plants, dome-shaped in form, with a hole near the top for the ingress and egress of the bird. The eggs are five or six in number, rather lengthened in form and of a beautiful flesh-white, eight and a half lines long by six and a half lines broad. It breeds from September to January, during which period two or three broods are reared. Its note is a single mournful sound emitted while perched on the low branches of the trees in the neighbourhood of its feeding-places.

The sexes present no external difference, and may be thus described:—circle surrounding the eyes, lores, and a line crossing the forehead black; all the upper surface, wings, and tail olive-brown, crossed by numerous narrow crescentic lines of black; rump and base of the tail-feathers shining scarlet; all the under surface grey, crossed by numerous narrow crescentic lines of black; centre of the abdomen and under tail-coverts black; tips of the primaries and tail-feathers brown without bars; bill crimson, becoming paler at the base of the upper mandible; irides very dark brown; eyelash beautiful light blue; feet flesh-colour.

Sp. 250.   ZONÆGINTHUS OCULEUS.
Red-eared Finch.

Fringilla oculea, Quoy et Gaim. Voy. de l'Astrolabe, Zool., part i. p. 211; Ois., pl. 18. fig. 2.
Zonaginthus oculeus, Cab. Mus. Hein., Theil i. p. 171, note.
Jed-ree, Aborigines of the lowland, and
Dwel-den-ngool-gnall-neer, Aborigines of the mountain districts of Western Australia.
Native Sparrow, Colonists of Swan River.

Estrelda oculea, Gould, Birds of Australia, fol., vol. iii. pl. 79.

This species is abundant in many parts of the colony of

Swan River. Like its near ally the *Zonæginthus bellus*, it inhabits open grassy glades studded with thickets, particularly in moist swampy districts and along the borders of lakes and rivers. Its food consists of small grass-seeds procured among the herbage. Gilbert states that "it is a solitary species and is generally found in the most retired spots in the thickets, where its mournful, slowly drawn-out note only serves to add to the loneliness of the place. Its powers of flight, although sometimes rapid, would seem to be feeble, as they are merely employed to remove it from tree to tree. The natives of the mountain districts of Western Australia have a tradition that the first bird of this species speared a dog and drank its blood, and thus obtained its red bill."

The sexes are so much alike that dissection is necessary to distinguish the male from the female. The beautiful patch of scarlet feathers behind the eye, together with the rich colouring of the bill, assists very materially in relieving the more sombre but delicate markings of the remainder of the body.

Lores, line over the bill and a narrow circle surrounding the eye black; behind the eye a small patch of shining scarlet; all the upper surface olive-brown, crossed by numerous fine irregular crescent-shaped bands of black, which are broadest and most conspicuous on the lower part of the back; wings and tail similarly marked, but with the black bands still broader and more distinct; rump and the margins of the base of the central tail-feathers shining scarlet; throat and breast light brown, crossed by numerous crescent-shaped bands of black; abdomen and under surface black, with a large spot of white near the tip of each feather; irides red; bill bright vermilion, the base of the upper mandible edged with pearl-grey; eyelash greenish blue; legs yellowish grey.

Genus STICTOPTERA, *Reichenbach*.

Two species of this form are found in Australia; one inhabits the south-eastern, the other the northern parts of the country. Both are distinguished by a conspicuous double zone of black on the breast; and their short and rather gibbose bills are doubtless admirably adapted for procuring some particular kind of food, probably the seeds of grasses.

Sp. 251.   STICTOPTERA BICHENOVII.

Bicheno's Finch.

Fringilla bichenovii, Vig. and Horsf. in Linn. Trans., vol. xv. p. 258.

Estralda Bichenovii, Gould, Birds of Australia, fol., vol. iii. pl. 80.

This beautiful little Finch inhabits the extensive plains of the interior, particularly such portions of them as are thinly intersected with low scrubby trees and bushes. My specimens were obtained on the Liverpool and Brezi Plains. As I have had occasion to remark with respect to other species, it will be impossible to determine the precise extent of its range until Australia has been more fully explored.

The Bicheno's Finch is very tame in its disposition, and is generally to be observed on the ground, occupied in procuring the seeds of the grasses and other small plants, which form its principal food. When I visited the interior in the month of December, it was assembled in small flocks of from four to eight in number; these, when flushed from among the grasses, would perch on the neighbouring bushes rather than fly off to any distance, and indeed the form of its wings and tail indicate that it possesses lesser powers of flight than many of the other Finches.

I was not fortunate enough to obtain its nest or eggs, neither did I ever hear it utter any kind of song; consequently I am unable to give any information on those points.

The male has the face, ear-coverts, and throat pure white, completely surrounded by a band of black, which is broadest on the forehead; crown of the head, nape of the neck, and back broccoli-brown, each feather crossed by numerous transverse lines of a lighter tint; upper part of the rump black; lower part of the rump and upper tail-coverts snow-white; wings black, all the feathers except the primaries beautifully spotted with white; chest greyish white, tinged with buff, bounded below by a broadish band of jet-black; abdomen and flanks buffy white; under tail-coverts and tail black; irides black, surrounded by a narrow black lash; bill beautiful pale blue.

The sexes, although having a similar character of marking, may be distinguished from each other by the male having the black bands of the chest and throat broader, and its plumage more brilliant. The young also at an early age possess the characteristic markings of the adult.

Sp. 252.   STICTOPTERA ANNULOSA, *Gould*.

BLACK-RUMPED FINCH.

Amadina annulosa, Gould in Proc. of Zool. Soc., part vii. p. 143.
Stictoptera annulosa, Reich. Sing.-Vögel.

Estrelda annulosa, Gould, Birds of Australia, fol. vol. iii. pl. 81.

This species was one of several, collected by the Officers of H.M.S.S. Beagle, and for the specimens from which my descriptions were taken I am indebted to Messrs. Bynoe and Dring. The bird has also been brought to England by Sir George Grey: all these specimens were collected on the north-west coast, and it is not unfrequently seen on the Coburg Peninsula, where it inhabits the grassy banks of running streams, in small families of from six to ten in number.

It differs from Bicheno's Finch in the spots and markings on the upper surface being rather less defined, and in the colouring of the rump, which in this species is black, while in the other it is white.

Face, ear-coverts, and throat white, surrounded by a jet-black band, which is broadest on the forehead; chest greyish white, bounded below by a conspicuous band of black; lower part of the abdomen white; crown of the head, back of the neck, and back greyish brown marked with numerous fine transverse lines of greyish white; rump, upper and under tail-coverts and tail black; wings blackish brown, the secondaries and coverts thickly dotted with fine markings of greyish white; bill and feet lead-colour.

Total length 4 inches; bill ¾; wing 2; tail 2½; tarsi ½.

Genus ÆGINTHA, *Cabanis*.

M. Cabanis has instituted this genus for *Fringilla temporalis* of Latham. It is the only species of this form yet discovered in Australia, and is nearly allied to *Estrelda*.

Sp. 253. ÆGINTHA TEMPORALIS.

RED-EYEBROWED FINCH.

Fringilla temporalis, Lat. Ind. Orn., Supp. p. xlviii.
Temporal Finch, Lat. Gen. Syn. Supp., vol. ii. p. 211.
Le Sénégali quinticolor, Vieill. Ois. Chant., p. 38, pl. 15.
Fringilla quinticolor, Vieill. 2nde édit. du Nouv. Dict. d'Hist. Nat., tom. xii. p. 183.
Amadina temporalis, Gray and Mitch, Gen. of Birds, vol. ii. p. 370, Amadina, sp. 25.
Ægintha temporalis, Cab. Mus. Hein., Theil i. p. 170.
Goo-lung-ag-ga, Aborigines of New South Wales.
Red-Bill of the Colonists.

Estrelda temporalis, Gould, Birds of Australia, fol., vol. iii. pl. 82.

This species of Finch is very generally spread over the gardens and all such open pasture lands of New South Wales and South Australia as abound in grasses and small plants, upon the seeds of which it chiefly subsists. It is particularly abundant in the neighbourhood of Sydney; even in the Botanic Garden numbers may always be seen flitting from border to border. It is easily domesticated, and it is of a lively disposition in captivity, even old birds becoming perfectly reconciled after a few days. In the autumn it is gregarious, and Mr. Calcy states it often assembles in large flocks; in the spring they are mostly seen in pairs, and then build their large and conspicuous nest, which is formed of dead grass, lined with thistle down, in any low bush adapted for a site, but in none more frequently than in the beautiful *Leptospermum squarrosum*.

The eggs are five or six in number, of a beautiful fleshy white, seven lines long by five and a half lines broad.

Crown of the head bluish grey; upper surface, wings, and tail olive-brown; under surface white; patch over the eye and rump crimson; irides brownish red; eyelash narrow, naked and black; bill fine blood-red, with the ridge of the upper and the lower part of the under mandible black; legs yellowish white.

Genus BATHILDA, *Reichenbach*.

This genus has been established for the beautiful *Estrelda ruficauda* of the folio edition of the Birds of Australia, a delicately coloured bird, rendered conspicuously different from other Finches by the spotted markings of the chest.

Sp. 254.   BATHILDA RUFICAUDA, *Gould.*
RED-TAILED FINCH.

Amadina ruficauda, Gould in Proc. of Zool. Soc., part iv. p. 106.
Bathilda ruficauda, Reich. Sing-Vögel.

Estrelda ruficauda, Gould, Birds of Australia, fol, vol. iii. pl. 84.

I observed this beautiful Finch rather thinly dispersed on

the sides of the river Namoi, particularly along the sloping banks covered with herbage, where it appeared to be feeding upon such grasses and other annuals as afforded seeds congenial to its taste; I also frequently observed it among the rushes which grow in the beds of mud along the sides of the water.

The adult male and female are scarcely to be distinguished by outward appearance; the female is, however, a trifle less than the male in size. The young, on the contrary, present a very different appearance; the whole of their plumage being of a uniform buffy brown; eye yellowish olive surrounded by a narrow olive lash; bill reddish brown; legs brownish yellow.

Face and cheeks scarlet, the latter covered with narrow feathers, which are finely spotted with white at the tip; upper surface and wings olive-brown; upper tail-coverts and tail deep crimson-brown, the former having a large spot of pinkish white near the tip of each feather; throat, chest, and flanks delicate olive-grey, each feather having a large oval white spot transversely disposed near the tip; centre of the abdomen and under tail-coverts dirty yellowish white; bill scarlet; irides orange slightly inclining to hazel, surrounded by a rather broad, naked, flesh-coloured lash; legs and feet rather darker than fine lemon-yellow.

Total length $4\frac{1}{2}$ inches; wing $2\frac{1}{8}$; tail $1\frac{2}{8}$; tarsi $\frac{1}{2}$.

Genus AIDEMOSYNE, *Reichenbach.*

The extreme modesty of its colouring and the jetty hue of its bill afford sufficient differential characters from the other Australian members of the family to warrant its separation into the type of a new genus. Without questioning the propriety of these numerous subdivisions, I must, in justice to those who make them, remark that they do differ very considerably, and that additional species of most of the forms occur in other countries.

Sp. 255. AIDEMOSYNE MODESTA, *Gould.*
PLAIN-COLOURED FINCH.

Amadina modesta, Gould in Proc. of Zool. Soc., part iv. p. 105.
Estrelda modesta, Gould, Birds of Australia, fol., vol. i. Introd. p. xlix.
Aidemosyne modesta, Sing-Vögel.

Amadina modesta, Gould, Birds of Australia, fol., vol. iii. pl. 85.

I found the Plain-coloured Finch tolerably abundant on the Liverpool Plains and on the banks of the Namoi, and Gilbert also mentions his having observed it on the low ranges to the northward of Moreton Bay. In its habits, actions, and economy no remarkable differences were observed from those of the other species of the genus.

It is usually seen in pairs or associated in small companies, feeding either on or near the ground; the seeds of grasses and other annuals forming its chief supply of food.

A nest found by Gilbert was of a domed form, composed of grasses, and contained five or six white eggs, about half an inch long by three-eighths broad.

The sexes may be distinguished by the absence of the black mark in the female, as shown in the accompanying plate.

The male has the fore-part of the head deep crimson-red; lores and a spot on the chin black; nape of the neck, mantle, and back brown; wings brown; tertials (which are very long in this species), together with the greater and lesser quill-feathers, having a spot of white at the tip; rump and upper tail-coverts alternately barred with lines of greyish white and brown; tail-feathers black, the two outer ones on each side tipped with white; under surface white, transversely barred with lines of brown, which are strongest on the flanks; middle of the abdomen and under tail-coverts white; bill black; irides reddish brown; eyes surrounded by a very narrow lash of blackish brown; legs flesh-white.

The female differs in having the colouring of the crown less extensive, and in wanting the black on the chin and lores.

Total length $4\frac{1}{2}$ inches; wing $2\frac{1}{4}$; tail 2; tarsi $\frac{3}{4}$.

Genus NEOCHMIA, *Hombron et Jacquinot*.

The bird to which the above generic title has been given, and by which it must hereafter be known, differs from all the other members of its family in the lengthened form of its tail, and in its peculiar red colouring.

Sp. 250.   NEOCHMIA PHAETON.

CRIMSON FINCH.

Fringilla phaeton, Homb. et Jacq. Ann. des Sci. Nat., tom. vi. p. 314.
Neochmia phaeton, G. R. Gray, Cat. of Gen. and Subgen. of Birds in Brit. Mus., p. 76.
Ing-a-dăm-oon, Aborigines of Port Essington.
Red Finch, Residents of Port Essington.

Estrelda phaeton, Gould, Birds of Australia, fol., vol. iii. pl. 83.

In a paper addressed by MM. Hombron and Jacquinot to the Académie des Sciences on the 9th of August 1841, entitled "*Description de plusieurs Oiseaux nouveaux ou peu connus, provenant de l'expédition autour du monde faite sur les corvettes l'Astrolabe et la Zélée*," I find the characters of a Finch, which, although the colouring does not quite agree with that of the bird here figured, I have little doubt is identical with it. I am the more inclined to consider this to be the case from the circumstance of MM. Hombron and Jacquinot's bird having been collected at Raffles' Bay, a locality closely bordering that in which Gilbert procured his specimens, and who states that "this bird is an inhabitant of moist grassy meadows, particularly where the *Pandanus* (Screw Pine) is abundant. It is generally found feeding among the grass, and when disturbed invariably takes to those trees. From July to November it is to be observed in large flocks, sometimes of several hundreds; but although great numbers were shot during this period, not more than three or four were obtained in the rich plumage. About the latter part of November they were either in pairs or in small companies, not exceeding six in number; the males decorated with their rich red and spotted dress."

The stomach is muscular, and the food consists of grass and other small seeds.

Crown of the head deep bluish black; lores, line over the eye, sides of the face, and ear-coverts rich crimson red; under surface crimson red, spotted on the flanks with white; centre of the abdomen and under tail-coverts black; back of the neck and rump dark brownish grey; back and wings brownish grey, each feather crossed near the extremity with a band of deep crimson red; upper tail-coverts and two centre tail-feathers deep red; the remainder deep red at the base, passing into brown at the tip; bill rich carmine, bounded at the base by a band of greyish white about one-tenth of an inch in breadth; hinder part of the tarsi and inside of the feet ochre-yellow; front of tarsi and upper surface of the feet ochre-yellow, strongly tinged with hyacinth-red.

The female, which is rather smaller than the male, is brown above, a few of the feathers on the back and the wing-coverts crossed with red as in the male; lores, line over the eye, sides of the face, chin, upper tail-coverts, and tail as in the male, but not quite so brilliant; breast and flanks greyish brown, the latter ornamented with a few small spots of white; centre of the abdomen buff.

Genus STAGONOPLEURA, *Cabanis.*

This genus has been instituted for the *Loxia guttata* of Shaw, and as yet Australia has given us but a single species; but when the interior and the northern coast line have been explored, others may be discovered there.

When fully adult, the sexes of this form are very similar; but the young birds are very different.

Sp. 257.   STAGONOPLEURA GUTTATA.

SPOTTED-SIDED FINCH.

Fringilla leucocephala, var., Lath. Ind. Orn., Supp. p. xlviii.
Loxia guttata, Shaw, Mus. Lev., pl.
Spotted Grosbeak, Lewin, Birds of New Holl., pl. 9.
White-headed Finch, Lath. Gen. Syn. Supp., vol. ii. p. 210, pl. 132.
Spotted-sided Grosbeak, Lath. Gen. Hist., vol. v. p. 248, pl. 89.
Fringilla lathami, Vig. and Horsf. in Linn. Trans., vol. xv. p. 256.
Amadina guttata, Gray and Mitch. Gen. of Birds, vol. ii. p. 370, *Amadina*, sp. 8.
Stagonopleura guttata, Cab. Mus. Hein., Theil i. p. 172.

Amadina Lathami, Gould, Birds of Australia, fol., vol. iii. pl. 86.

I found this species plentiful in South Australia and in every part of New South Wales that I visited; and it was equally numerous on the Liverpool Plains, the sides of the River Mokai, Namoi, &c. It is a showy attractive species, and passes much of its time on the ground, where it procures its food, which consists of the seeds of various kinds of grasses, &c.

The nest is frequently built among the large sticks forming the under surface of the nest of the smaller species of Eagles, and that too during the time the Eagle is incubating, both species hatching and rearing their progeny in harmony; this I have witnessed in several instances, and have taken the eggs of the Eagle and of the Finch at the same time, as mentioned in the following extract from my journal:—" Oct. 23. Found the nest of the Spotted-sided Finch placed under and among the sticks of a Whistling Eagle's (*Haliastur? sphenurus*) nest, in which latter the old bird was then sitting. My black companion Natty ascended the tree, a high swamp oak (*Casuarina*) on the bank of the Dartbrook, and brought down the eggs of both birds. The little Finches were sitting on the small twigs close to their rapacious but friendly neighbour." At other times the nest of this Finch is placed on the leafy branch of a gum- or apple-tree. It is of a large size, and is constructed of grasses of various kinds; in form it is nearly spherical, with a short pendant spout on one side, through which the bird obtains access to the interior; the eggs are white, rather long in shape, and five or six in number.

The sexes offer little or no difference in the markings of their plumage.

Crown of the head and back of the neck brownish grey; back and wings brown, becoming deeper on the tips of the primaries; lores, a broad band across the breast, flanks, and tail deep black; each feather of the flanks with a large spot of white near the tip; rump and upper tail-coverts shining scarlet; throat, abdomen, and under tail-coverts white; irides red, surrounded by a narrow, naked, lilac-red lash; bill blood-red, passing into lilac at the base and on the culmen; feet purplish brown.

The young for the first year has the bill black, except at the base, where it is flesh-colour; the band across the breast and the flanks greyish brown, the latter being barred indistinctly with black and greyish white; in other respects the plumage nearly resembles the adult.

Genus TÆNIOPYGIA, *Reichenbach*.

The Berlin Professor, Cabanis, established the genus *Staganopleura* for the *Loxia guttata* of Shaw, and associated therewith the bird described by me as *Amadina castanotis*; the Dresden Professor, Reichenbach, has, however, gone further still, for he has separated the latter bird from the former under the generic name of *Tænioptera*. To this form must be added the *Loxia guttata* of Vieillot, not of Shaw. Probably the Timor bird recently described by Mr. Wallace as *Amadina insularis* is Vieillot's bird; there are, therefore, two, if not three, distinct species of this form.

Sp. 258. TÆNIOPYGIA CASTANOTIS, *Gould*.

CHESTNUT-EARED FINCH.

Amadina castanotis, Gould in Proc. of Zool. Soc., part iv. p. 105.
Stagonopleura castanotis, Cab. Mus. Hein., Theil i. p. 172.
Tæniopygia castanotis, Reich. Sing. Vögel.

Amadina castanotis, Gould, Birds of Australia, fol., vol. iii. pl. 87.

This bird appears to be almost peculiar to the interior of Australia; among other places it inhabits the large plains to the north of the Liverpool range and is particularly abundant about Brezi and the banks of the river Mokai; but that it sometimes occurs on the southern side of the range is proved by my having killed five specimens on the Upper Hunter. It has also been found, though very sparingly, at Swan River, and a specimen is contained in the collection formed by Mr. Bynoe at Port Essington. It passes much of its time on the ground, and feeds upon the seeds of various kinds of grasses. On the plains it congregates in small flocks, and evinces a decided preference to those spots where the trees are thinly dispersed and grasses abundant.

The Chestnut-eared Finch is one of the smallest of the genus yet discovered in Australia; it is also one of the most beautiful, and in the chasteness of its colouring can scarcely be excelled.

The two sexes differ very considerably in their markings, and may be thus described:—

The male has the crown of the head, nape, and back brownish grey; wings brown; rump white; upper tail-coverts jet black, each feather having three large and conspicuous oval spots of white; tail-feathers blackish brown, slightly tinged with white at their tips; cheeks and ear-coverts reddish chestnut, separated from the bill by a narrow transverse line of white, which white line is bounded on each side by a still finer line of black; throat and chest grey, the feathers transversely marked with fine lines of black; a small black patch on the middle of the chest; abdomen white; under tail-coverts buffy white; flanks chestnut, each feather marked near the tip with two small oval spots of white; bill reddish orange; feet reddish orange, rather lighter than the bill; irides red.

The female has the transverse lines on the face, upper tail-coverts, and feet as in the male; upper surface, ear-coverts, wings, tail, and flanks greyish brown; throat and chest grey, slightly tinged with brown; abdomen yellowish brown; bill reddish orange.

Total length 4¼ inches; wing 2⅛; tail 1½; tarsi ½.

Genus POËPHILA, *Gould*.

A generic division proposed for a number of Grass-Finches distinguished for the beauty of their plumage and the elegance of their form; they principally inhabit the plains of the northern portions of Australia.

Sp. 250.   POËPHILA GOULDIÆ, *Gould*.

GOULDIAN GRASS-FINCH.

Amadina gouldiæ, Gould in Proc. of Zool. Soc., part xii. p. 6.
Poëphila gouldiæ, Gould, Birds of Australia, fol., vol. i. Introd. p. xlix.
Chloëbia gouldiæ, Reich. Sing. Vögel.

Amadina Gouldiæ, Gould, Birds of Australia, fol., vol. iii. pl. 88.

It was with feelings of the purest affection that I ventured, in the folio edition, to dedicate this lovely bird to the memory of my late wife, who for many years laboriously assisted me with her pencil, accompanied me to Australia, and cheerfully interested herself in all my pursuits. The dedication of this bird to Mrs. Gould's memory will surely then receive the sanction of every scientific ornithologist.

The *Poëphila gouldiæ* was discovered by Gilbert on Green-bill Island at the head of Van Diemen's Gulf, "where it inhabited the edges of the mangroves and thickets: when disturbed it invariably flew to the topmost branches of the loftiest gums, a habit I have not before observed in any other member of the genus. Its note is a very mournful sound added to a double twit. Those I observed were feeding among the high grass in small families of from four to seven in number, and were very shy. The stomach is tolerably muscular, and the food consists of grass and other seeds."

More recently the late Mr. Elsey observed it in great abundance on the Victoria River.

The adult has the forehead, lores, ear-coverts, and throat deep velvety black; from behind the eye, round the occiput, and down the sides of the neck a mark of verdigris-green, gradually blending into the yellowish green of the upper surface and wings; across the breast a broad band of shining lilac-purple, below which all the under surface is shining wax-yellow; tail black; bill flesh-white at the base, tipped with blood-red; feet flesh-colour.

Total length 3¾ inches; bill ⅜; wing 2¼; tail 2½; tarsi ½.

The young bird has the head grey; upper surface light olive; under surface pale buff; chin white; primaries and tail brown; irides dark brown.

Sp. 260. POËPHILA MIRABILIS, *Homb. et Jacq.*

BEAUTIFUL GRASS-FINCH.

Poëphila mirabilis, Homb. et Jacq. Voy. au Pôle Sud., tab. 22. fig. 2.

Poephila mirabilis, Gould, Birds of Australia, fol., vol. iii. pl. 89.

Some ornithologists have entertained the opinion that the *P. mirabilis* and the *P. gouldiæ* were one and the same species; but that such is not the case has been proved by the researches of the late Mr. Elsey, who lived for some time at the Victoria River, surrounded by hundreds of both these birds. This gentleman found them breeding, and collected many examples, which, all carefully labelled, are now in the national collection. Some of the black-headed ones, or *P. gouldiæ*, are labelled "adult male;" one is marked "female obtained from the nest;" there are also red-headed specimens labelled "adult male" and "adult female," and young birds which are totally different in colouring, being nearly uniform olive, without markings of any kind. Mr. Elsey informed me that he often saw the two species associated in large flocks.

Crown of the head and cheeks of a beautiful carmine, bounded posteriorly by a narrow line of black; throat black; to this succeeds a band of pale blue, narrow on the throat and broad on the back of the neck; back and wings green, passing into yellow at the nape of the neck; breast crossed by a broad band of lilac, separated from the yellow of the abdomen by a narrow line of orange; rump and upper tail-coverts pale blue; quills brown; tail black; bill fleshy white, becoming redder at the tip; feet flesh-colour.

Sp. 261. POËPHILA ACUTICAUDA, *Gould*.

LONG-TAILED GRASS-FINCH.

Amadina acuticauda, Gould in Proc. of Zool. Soc., part vii. p. 143.

Poephila acuticauda, Gould, Birds of Australia, fol., vol. iii. pl. 80.

The specimens from which my description of this bird was taken are from the interesting collection placed in my hands by the late Mr. Bynoe, whose great perseverance and assiduity have enabled me to add many species to the fauna of Australia. Indeed many of the officers of the 'Beagle' will have their names handed down to posterity in consequence of the attention they paid to this branch of science, independently of the legitimate objects of their various expeditions; among others I may particularly allude to Mr. Charles Darwin, Captain Wickham, Captain Stokes, Mr. Dring, &c. Since the arrival of Bynoe's birds I have also received specimens from Port Essington, which, like their analogue the *Poëphila cincta* of the eastern coast, inhabit the open plains bordering streams, and feed on the seeds of various grasses and other plants.

I regret that so little information has been transmitted to me respecting the habits and economy of this beautiful species.

The sexes differ but little in outward appearance; the female is, however, rather less in size, is less strikingly marked, and has the two middle tail-feathers shorter than her mate.

Crown of the head and cheeks grey; upper and under surface of the body fawn-colour, becoming more delicate, and assuming a pinky hue on the abdomen; lores, throat, band across the rump, and tail jet-black; upper and under tail-coverts and thighs white; wings fawn-grey; bill and feet yellow.

Total length 5¾ inches; bill ⅜; wing 2⅝; tail 3¼; tarsi ½.

There are magnificent specimens in the British Museum—one a male, having the centre tail-feathers 6¼ inches in length.

Sp. 202.   POËPHILA PERSONATA, *Gould*.

MASKED GRASS-FINCH.

Poëphila personata, Gould in Proc. of Zool. Soc., part x. p. 18.

Poëphila personata, Gould, Birds of Australia, fol., vol. iii. pl. 91.

This beautiful and well-marked species of Grass-Finch is a native of the north-west coast of Australia, where several specimens were shot by Gilbert during an excursion from Port Essington towards the interior of the country, who states that it inhabits grassy meadows near streams, feeding on grass-seeds, &c. It was tolerably abundant, being congregated in flocks of from twenty to forty. When on the wing it utters a very feeble cry of *twit, twit, twit,* but at other times pours forth a drawn-out mournful note, like that of some of the other Grass-Finches.

The sexes are scarcely to be distinguished by their outward

appearance, both possessing the masked face; the female is, however, rather less in size, and her markings are not quite so brilliant or decided as those of the male.

Base of the bill surrounded by an irregular ring of deep velvety black; crown of the head, upper surface, and wings light cinnamon-brown; lower part of the abdomen banded with deep velvety black; lower part of the rump and under tail-coverts white; upper tail-coverts white, striped longitudinally with black on the outer side; tail deep blackish brown; irides of the old birds red, of the young birds dark brown; bill bright orange; legs and feet fleshy red.

Total length $3\frac{1}{2}$ inches; bill $\frac{3}{8}$; wing $2\frac{1}{4}$; tail 2; tarsi $\frac{9}{16}$.

In some specimens the upper and lower ridges of the bill are black, while in others the basal half only is orange, the remaining portion being brown.

Sp. 203.   POËPHILA LEUCOTIS, *Gould*.

WHITE-EARED GRASS-FINCH.

Poëphila leucotis, Gould in Proc. of Zool. Soc., part xiv. p. 106.

Poëphila leucotis, Gould, Birds of Australia, fol., vol. iii. pl. 92.

The present beautiful species of *Poëphila* is one of the novelties discovered during Dr. Leichardt's expedition from Moreton Bay to Port Essington; it was killed in the neighbourhood of the river Lynd by Gilbert, in whose Journal, under the date of June 3, 1845, I find the following remark:—"The most interesting circumstance that occurred to me to-day was the discovery of a new species of *Poëphila*, which is very nearly allied to the one from Port Essington (*P. personata*), but which differs from that bird in having the bill light yellowish horn-colour instead of orange, the irides dark brown, and the legs red; it is in every respect a true *Poëphila*, having the black face and throat, the black marks on the flanks, the lengthened tail-feathers, and the general plumage of a light brown; like the other members of the genus, it inhabits the open spots of country, and feeds on grass-seeds."

In addition to the differences pointed out by Gilbert, I may mention that it may also be distinguished from the *P. personata* by its white ear-coverts and by the black of the throat being bounded below, and the black marks on the flanks anteriorly, with white; the colouring of the upper surface is also a somewhat richer brown.

As is the case with the other members of the genus, the sexes of this species differ but little from each other.

Band crossing the forehead, lores, throat, and a large patch on each flank deep velvety black; ear-coverts, a narrow line beneath the black of the throat, and a space surrounding the black patch on the flanks white; crown of the head deep reddish chestnut; all the upper surface and wings dark cinnamon-brown; chest and abdomen pale vinous brown; upper and under tail-coverts white, the former margined externally with deep black; tail black; irides dark brown; feet red; bill yellowish horn-colour.

Total length $4\frac{3}{4}$ inches; bill $\frac{2}{8}$; wing $2\frac{1}{4}$; tail $2\frac{1}{4}$; tarsi $\frac{5}{8}$.

Sp. 204.   POËPHILA CINCTA, *Gould.*

Banded Grass-Finch.

Amadina cincta, Gould in Proc. of Zool. Soc., part iv. p. 105.

Poëphila cincta, Gould, Birds of Australia, fol., vol. iii. pl. 93.

This species is tolerably abundant on the Liverpool Plains, and the open country to the northward towards the interior. It occurs so rarely on the sea side of the ranges, that I only once met with it during my sojourn in New South Wales. It is doubtless a native of the great basin of the interior, where, like the *P. acuticauda*, *P. personata*, and *P. leucotis*, it frequents those parts of the open plains which abound in grasses, upon the seeds of which and other plants it mostly subsists. The range of this species is entirely unknown; I have never seen a specimen except from the localities above mentioned.

Crown of the head and back of the neck grey; ear-coverts and sides of the neck silvery grey; throat and lores black; back, chest, and abdomen chestnut-brown; wings the same, but darker; lower part of the body surrounded by a black band; tail-coverts white; tail, which is short when compared with other species of the genus, black; bill black; irides reddish brown; eyelash blackish brown; feet pink-red.

Total length 4½ inches; bill ⅜; wing 2⅛; tail 2¼; tarsi ⅝.

The female differs from her mate by all her markings being more obscurely defined.

Genus DONACOLA, *Gould*.

When the habits of the Australian Finches become fully known, I have no doubt that they will be found to differ considerably, and that the members of each division of them will exhibit as marked a difference in their economy as they do in their structure and markings.

The late Mr. Elsey informed me that the *Donacolæ* build, in low tea-trees overhanging water, a large spouted nest, with a small cavity, of dry bark of those trees and of *Pandanus*.

Sp. 205. **DONACOLA CASTANEOTHORAX**, *Gould*.

Chestnut-breasted Finch.

Amadina castaneothorax, Gould in Syn. Birds of Australia, part ii.

Donacola castaneothorax, Gould, Birds of Australia, fol., vol. iii. pl. 94.

I had not the good fortune to meet with this bird in a state of nature, but I have been informed that it frequents reed-beds bordering the banks of the rivers and lagoons of the eastern coast, and that it much resembles the Bearded Tit

(*Calamophilus biarmicus*) of Europe, in the alertness with which it passes up and down the upright stems of the reeds, from the lower part to the very top, a habit for which the lengthened and curved form of its claws seems well adapted.

The sexes appear to differ but little in colouring; in some individuals, however, the cheeks and throat are black instead of brown, a character doubtless dependent on age or season.

I have not as yet seen this bird from the northern or western coast.

Crown of the head and back of the neck grey, the centre of each feather being brown; cheeks, throat, and ear-coverts blackish brown in some specimens, each feather slightly tipped with pale buff; upper surface and wings reddish brown; upper tail-coverts orange; tail brown, margined with paler brown; across the chest a broad band of pale chestnut, bounded below by a line of black, which gradually widens towards the flanks, along which it is continued for some distance; the remainder of the feathers on this part white, with a spot of blackish brown at the extremity of each; abdomen white; thighs black; under tail-coverts white, with a spot of blackish brown at the extremity of each; bill black; feet brown.

Total length 4 inches; bill $\frac{1}{4}$; wing 2$\frac{1}{4}$; tail 1$\frac{1}{4}$; tarsi $\frac{1}{2}$.

Sp. 266.   DONACOLA PECTORALIS, *Gould*.

WHITE-BREASTED FINCH.

Amadina pectoralis, Gould in Proc. of Zool. Soc., part viii. p. 127.

Donacola pectoralis, Gould, Birds of Australia, fol., vol. iii. pl. 95.

For two beautiful specimens of this entirely new Finch I am indebted to E. Dring, Esq., of the Beagle, who procured them on the north-west coast of Australia: no notes of their habits or economy having been forwarded with the specimens, I am unable to give any particulars respecting them.

In structure and in the general disposition of its markings, the White-breasted Finch offers a considerable resemblance to the *Donacola castaneothorax* of the eastern coast, and in all probability they are analogues of each other, in accordance with a law which appears very generally to prevail among the birds of Australia; each great division of this vast country having its own peculiar species.

Crown of the head, all the upper surface, and wings delicate greyish brown; the tips of the wing-coverts very minutely spotted with white; tail blackish brown; throat and ear-coverts glossy blackish purple; chest crossed by a band of feathers, black at the base, largely tipped with white; abdomen and under tail-coverts vinous grey; flanks ornamented with a few feathers similar to those crossing the breast; bill bluish horn-colour; feet flesh-colour.

Total length 4½ inches; bill ½; wing 2¼; tail 1¾; tarsi ½.

Genus MUNIA, *Hodgson*.

This genus has been established for the *Loxia malacca* of Linnæus, to which may be added the *Loxia ferruginea* of Sparrmann, and the Australian bird to which I gave the name of *Donacola flaviprymna*.

Sp. 207.   MUNIA FLAVIPRYMNA, *Gould*.

YELLOW-RUMPED FINCH.

Donacola flaviprymna, Gould in Proc. of Zool. Soc., part xiii. p. 80.
Dermophrys flaviprymnus, Cab. Mus. Hein., Theil i. p. 174, note.

Donacola flaviprymna, Gould, Birds of Australia, fol., vol. III. pl. 98.

A single specimen, and the only one I have ever seen of this pretty Finch, was presented to me by the late Mr. Bynoe, who procured it on the banks of the Victoria River during the late surveying voyage of H.M.S. Ship Beagle. It is very nearly allied to the *Donacola castaneothorax*, but is specifically distinct from that as well as from every other known species of this now numerous tribe of birds. I regret to add that nothing whatever is known of its habits or mode of life; but in these respects it doubtless as closely assimilates to its congeners as it does in form.

Head pale fawn-colour; back and wings light chestnut-brown; under surface buff; upper tail-coverts wax-yellow; under tail-coverts black; tail brown.

Total length 4½ inches; bill ¼; wing 2¼; tail 1¾; tarsi ¾.

Genus EMBLEMA, *Gould*.

The bird to which I have assigned the above generic designation differs from all the other Finches in its lengthened and pointed bill, and in the character and disposition of its markings.

Sp. 268.   EMBLEMA PICTA, *Gould*.

Painted Finch.

Emblema picta, Gould in Proc. of Zool. Soc., part x. p. 17.

Emblema picta, Gould, Birds of Australia, fol., vol. iii. pl. 87.

This beautiful Finch is a native of the north-west coast of Australia, where it was procured by the late Mr. Bynoe. The single individual sent me by that gentleman was unaccompanied by any account whatever of its habits and economy; but we may reasonably infer from the lengthened and pointed form of its bill, that the kind of food upon which it subsists will be somewhat different from that of the other Australian Finches. The disposition of the colouring of the present bird is very singular, the under parts being extremely beautiful, while on the upper, which is generally the most highly ornamented, a more than ordinary degree of plainness prevails.

The example of this beautiful bird above mentioned, which was presented to me by Bynoe, is, I believe, all that has ever been seen; I regret to say it no longer graces my collection, having been stolen therefrom, together with some other valuable birds, in the year 1846; and up to the present time the bird has not been again discovered.

Face and throat deep vermilion red; the base of all the feathers of the throat black, giving that part a mingled appearance of black and red; crown of the head, all the upper surface, and wings brown; rump deep vermilion red; tail dark brown; chest and all the under surface jet-black, the flanks numerously spotted with white, and the centre of the abdomen dashed with deep vermilion red; feet light red; upper mandible black, under mandible scarlet, with a triangular patch of black at the base.

Total length $3\frac{1}{4}$ inches; bill $\frac{9}{16}$; wing $2\frac{1}{4}$; tail $1\frac{3}{8}$; tarsi $\frac{3}{4}$.

Family MERULIDÆ.

Genus PITTA, *Vieillot*.

The members of this genus extend from India, throughout the islands of the Indian Archipelago to New Guinea and Australia: one species also occurs in Africa.

Sp. 269.   PITTA STREPITANS, *Temm.*

Noisy Pitta.

Pitta strepitans, Temm. Pl. Col., 333.
—— *versicolor*, Swains. in Zool. Journ., vol. i. p. 468.
Brachyurus strepitans, Bonap. Consp. Gen. Av., tom. 1. p. 254, *Brachyurus*, sp. 5.
Coloburis strepitans, Cab. et Hein. Mus. Hein., Theil ii. p. 3.

Pitta strepitans, Gould, Birds of Australia, fol., vol. iv. pl. 1.

This species inhabits the eastern coast of Australia, and is tolerably abundant between the river Macquarrie and Moreton Bay. Specimens from Cape York are smaller in all their admeasurements; but the differences, I think, are too trivial to be regarded as specific. It is said to be very Thrush-like in its habits and disposition, and, as its long legs would lead us to suppose, to resort much to the ground, but to take readily to the branches of trees when its haunts are intruded upon. Its food consists of insects, and probably berries, fruits, and snails.

Since my account of this species was printed in the folio edition, I have received its eggs, accompanied by the following notes from the late F. Strange of Sydney:—

"I never saw any bird whose actions are more graceful than those of the *Pitta strepitans*, when seen in its native brushes, where its presence is indicated by its singular call, resembling the words '*want a watch*,' by imitating which you can call it close to the muzzle of your gun; no sooner, however, does it commence breeding, than it becomes shy and retiring, keeping out of sight in the most artful manner, moving about from place to place, and occasionally uttering its cry until it has drawn you away from the nest. The nests I have seen were generally placed in the spur of a fig-tree, sometimes near the ground, and were outwardly constructed of sticks and lined with moss, leaves, and fine pieces of bark; the eggs are four in number," of a pale creamy-white, marked all over with irregularly-shaped blotches of brown and deep vinous grey, the latter appearing as if beneath the surface of the shell; they are one inch and a quarter in length by seven-eighths of an inch in breadth.

The sexes present but little differences either in colour or size; some specimens, which I take to be males, however, have the tail-feathers more largely tipped with green than others.

Crown deep ferruginous, with a narrow stripe of black down the centre; on the chin a large spot of black, terminating in a point on the front of the neck, and uniting to a broad band on each side of the head, encircles the crown, and terminates in a point at the back of the neck; back and wings pure olive-green; shoulders and lesser wing-coverts bright metallic cærulean blue; across the rump a band of the same colour; upper tail-coverts and tail black, the latter tipped with olive-green; primaries black, becoming paler at the tips; at the base of the fourth, fifth, and sixth a small white spot; sides of the neck, throat, breast, and flanks buff; in the centre of the abdomen a patch of black; vent and under tail-coverts scarlet; irides dark brown; bill brown; feet flesh-colour.

The young, like those of the Kingfisher, assume the characteristic plumage of the adult from the time they leave the nest.

Sp. 270. PITTA IRIS, *Gould*.
RAINBOW PITTA.

Pitta iris, Gould in Proc. of Zool. Soc., part x. p. 17.

Pitta iris, Gould, Birds of Australia, fol., vol. iv. pl. 3.

The Rainbow Pitta inhabits the Cobourg Peninsula, and will doubtless, hereafter, be found to range over a great portion of the northern part of the country. No further account of this fine bird has been received than that it frequents the thick "cane-beds" near the coast, through which it runs with great facility, and that the boldness and richness of its markings render it a most attractive object in the bush.

Head, neck, breast, abdomen, flanks, and thighs deep velvety black; over the eye, extending to the occiput, a band of ferruginous brown; upper surface and wings golden green; shoulders bright metallic cærulean blue, bordered below with lazuline blue; primaries black, passing into olive-brown at their tips, the third, fourth, fifth, and sixth having a spot about the centre of the feather; tail black at the base, green at the tip, the former colour running on the inner web nearly to the tip; rump-feathers tinged with cærulean blue; lower part of the abdomen and under tail-coverts bright scarlet, separated from the black of the abdomen by yellowish brown; irides dark brown; bill black; feet flesh-colour.

Total length 7 inches; bill 1⅛; wing 4; tail 1¾; tarsi 1½.

Genus CINCLOSOMA, *Vigors* and *Horsf.*

Among the novelties comprised in the present work, there are none more important than the additional members of this genus; four well-defined species being described, of which only one was previously known. The form is peculiar to Australia.

Sp. 271. CINCLOSOMA PUNCTATUM, *Vig. and Horsf.*

SPOTTED GROUND-THRUSH.

Turdus punctatus, Lath. Ind. Orn., Supp. p. xliv.
Punctated Thrush, Lath. Gen. Syn. Supp., vol. ii. p. 187.
Cinclosoma punctatum, Vig. and Horsf. in Linn. Trans., vol. xv. p. 220.

Cinclosoma punctatum, Gould, Birds of Australia, fol., vol. iv. pl. 4.

The *Cinclosoma punctatum* is a stationary species, and is distributed over the whole of Tasmania and the eastern portion of Australia, from Moreton Bay to Spencer's Gulf.

It gives a decided preference to the summits of low stony hills and rocky gullies, particularly those covered with scrubs and grasses. Its flight is very limited, and this power is rarely employed, except for the purpose of crossing a gully or passing to a neighbouring scrub; it readily eludes pursuit by the facility with which it runs over the stony surface and conceals itself among the underwood. When suddenly flushed it rises with a loud burring noise, like a Quail or Partridge. Its short flight is performed by a succession of undulations, and is terminated by the bird pitching abruptly to the ground almost at right angles.

It seldom perches on the smaller branches of trees, but may

be frequently seen to run along the fallen trunks so common in the Australian forests.

Unlike many others of the Thrush family which are celebrated for their song, the note of this species merely consists of a low piping whistle, frequently repeated while among the underwood, and by which its presence is often indicated.

In Hobart Town it is frequently exposed for sale in the markets with Bronzewing Pigeons and Wattlebirds, where it is known by the name of Ground-Dove, an appellation which has doubtless been given both from its habit of running and feeding upon the ground like the Pigeons, and the circumstance of its flesh being very delicate eating; to its excellence in this respect I can bear testimony. The pectoral muscles are very largely developed, and the body, when plucked, has much the contour of a Quail.

The duty of incubation is performed in October and the three following months, during which period two and often three broods are produced. The nest is a slight and rather careless structure, composed of leaves and the inner bark of trees, and is of a round open form; it is always placed on the ground, under the shelter of a large stone, stump of a tree, or a tuft of grass. The eggs are two, and sometimes three, in number, one inch and three lines long, and are white, blotched with large marks of olive-brown, particularly at the larger end, some of the spots appearing as if on the inner surface of the shell. The young, which at two or three days old are thickly clothed with long black downy feathers, soon acquire the power of running, and at an early age assume the plumage of the adult, after which they are subject to no periodical change in their plumage. The stomach is very muscular, and in those dissected were found the remains of seeds and caterpillars mingled with sand.

Adult males have the forehead and chest ash-grey; crown of the head, back, rump, and the middle tail-feathers rufous-brown, each feather of the back having a broad longitudinal stripe of black down the centre; shoulders and wing-coverts steel-black, each feather having a spot of white at the extreme tip; primaries blackish-brown, margined on their outer edges with lighter brown; throat and a narrow band across the chest steel-black; stripe over the eye, a nearly circular spot on the side of the neck, and the centre of the abdomen white; flanks and under tail-coverts reddish-buff, with a large oblong stripe of black down the centre of each feather; lateral tail-feathers black, broadly margined with grey on their inner webs, and largely tipped with white; bill black; legs fleshy-white; feet darker; eyes very dark lead colour, with a naked blackish-brown eyelash. The female differs from the male in having all the upper surface of a lighter hue; the throat greyish-white instead of black; the spot on the neck rufous instead of white, and in being destitute of the black pectoral band.

Sp. 272. CINCLOSOMA CASTANEONOTUM, *Gould*.

CHESTNUT-BACKED GROUND-THRUSH.

Cinclosoma castanotus, Gould in Proc. of Zool. Soc., part viii. p. 113.
—— *castanotum*, Cab. Mus. Hein., Theil i. p. 85.
Boent-Yong, Aborigines of the mountain districts of Western Australia.

Cinclosoma castanotus, Gould, Birds of Australia, fol., vol. iv. pl. 5.

The habits and economy of the present bird closely resemble those of the Spotted Ground-Thrush; but the more level plains, particularly those that are studded with clumps of dwarf trees and scrubs, would appear to be the situations for which it is more peculiarly adapted, at least such was the character of the country in the Belts of the Murray where I discovered it. On the other hand, it is stated in the notes accompanying specimens received from Swan River, that "it is rarely seen in any but the most barren and rocky places. The white gum forests, here and there studded with small patches of scrub, are its favourite haunts. It is only found in the interior; the part nearest to the coast, where it has been observed, being Dank's Hutts on the York Road about fifty-three miles from Fremantle."

Its disposition is naturally shy and wary, a circumstance which cannot be attributed to any dread of man as an enemy, since it inhabits parts scarcely ever visited either by the natives or Europeans. Few persons, I may safely say, had ever discharged a gun in that rich arboretum, the Belts of the Murray, before the period of my being there; still the bird was so difficult of approach, that it required the utmost exertion to procure specimens. They were generally observed in small troops of four or six in number, running through the scrub one after another in a line, and resorting to a short low flight, when crossing the small intervening plains. It runs over the surface of the ground with even greater facility than *C. punctatum*.

In its mode of flight and nidification it assimilates so closely to the Spotted Ground-Thrush, as to render a separate description superfluous.

The stomach is extremely muscular, and the food consists of seeds and the smaller kind of *Coleoptera*.

The male has the crown of the head, ear-coverts, back of the neck, upper part of the back, upper tail-coverts, and two central tail-feathers brown; stripe over the eye and another from the base of the lower mandible down the side of the neck white; scapularies and lower part of the back rich chestnut; shoulders and wing-coverts black, each feather having a spot of white at the tip; primaries and secondaries dark brown, margined with lighter brown; lateral tail-feathers black, largely tipped with white; chin, throat, and centre of the breast steel black; sides of the chest and flanks brownish grey, the latter blotched with black; centre of the abdomen and under tail-coverts white; bill black; base of the under mandible lead colour; irides reddish hazel; legs blackish brown. The female differs in having the whole of the plumage much lighter, and with only a slight tinge of chestnut on the rump; the stripes of white over the eye and down the sides of the neck less distinctly marked; the chin, throat, and breast grey instead of black; the irides hazel, and the feet leaden brown.

Total length 9 inches; bill 1; wing 4¼; tail 4½; tarsi 1¼.

Sp. 273. CINCLOSOMA CINNAMOMEUM, *Gould*.

CINNAMON-COLOURED CINCLOSOMA.

Cinclosoma cinnamomeus, Gould in Proc. of Zool. Soc., part xiv. p. 68.

Cinclosoma cinnamomeus, Gould, Birds of Australia, fol., vol. iv. pl. 6.

For our knowledge of this new *Cinclosoma* we are indebted to the researches of that enterprising traveller Captain Sturt, who procured a single specimen during his lengthened sojourn at the Depôt in that sterile and inhospitable region, the interior of South Australia. Since that date many other examples have been sent to Europe, which have been collected in other parts of the country.

It is considerably smaller than either of its congeners, the *C. castanonotum*, *C. punctatum*, and *C. castaneothorax*, and, moreover, differs from them in the cinnamon colouring of the greater portion of its plumage.

The female differs from the opposite sex in the absence of the black markings of the throat, breast, and wings, those parts being brownish grey.

The whole of the upper surface, scapularies, two central tail-feathers, sides of the breast, and flanks cinnamon-brown; wing-coverts jet-black, each feather largely tipped with white; above the eye a faint stripe of white; lores and throat glossy black, with a large oval patch of white seated within the black, beneath the eye; under surface white, with a large arrow-shaped patch of glossy black on the breast; feathers on the sides of the abdomen with a broad stripe of black down the centre; lateral tail-feathers jet-black, largely tipped with pure white; under tail-coverts black for four-fifths of their length on the outer web, their inner webs and tips white; eyes brown; tarsi olive; toes black.

Total length 7½ inches; bill ⅞; wing 3¾; tail 3¼; tarsi 1⅜.

Sp. 274. CINCLOSOMA CASTANEOTHORAX, *Gould*.

CHESTNUT-BREASTED GROUND THRUSH.

Cinclosoma castaneothorax, Gould in Proc. of Zool. Soc., 1848, p. 139, Aves, pl. 6.

—— *castaneithorax*, Bonap. Consp. Gen. Av., p. 278, *Cinclosoma*, sp. 4.

Cinclosoma castaneothorax, Gould, Birds of Australia, fol. vol. Supplement, pl.

For a knowledge of this richly coloured and very distinct species of Ground-Thrush science is indebted to Charles Coxen, Esq., of Brisbane, who discovered it in the scrubby belts of trees growing on the table-land to the northward of the Darling Downs. In size it nearly equals the *Cinclosoma castaneonotum*, but differs from that bird in the buffy stripe over the eye, in the colouring of the back, and in the band of chestnut-brown which crosses the breast. To my regret, only a single male specimen has yet been forwarded to me; I trust, however, that through Mr. Coxen or some other lover of ornithology I may ere long be favoured with an example of the female.

Crown of the head, ear-coverts, back of the neck, and upper tail-coverts brown; stripe over the eye and another from the base of the lower mandible, down the side of the neck, white; shoulders and wing-coverts black, each feather with a spot of white at the tip; all the upper surface, the outer margins of the scapularies and a broad longitudinal stripe on their inner webs next the shaft deep rust-red; primaries, secondaries, and the central portion of the scapularies dark brown; tail black, all but the two central feathers largely tipped with white; chin and throat black; chest crossed by a band of rich rust-red; sides of the chest and flanks brownish-grey, the latter blotched with black; centre of the abdomen white; under tail-coverts brown, deepening into black near the tip, and margined with white; bill and feet black.

Total length 8½ inches; bill 1; wing 4; tail 4½; tarsi 1.

Genus OREOCINCLA, *Gould*.

Species of this genus inhabit India, the Indian Islands, and Australia, in which latter country, although much difference in size is observable in specimens from different localities, I believe only one exists. It is decidedly a brush bird, and has many habits in common with the typical Thrushes, but is more shy and retiring.

Sp. 275. OREOCINCLA LUNULATA.

MOUNTAIN-THRUSH.

Turdus lunulatus, Lath. Ind. Orn., Supp. p. xlii.
Lunulated Thrush, Lath. Gen. Syn. Supp., vol. ii. p. 184.
—— *Honey-eater*, Lath. Gen. Hist., vol. iv. p. 180.
Turdus varius, Vig. and Horsf. in Linn. Trans., vol. xv. p. 218.
Oreocincla novæ-hollandiæ et *O. macrorhyncha*, Gould in Proc. of Zool. Soc., part v. p. 145.
Mountain Thrush, Colonists of Tasmania.

Oreocincla lunulata, Gould, Birds of Australia, fol., vol. iv. pl. 7.

In all localities suitable to its habits and mode of life this species is tolerably abundant, both in Tasmania and in New South Wales; it has also been observed in South Australia, where however it is rare. From what I saw of it personally, I am led to infer that it gives a decided preference to thick mountain forests, where large boulder stones occur covered with green moss and lichens, particularly if there be much humidity; rocky gulleys and the sides of water-courses are also among its favourite places of resort. In Tasmania, the slopes of Mount Wellington, and other similar bold elevations are situations in which it may always be seen if closely looked for. During the summer it ascends high up the mountain sides, but in winter it descends to the lower districts, the outskirts of the forests, and occasionally visits the gardens of the settlers. In New South Wales, the Cedar Brushes of the Liverpool range and all similar situations are frequented by it; I also observed it on the islands at the mouth of the Hunter; and I possess specimens from the north shore near Sydney and the banks of the Clarence. Its chief food is helices and other mollusks, to which insects of many kinds are added; and it is most likely that fruits and berries occasionally form a part of its diet. It is a solitary species, more than two being rarely observed together, and frequently a single individual only is to be seen, noiselessly hopping over the rugged ground in search of food. Its powers of flight are seldom exercised, and so far as I am aware it has no song. Considerable variation exists in the size and colouring of individuals from different districts. The Tasmanian specimens are larger, and have the bill more robust, than those from New South Wales; considerable difference also exists in the lunations at the tip of the feathers, some being much darker and more distinctly defined than others. The young assume the plumage of the adults from the nest, but have the lunations paler and the centre of the feathers of the back bright tawny instead of olive-brown.

The Mountain-Thrush breeds in many of the localities above-mentioned during the months of August, September, and October, the nest being placed on the low branches of the trees, often within reach of the hand; those I saw were outwardly formed of green moss and lined with fine crooked black fibrous roots, and were about seven inches in diameter by three inches in depth; the eggs are of a buffy white or stone-colour, minutely freckled all over with reddish brown, about one inch and three-eighths long by seven-eighths broad.

The sexes are alike in plumage, and may be thus described:
The whole of the upper surface olive-brown, each feather with a lunar-shaped mark of black at the tip; wings and tail olive-brown, the former fringed with yellowish olive and the outer feather of the latter tipped with white; under surface white, stained with buff on the breast and flanks, each feather, with the exception of those of the centre of the abdomen and the under tail-coverts, with a lunar-shaped mark of black at the tip, narrow on the breast and abdomen, and broad on the sides and flanks; irides very dark brown; bill horn-colour, becoming yellow on the base of the lower mandible; feet horn-colour.

Family PARADISEIDÆ?

I certainly consider the following accounts of the extraordinary habits of the *Ptilonorhynchi* and *Chlamyderæ* as some of the most valuable and interesting portions of my work; and, however incredible they may appear, they have been fully confirmed by specimens of the *Ptilonorhynchus holosericeus* having constructed their bowers in the Gardens of the Zoological Society of London, and by the observation of other persons in Australia. These, with the genera *Ailuroedus* and *Sericulus*, appear to me to constitute a very natural group, and to be nearly allied to the *Paradiseidæ*.

Genus PTILONORHYNCHUS, *Kuhl.*

Of this genus I am acquainted with only a single species, the well-known Satin-Bird of the colonists.

Sp. 270. PTILONORHYNCHUS HOLOSERICEUS, *Kuhl.*

SATIN BOWER-BIRD.

Ptilonorhynchus holosericeus, Kuhl, Beytr. zur Zool., S. 150.
Pyrrhocorax violaceus, Vieill. Nouv. Dict. d'Hist. Nat., tom. vi. p. 569.
Kitta holosericea, Temm. Pl. Col., 395 and 422.
Satin Grakle, Lath. Gen. Hist., vol. iii. p. 171.
Ptilonorhynchus macleayii, Lath. MSS., Vig. and Horsf. in Linn. Trans., vol. xv. p. 263.
Corvus squamulosus, Ill., female or young?
Ptilonorhynchus squamulosus, Wagl. Syst. Av., sp. 2, female or young?
Ptilorhynchus holosericeus, Cab. Mus. Hein., Theil i. p. 213.
Satin Bird of the Colonists of New South Wales.
Cowry of the Aborigines of the coast of New South Wales.

Ptilonorhynchus holosericeus, Gould, Birds of Australia, fol., vol. iv. pl. 10.

Although this species had been long known to ornithologists and to the colonists of New South Wales, its extraordinary habits had never been brought before the scientific world until I had the gratification of publishing an account of them after my return from Australia.

The localities frequented by the Satin Bower-bird are the luxuriant and thickly-foliaged brushes stretching along the coast from Port Philip to Moreton Bay, and the cedar-brushes of the Liverpool range. So far as is at present known, it is restricted to New South Wales; certainly it is not found so far to the westward as South Australia, and I am not aware of its having been seen on the north coast; but its range in that direction can only be determined by future research.

It is a stationary species, but appears to roam from one part of a district to another, either for the purpose of varying the nature, or of obtaining a more abundant supply of food. Judging from the contents of the stomachs of the many specimens I dissected, it would seem that it is altogether frugivorous, or if not exclusively so, that insects form but a small portion of its diet. Independently of numerous berry-bearing plants and shrubs, the brushes it inhabits are studded with enormous fig-trees, to the fruit of which it is especially partial. It appears to have particular times in the day for feeding, and when thus engaged among the low shrub-like trees, I have approached within a few feet without creating alarm; but at other times the bird was extremely shy and watchful, especially the old males, which not unfrequently perch on the topmost branch or dead limb of the loftiest tree in the forest, whence they can survey all round, and watch the movements of their females and young in the brush below.

In the autumn they associate in small flocks, and may often be seen on the ground near the sides of rivers, particularly where the brush descends in a steep bank to the water's edge.

The extraordinary bower-like structure, alluded to in my remarks on the genus, first came under my notice in the Sydney Museum, to which an example had been presented by Charles Coxen, Esq., of Brisbane, as the work of the Satin Bower-bird. This so much interested me that I determined to leave no means untried for ascertaining every particular relating to this peculiar feature in the bird's economy; and on visiting the cedar-brushes of the Liverpool range, I discovered several of these bowers or playing-places on the ground, under the shelter of the branches of overhanging trees, in the most retired part of the forest: they differed considerably in size, some being a third larger than others. The base consists of an extensive and rather convex platform of sticks firmly interwoven, on the centre of which the bower itself is built: this, like the platform on which it is placed, and with which it is interwoven, is formed of sticks and twigs, but of a more slender and flexible description, the tips of the twigs being so arranged as to curve inwards and nearly meet at the top: in the interior the materials are so placed that the forks of the twigs are always presented outwards, by which arrangement not the slightest obstruction is offered to the passage of the birds. The interest of this curious bower is much enhanced by the manner in which it is decorated with the most gaily-coloured articles that can be collected, such as the blue tail-feathers of the Rose-hill and Pennantian Parrakeets, bleached bones, the shells of snails, &c.; some of the feathers are inserted among the twigs, while others with the bones and shells are strewed about near the entrances. The propensity of these birds to fly off with any attractive object, is so well known to the natives, that they always search the runs for any small missing article that may have been accidentally dropped in the brush. I myself found at the entrance of one of them a small neatly-worked stone tomahawk, of an inch and a half in length, together with some slips of blue cotton rags, which the birds had doubtless picked up at a deserted encampment of the natives.

It has now been clearly ascertained that these curious bowers are merely sporting-places in which the sexes meet, and the males display their finery, and exhibit many remarkable actions; and so inherent is this habit, that the living examples, which have from time to time been sent to this country, continue it even in captivity. Those belonging to the Zoological Society have constructed their bowers, decorated and kept them in repair, for several successive years.

In a letter received from the late F. Strange, he says—

"My aviary is now tenanted by a pair of Satin-birds, which for the last two months have been constantly engaged in constructing bowers. Both sexes assist in their erection, but the male is the principal workman. At times the male will chase the female all over the aviary, then go to the bower, pick up a gay feather or a large leaf, utter a curious kind of note, set all his feathers erect, run round the bower, and become so excited that his eyes appear ready to start from his head, and he continues opening first one wing and then the other, uttering a low whistling note, and, like the domestic Cock, seems to be picking up something from the ground, until at last the female goes gently towards him, when, after two turns round her, he suddenly makes a dash, and the scene ends."

I regret to state, that although I have used my utmost endeavours, I could never discover the nest and eggs of this species, neither could I obtain any authentic information respecting them, either from the natives or the colonists.

The adult male has the whole of the plumage of a deep shining blue-black, closely resembling satin, with the exception of the primary wing-feathers, which are of a deep velvety black, and the wing-coverts, secondaries, and tail-feathers, which are also of a velvety black, tipped with the shining blue-black lustre; irides beautiful light blue with a circle of red round the pupil; bill bluish horn, passing into yellow at the tip; legs and feet yellowish white.

The female has the head and all the upper surface greyish green; wings and tail dark sulphur-brown, the inner webs of the primaries being the darkest; under surface containing the same tints as the upper, but very much lighter, and with a wash of yellow; each feather of the under surface also has a crescent-shaped mark of dark brown near the extremity, giving the whole a scaly appearance; irides of a deeper blue than in the male, and with only an indication of the red ring; bill dark horn-colour; feet yellowish white tinged with olive.

Young males closely resemble the females, but differ in having the under surface of a more greenish-yellow hue, and the crescent-shaped markings more numerous; irides dark blue; feet olive brown; bill blackish olive.

Genus AILURŒDUS, *Cabanis.*

I quite agree with Dr. Cabanis in the propriety of instituting a new genus for the reception of the Cat-bird of Australia, inasmuch as it certainly differs from the Satin-bird in the structure of its bill and in the character and colouring of its plumage. A single species only inhabits Australia.

Sp. 277. AILURŒDUS SMITHII, *Vig. and Horsf.*
CAT-BIRD.

Varied Roller, Lath. Gen. Hist., vol. iii. p. 86.
Ptilonorhynchus smithii, Lath. MSS. Vig. and Horsf. in Linn. Trans., vol. xv. p. 264.
—— *viridis*, Wagl. Syst. Av., sp. 3.
Kitta virescens, Temm. Pl. Col., 396.
Ailurœdus smithi, Cab. Mus. Hein., Theil i. p. 213.
Cat-Bird of the Colonists of New South Wales.

Ptilonorhynchus smithii, Gould, Birds of Australia, fol., vol. iv. pl. 11.

So far as our knowledge extends, this species is only found in New South Wales, where it inhabits the luxuriant forests that extend along the eastern coast between the mountain ranges and the sea; those of Illawarra, the Hunter, the MacLeay, and the Clarence and the cedar brushes of the Liverpool range being, among many others, localities in which it may be found: situations suitable to the Regent- and Satin-Birds are equally adapted to the habits of the Cat-Bird, and I have not unfrequently seen them all three feeding together on the same tree. The wild fig, and the native cherry, when in season, afford an abundant supply. So rarely does it take insects, that I do not recollect ever finding any remains in the stomachs of those specimens I dissected. In its disposition it is neither a shy nor a wary bird, little caution being required to approach it, either when feeding or while quietly perched upon the lofty branches of the trees. It is at such times that its loud, harsh and extraordinary note is heard; a note which differs so much from that of all other birds, that having been once heard it can never be mistaken. In comparing it to the nightly concerts of the domestic cat, I conceive that I am conveying to my readers a more perfect idea of the note of this species than could be given by pages of description. This concert, is performed either by a pair or several individuals, and nothing more is required than for the hearer to shut his eyes to the neighbouring foliage to fancy himself surrounded by London grimalkins of house-top celebrity.

While in the district in which this bird is found, my attention was directed to the acquisition of all the information I could obtain respecting its habits, as I considered it very probable that it might construct a bower similar to that of the Satin-Bird; but I could not satisfy myself that it does, nor could I discover its nest, or the situation in which it breeds; it is doubtless, however, among the branches of the trees of the forest in which it lives.

The sexes do not offer the slightest difference in plumage, or any external character by which the male may be distinguished from the female; she is, however, rather less brilliant in her markings, and somewhat smaller in size.

Head and back of the neck olive-green, with a narrow line of white down each of the feathers of the latter; back, wings, and tail grass-green, with a tinge of blue on the margins of the back-feathers; the wing-coverts and secondaries with a spot of white at the extremity of their outer web; primaries black, their external webs grass-green at the base and bluish green for the remainder of their length; all but the two central tail-feathers tipped with white; all the under surface yellowish green, with a spatulate mark of yellowish white down the centre of each feather; bill light horn-colour; irides brownish red; feet whitish.

Genus CHLAMYDODERA, *Gould.*

Of this well-defined genus four very distinct species are now known: viz., *C. nuchalis* which frequents the northern parts of the country, *C. maculata* of the east coast, *C. cerviniventris* of Cape York, and *C. guttata* of the north-western districts.

Some parts of their economy are more astonishing than

those of *Ptilonorhynchus*. I allude more particularly to their bowers or playing-places, which are of no great size in the former case, but here attain their maximum so far as is known. These extraordinary playing-places have been a source of much speculation, and by some persons have been considered to be made by the Aborigines as cradles for their children; but it is now known that they are places of resort for both sexes of these birds at that season of the year when nature prompts them to reproduce their kind. Here the males meet and contend with each other for the favours of the females, and here also the latter assemble and coquet with the males. These highly decorated halls of assembly must, therefore, be regarded as the most wonderful instances of bird-architecture yet discovered. Those of my readers who are not acquainted with these curious structures will do well to refer to the drawings of them in the folio edition, for no description, however accurate, can convey an adequate idea of them. The bowers must not be confounded with their nests, which are made in the ordinary way among the branches of trees, and, as far as we yet know, assimilate very closely in size and form to that of the Jay of Europe, *Garrulus glandarius*.

Sp. 278.   CHLAMYDODERA NUCHALIS.

Great Bower-bird.

Ptilonorhynchus nuchalis, Jard. and Selb. Ill. Orn., vol. ii. pl. 103.
Calodera nuchalis, Gould, Syn. Birds of Australia, part i.
Chlamydera nuchalis, Gould, Birds of Australia, 1837, part i. cancelled.
Chlamydodera nuchalis, Cab. Mus. Hein., Theil i. p. 212.

Chlamydera nuchalis, Gould, Birds of Australia, fol., vol. iv. pl. 9.

This fine species was first described and figured in the "Illustrations of Ornithology," by Sir William Jardine and Mr. Selby, from the then unique specimens in the collection of the Linnean Society; but neither the part of Australia of which it is a native, nor any particulars relative to its habits were known to those gentlemen: it is now clearly ascertained that it is an inhabitant of the north-west coast, a portion of the Australian continent that has, as yet, been but little visited. I am indebted for individuals of both sexes to two of the officers of the 'Beagle,' Messrs. Dynoe and Dring; but neither of these gentlemen furnished me with any account of its economy.

The following passage from Captain Stokes's 'Discoveries in Australia,' vol. ii, p. 97, comprises all that has, as yet, been recorded respecting the curious bower constructed by this bird.

"I found matter for conjecture in noticing a number of twigs with their ends stuck in the ground, which was strewed over with shells, and their tops brought together so as to form a small bower; this was $2\frac{1}{2}$ feet long, $1\frac{1}{2}$ foot wide at either end. It was not until my next visit to Port Essington that I thought this anything but some Australian mother's toy to amuse her child; upon being asked, one day, to go and see the 'birds' playhouse,' I immediately recognized the same kind of construction I had seen at the Victoria River, and found the bird amusing itself by flying backwards and forwards, taking a shell alternately from each side, and carrying it through the archway in its mouth."

Head and all the upper surface greyish brown, the feathers of the former with a shining or satiny lustre; the feathers of the back, wing-coverts, scapulars, quills, and tail tipped with greyish white; on the nape of the neck a beautiful rose-pink fascia, consisting of narrow feathers, partly encircled by a ruff of satin-like plumes, the tips distinct, rounded, and turning inwards; under surface yellowish grey, the flanks tinged with brown; irides, bill, and legs brownish black.

In one of the specimens I possess no trace of the nuchal ornament is observable, a circumstance I conceive to be indicative of the female.

Sp. 279. CHLAMYDODERA MACULATA, *Gould*.

Spotted Bower-bird.

Calodera maculata, Gould in Proc. of Zool. Soc., part iv. p. 106.
Chlamydera maculata, Gould, Birds of Australia, 1837, part i. cancelled.
Chlamydodera maculata, Cab. Mus. Hein., Theil i. p. 212.

Chlamydera maculata, Gould, Birds of Australia, fol., vol. iv. pl. 8.

During my journey into the interior of New South Wales, I observed this bird to be tolerably abundant at Brezi on the river Mokai to the northward of the Liverpool Plains: it is also equally numerous in all the low scrubby ranges in the neighbourhood of the Namoi, as well as in the open brushes which intersect the plains on its borders; and collections from Moreton Bay generally contain examples; still from the extreme shyness of its disposition, the bird is seldom seen by ordinary travellers, and it must be under very peculiar circumstances that it can be approached sufficiently close to observe its colours. The Spotted Bower-bird has a harsh, grating, scolding note, which is generally uttered when its haunts are intruded on, and by which means its presence is detected when it would otherwise escape observation: when disturbed it takes to the topmost branches of the loftiest trees, and frequently flies off to another neighbourhood.

In many of its actions and in the greater part of its economy much similarity exists between this species and the Satin Bower-bird, particularly in the curious habit of constructing an artificial bower or playing-place. I was so far fortunate as to discover several of these bowers during my journey in the interior, the finest of which I succeeded in bringing to England; it is now in the British Museum. The situations of these runs or bowers are much varied; I found them both on the plains studded with Myalls (*Acacia pendula*) and other small trees, and in the brushes clothing the lower hills. They are considerably longer and more avenue-like than those of the Satin Bower-bird, being in many instances three feet in length. They are outwardly built of twigs, and beautifully lined with tall grasses, so disposed that their heads nearly meet; the decorations are very profuse, and consist of bivalve shells, crania of small mammalia and other bones bleached by exposure to the rays of the sun or from the camp-fires of the natives. Evident indications of high instinct are manifest throughout the whole of the bower and decorations formed by this species, particularly in the manner in which the stones are placed within the bower, apparently to keep the grasses with which it is lined fixed firmly in their places: these stones diverge from the mouth of the run on each side so as to form little paths, while the immense collection of decorative materials are placed in a heap before the entrance of the avenue, the arrangement being the same at both ends. In some of the larger bowers, which had evidently been resorted to for many years, I have seen half a bushel of bones, shells, &c., at each of the entrances. I frequently found these structures at a considerable distance from the rivers, from the borders of which they could alone have procured the shells and small round pebbly stones; their collection and transportation must therefore be a task of great labour. I fully ascertained that these runs, like those of the Satin Bowerbird, formed the rendezvous of many individuals; for, after secreting myself for a short space of time near one of them, I killed two males which I had previously seen running through the avenue.

The natives unhesitatingly state that the bird makes its nest in the high gum trees, and Mr. Charles Coxen of Brisbane found a nest of the *Chlamydodera maculata* with young birds in it some years ago on Oaky Creek near the present Jondaryan head station, on the Darling Downs; the nest was built in one of the *Myrtaceæ* overhanging a waterhole, near a scrub, on which a bower was built; and was in form very similar to that of the Common Thrush of Europe, being of a cup-shape, constructed of dried sticks with a slight lining of feathers and fine grass. The eggs are still unknown.

Crown of the head, ear-coverts, and throat rich brown, each feather surrounded with a narrow line of black; feathers on the crown small, and tipped with silvery grey; a beautiful band of elongated feathers of light rose-pink crosses the back of the neck, forming a broad, fan-like, occipital crest; all the upper surface, wings, and tail of a deep brown; every feather of the back, rump, scapularies, and secondaries tipped with a large round spot of rich buff; primaries slightly tipped with white; all the tail-feathers terminated with buffy white; under surface greyish white; feathers of the flanks marked with faint, transverse, zigzag lines of light brown; bill and feet dusky brown; irides dark brown; bare skin at the corner of the mouth thick, fleshy, prominent, and of a pinky flesh-colour.

I am in some doubt as to whether the female ever acquires the lilaceous mark at the back of the neck: for the first and perhaps the second year, she is certainly without it.

Total length $11\frac{1}{4}$ inches; bill $1\frac{1}{4}$; wing 6; tail $4\frac{3}{8}$; tarsi $1\frac{1}{8}$.

Sp. 280. CHLAMYDODERA GUTTATA, *Gould*.

GUTTATED BOWER-BIRD.

Chlamydera guttata, Gould in Proc. of Zool. Soc., 1862, p. 161.

I am indebted to the researches of T. F. Gregory, Esq., the West Australian explorer, for a knowledge of this new species. It was collected in North-western Australia, and is doubtless the bird which constructs the bowers described by Captain (now Sir George) Grey in his "Travels," vol. i. pp. 196 and 245, where he states, that on gaining the summit of one of the sandstone-ranges forming the watershed of the streams flowing into the Glenelg and Prince Regent's Rivers, "We fell in with a very remarkable nest, or what appeared to me to be such. We had previously seen several of them, and they had always afforded us food for conjecture as to the agent and purpose of such singular structures. This very curious sort of nest, which was frequently found by myself and other individuals of the party, not only along the sea-shore, but in some instances at a distance of six or seven miles from it, I once conceived must have belonged to a Kangaroo, until I was informed that it was the run or playing-place of a species of *Chlamydodera*. These structures were formed of dead grass and parts of bushes, sunk a slight depth into two parallel furrows in sandy soil, and then nicely arched above. But the most remarkable fact connected with them was, that they were always full of broken sea-shells, large heaps of which protruded from each extremity. In one instance, in a bower the most remote from the sea that we discovered, one of the men of the party found and brought to me the stone of some fruit which had evidently been rolled in the sea; these stones he found lying in a heap in the nest, and they are now in my possession."

The bird sent to me by Mr. Gregory is rather larger, but bears a very general resemblance to the *Chlamydodera maculata*, being spotted all over like that species; but it differs in the guttations of the upper surface being of a larger size and much more distinct, in the abdomen being buff, and in the shafts of the primaries being of a richer yellow. In all probability, the specimen is a female, since there is no trace of the beautiful lilaceous nuchal mark seen in the males only of *C. maculata* and *C. nuchalis*. Since Mr. Gregory discovered this interesting bird, Mr. Stuart, as all the world knows, has crossed the continent of Australia from Adelaide to the Victoria River; and that he met with this bird in some part of his journey is shown by his having kindly left at my house the head of a male adorned with fine lilaceous feathers at the back of the neck like *C. nuchalis* and *C. maculata*.

General tint of the upper surface and wings deep brownish black, with a spot of rich buff at the tip of each feather, those of the head and nape being very small, while those on the body and wings are of a large size, accordant in fact with the increased size of the feathers; the spots on the tips of the wing-feathers are not so round as those on the back; the primaries are very pale brown, fading into white on the basal portion of their inner webs, which is yellow on their under surface; their shafts straw-yellow; these feathers are much worn, and are doubtless tipped with white in fresh moulted specimens; tail-feathers pale brown, with buff shafts and white tips; throat-feathers brown at the base, with an arrow-head-shaped mark of pale buff at the tip of each, the buff tips becoming much larger on the chest; centre of the abdomen pale buff; flanks, thighs, and under tail-coverts buff, barred with light brown; bill black; gape rich yellow; feet apparently very dark olive.

Total length 11¼ inches; bill 1½; wing 6; tail 4¼; tarsi 1¾.

Sp. 281. CHLAMYDODERA CERVINIVENTRIS, *Gould*.

Fawn-breasted Bower-bird.

Chlamydera cerviniventris, Gould in Proc. of Zool. Soc., part xviii. p. 201.

Chlamydera cerviniventris, Gould, Birds of Australia, fol., Supplement, pl.

The discovery of the present species is due Mr. Macgillivray, who procured a specimen at Cape York, which with its curious bower he transmitted to the British Museum. Other examples have since been procured, but none are adorned with the lovely frill of liliaceous feathers at the nape of the neck although I believe some of them are very old birds. In size this species is rather larger than *C. maculata*, or almost intermediate between that species and *C. nuchalis*; its distinguishing feature is its rich, uniformly-coloured, buff under surface. Its bower differs from those of the other species; its walls, which are very thick, being nearly upright, or but little inclining towards each other at the top, so that the passage through is very narrow; it is formed of fine twigs, is placed on a very thick platform of thicker twigs, is nearly 4 feet in length and almost as much in breadth, and has here and there a small snail-shell or berry dropped in as a decoration.

The following note relative to this bird is extracted from Mr. Macgillivray's "Narrative of the Voyage of H.M.S. Rattlesnake:"—

"Two days before we left Cape York, I was told that some Bower-birds had been seen in a thicket or patch of low scrub, half a mile from the beach; and after a long search I found a recently-constructed bower, 4 feet long and 18 inches high, with some fresh berries lying upon it. The bower was situated near the border of the thicket, the bushes composing which were seldom more than 10 feet high, growing in smooth sandy soil without grass.

"Next morning I was landed before daylight, and proceeded to the place in company with Paida, taking with us a large board on which to carry off the bower as a specimen. I had great difficulty in inducing my friend to accompany me, as he was afraid of a war party of Gomokudins, which tribe had lately given notice that they were coming to fight the Evans Bay people. However, I promised to protect him, and loaded one barrel with ball, which gave him increased confidence; still he insisted upon carrying a large bundle of spears and a throwing-stick.

"While watching in the scrub, I caught several glimpses of the *tewinga* (its native name) as it darted through the bushes in the neighbourhood of the bower, announcing its presence by an occasional loud *churr-r-r*, and imitating the notes of various other birds, especially the *Tropidorhynchus*. I never before met with a more wary bird; and, for a long time, it enticed me to follow it to a short distance, then flying off and alighting on the bower it would deposit a berry or two, run through and be off again before I could reach the spot. All this time it was impossible to get a shot. At length, just as my patience was becoming exhausted, I saw the bird enter the bower and disappear, when I fired at random through the twigs, fortunately with effect. So closely had we concealed ourselves latterly, and so silent had we been, that a kangaroo, while feeding, actually hopped up within fifteen yards, unconscious of our presence until fired at."

Upper surface brown, each feather of the back and wings margined and marked at the tip with buffy white; throat striated with greyish brown and buff; under surface of the shoulder, abdomen, thighs, and under tail-coverts light pure fawn-colour.

Total length 11½ inches; bill 1½; wing 5¾; tail 5; tarsi 1⅜.

Genus SERICULUS, *Swainson*.

Of this genus only a single species is known; and that this bird has many characters in common with the *Paradiseidæ* will, I think, be evident to every one who will compare it with those birds. In my opinion much has been added to the interest of the Regent Bird by Mr. Coxen's discovery that it constructs a bower or playing-place like the members of the genera *Chlamydodera* and *Ptilinorhynchus*.

Sp. 282.   SERICULUS MELINUS.
REGENT-BIRD.

Turdus melinus, Lath. Ind. Orn. Supp., p. xliv.
Meliphaga chrysocephala, Lewin, Birds of New Holl., pl. 1.
Golden-crowned Honey-eater, Lath. Gen. Hist., vol. iv. p. 184.
Oriolus regens, Wagl. Syst. Av., *Oriolus*, sp. 2.
—— *regius*, Temm. Pl. Col., 320.
Sericulus chrysocephalus, Swains. in Zool. Journ., vol. i. p. 478.
—— *regens*, Less. Man. d'Orn., tom. i. p. 256.
—— *magnirostris*, Gould in Proc. of Zool. Soc., part v. p. 145.
—— *melinus*, Gray, Gen. of Birds, vol. i. p. 233, *Sericulus*, sp. 1.

Sericulus chrysocephalus, Gould, Birds of Australia, fol., vol. iv. pl. 12.

This beautiful species, one of the finest birds of the

Australian Fauna, is, I believe, exclusively confined to the eastern portion of the country; it is occasionally seen in the neighbourhood of Sydney, which appears to be the extent of its range to the southward and westward. I met with it in the brushes at Maitland in company, and feeding on the same trees, with the Satin- and Cat-Birds and the *Mimeta viridis*; it is still more abundant on the Manning, at Port Macquarrie, and at Moreton Bay; I sought for and made every inquiry respecting it at Illawarra, but did not meet with it, and was informed that it is never seen there, yet the district is precisely similar in character to those in which it is abundant, about two degrees to the eastward: while encamped on Mosquito Island, near the mouth of the River Hunter, I shot several, and observed it to be numerous on the neighbouring islands, particularly Baker's Island, where there is a fine garden, and where it commits serious injury to the fruit crops.

Although I have spoken of this bird as abundant in the various localities referred to, I must mention that at least fifty out of colour may be observed to one fully-plumaged male, which, when adorned in its gorgeous livery of golden yellow and deep velvety black, exhibits an extreme shyness of disposition, as if conscious that its beauty, rendering it a conspicuous object, might lead to its destruction; it is usually therefore very quiet in its actions, and mostly resorts to the topmost branches of the trees; but when two gay-coloured males meet, conflicts frequently take place. To obtain specimens in their full dress, considerable caution is necessary; on the other hand, females and immature males are very tame, and, when feeding among the foliage, appear to be so intent upon their occupation as not to heed the approach of an intruder; and I have occasionally stood beneath a low tree, not more than fifteen feet high, with at least ten feeding voraciously above me.

I did not succeed in discovering the nest; but the late

F. Strange, writing from Moreton Bay, informed me that it "is rudely constructed of sticks; no other material being employed, not even a few roots as a lining. On the 4th of November I observed one building, and, as I was leaving for the Richmond the next day, I gave instructions that it should be taken fifteen days after; when the time arrived, however, no native could be got to secure it, and it remained till my return on the 4th of December. I then sent a native up, and he brought me the nest, with two young ones covered with down, except the wings, which were feathered. As the two birds quite filled the nest, and I have heard of other nests being taken with the same number of birds in them, I am inclined to believe that two is the normal number of eggs laid. After taking the young, I wounded and succeeded in capturing the old bird; which, after being two days in confinement, became reconciled to captivity, attended to her progeny, fed them, and removed the dirt that accumulated in the nest."

The eggs are still a desideratum, and their acquisition would be a source of much gratification to me.

The following extracts from a paper on the habits of this fine bird, by C. Coxen, Esq., of Brisbane, read at a meeting of the Queensland Philosophical Society on the 23rd of May 1864, I consider to be of high interest, as affording a clue to the position the bird should occupy in our systems:—

"Although the Regent-bird has been known to ornithologists for many years, very little of its habits has become known, and it has been left for me to bring under your notice the very peculiar and curious habit it enjoys in common with the Satin-bird (*Ptilonorhynchus holosericeus*) and the Spotted Bower-bird (*Chlamydodera maculata*). My attention was called to this peculiarity in August last, by Mr. Waller, taxidermist, of Edward-street, in this city, to whose untiring energy and ability as a collector I must always bear testimony. Mr. Waller informed me that, while shooting in a scrub on the banks of the Brisbane River, he saw a male Regent-bird playing on the ground, jumping up and down, puffing out its feathers, and rolling about in a very odd manner, which occasioned much surprise, never having seen the bird on the ground before. The spot where it was playing was thickly covered with small shrubs; not wishing to lose the opportunity of procuring a specimen, he fired, but only succeeded in wounding it: and on searching the spot, he found a bower formed between, and supported by, two small brush plants, and surrounded by small shrubs, so much so, that he had to creep on his hands and knees to get to it; while doing so, the female bird came down from a lofty tree, uttered her peculiar note, and lit on a branch immediately over the bower, apparently with the intention of alighting in front of it, but was scared away on seeing Mr. Waller so close to her. She continued flitting over the place, and calling for her mate so long as he was in the neighbourhood. Mr. Waller believes that the male bird, after being wounded, fluttered to some distance from the bower, and died, as a male Regent-bird was found dead two days afterwards in a more open part of the brush. On visiting the scrub on the following and several successive days, the female bird was seen in the locality of the bower, and by her constant calling was apparently lamenting the loss, or what might seem to her the inconstancy of her mate. The ground around the bower was clear of leaves for some twelve or eighteen inches, and had the appearance of having been swept, the only objects in its immediate vicinity being a small specimen of helix. The structure was alike at both ends, but the part designated as the front was more easy of approach, and had the principal decorations; the approach to the back being more closed by scrub. Mr. Waller being desirous that this curious habit of the Regent-bird should be verified, determined to leave the bower untouched until he had acquainted me with his discovery. Circumstances occurred to prevent me from accompanying him to its whereabouts until the following November, when we found the bower in good preservation. Previous to my seeing and examining the structure, I must confess to having had considerable doubts as to whether it would not prove to be a bower of the Satin-bird, but these doubts were dissipated at the first glance, the formation of the structure differing considerably, and the decoration more so. With Mr. Waller's assistance I removed the building without injuring or in any way defacing its architectural style. It may not be inopportune for me to state that I was the first to discover the bower and habits of the Satin-bird, and, also, among the first discoverers of the bower of the Spotted Bower-bird, that I have had frequent opportunities of seeing them in the New South Wales brushes and the myall scrubs to the westward, and am consequently conversant with their peculiarities. The bower of the Regent-bird differs from the Satin-bird's in being less dome-shaped, straighter in the sides, platform much less, being only ten inches by ten, but thicker in proportion to its area, twigs smaller and not so arched, and the inside of the bower smaller; indeed, I believe, too small to admit an adult Satin-bird without injury to its architecture. The decorations of the bower are uniform, consisting only of a small species of helix, herein forming a marked distinction from the Satin-bird. Mr. Gould had shown his usual power of observation and knowledge of generic distinctions, in having placed the Regent-bird next in order to the Satin Bower-bird, without having any knowledge of its peculiar building-instincts. The Regent-bird frequents our river scrubs during the winter months, from the beginning of May to the end of September, coming from the south, whither he repairs during the summer. Its food consists of berries, wild fruits, and insects. In confinement it greedily disposes of house-flies, cockroaches, and small insects, showing great activity in their capture; but its principal food is the banana, of which it eats largely. It is very bold and pugnacious, the young males particularly so. In confinement several cases have occurred of one having killed the other. The young males closely resemble the females in plumage during their first year, in the second they partially assume the gay plumage of their sire, and in their third year they put on the full livery of the adult male."

The male has the head and back of the neck, running in a rounded point towards the breast, rich bright gamboge-yellow, tinged with orange, particularly on the centre of the forehead; the remainder of the plumage, with the exception of the secondaries and inner webs of all but the first primary, deep velvety black; the secondaries bright gamboge-yellow, with a narrow edging of black along the inner webs; the first primary is entirely black, the next have the tips and outer webs black —the half of the inner web and that part of the shaft not running through the black tip are yellow; as the primaries approach the secondaries, the yellow of the inner web extends across the shaft, leaving only a black edge on the outer web, which gradually narrows until the tips only of both webs remain black; bill yellow; irides pale yellow; legs and feet black.

The female has the head and throat dull brownish white, with a large patch of deep black on the crown; all the upper surface, wings, and tail pale olive-brown, the feathers of the back with a triangular-shaped mark of brownish white near the tip; the under surface is similar, but here, except on the breast, the white markings increase so much in size as to become the predominant hue; irides brown; bill and feet black.

Genus MIMETA, *Vigors and Horsfield.*

This form is merely an offshoot from *Oriolus*, from which it is distinguished by the absence of any gay colouring in the plumage of its members. Three species inhabit Australia, and others are found in the islands immediately to the northward of that country.

Sp. 293. MIMETA VIRIDIS.

NEW SOUTH WALES ORIOLE.

Gracula viridis, Lath. Ind. Orn., Supp. p. xxviii.
Green Grakle, Lath. Gen. Syn. Supp., vol. ii. p. 129.
Coracias sagittata, Lath. Ind. Orn., Supp. p. xxvi.
Striated Roller, Lath. Gen. Syn., Supp. vol. ii. p. 122.
Streaked Roller, Lath. Gen. Hist., vol. iii. p. 84, young.
Oriolus viridis, Vieill. 2nd edit. du Nouv. Dict. d'Hist. Nat., tom. xviii. p. 197.
—— *variegatus*, Vieill. Ib., tom. xviii. p. 196.
Mimetes viridis, King, Survey of Intertropical Coast of Australia, vol. ii. p. 419.
Mimeta viridis, Vig. and Horsf. in Linn. Trans., vol. xv. p. 326.
—— *maruloides*, Vig. and Horsf. Ib., vol. xv. p. 327, young.

Oriolus viridis, Gould, Birds of Australia, fol. vol. iv. pl. 13.

The true and probably the restricted habitat of this species is New South Wales, where in the months of summer it is tolerably plentiful in every part of the colony. I frequently observed it in the Botanic Garden at Sydney, and in all the gardens of the settlers where there were trees of sufficient size to afford it shelter; the brushes of the country, the sides of brooks, and all similar situations are equally inhabited by it. I did not find it in South Australia, neither has it been observed to the westward of that part of the country. That its range extends pretty far to the northward I have no doubt, as its numbers rather increased than diminished in the neighbourhood of the rivers Peel and Namoi.

The following notes respecting this species, by Mr. E. P. Ramsay, are extracted from the 'Ibis' for 1863:—

"During the winter months these birds may be found in flocks of from five to twenty in number, feeding upon various cultivated and wild fruits, and often in company with the

Fruit-eating Magpie, the note of which they often imitate. They frequent nearly all the orchards and gardens about Sydney, especially if they contain any of the native olive- or Moreton Bay fig-trees in fruit, to which they are very partial. I have known them, though seemingly with great reluctance, eat the berries of the white cedar. Towards the beginning of September those near Sydney pair, and seek for breeding-places, each couple selecting a distinct locality, where they remain during the whole of the season; even if the nest be taken, they will, like the *Grallina australis*, continue building near the same place until the season has expired.

"The nest is cup-shaped, and composed of shreds of the bark of the stringy-bark tree, a species of *Eucalyptus*, strongly interwoven, with the inside made thick and more compact by the addition of the white paper-like bark of the tea-tree, or any other material adapted for the purpose; and lined with the narrow leaves of the native oaks, or with grass and hair. It is from four to five inches in diameter, three to four inches wide inside, about three and a half inches deep, and is usually suspended between a fork at the extreme end of a horizontal bough of a gum-, tea-, or turpentine-tree, &c., and often in very exposed situations.

"The eggs are two or three in number, usually the latter; but in two instances I have found four. They are from one inch and two lines to one inch and four lines in length by from nine lines to one inch in breadth. Their ground-colour varies from a rich cream to a dull white or very light brown, minutely dotted and blotched with umber and blackish brown, and instances with faint lilac spots which appear beneath the surface, all over in some instances, but generally the spots are more numerous at the larger end, where they form an indistinct band.

"The note of this Oriole is very melodious and varied. It may often be seen perched on some shady tree, with its head thrown back, showing to perfection its mottled breast, singing in a low tone and imitating the notes of many birds, including the *Zosterops*, and particularly the Black or Fruit-eating Magpie. While feeding, it frequently utters a harsh guttural sort of squeak. During the breeding-season, which commences at the end of September and ends in January, it confines itself to a very monotonous although melodious cry, the first part of which is quickly repeated, and ends in a lower note."

The bird as observed by me in New South Wales was bold and active, and was often seen in company with the Regent-, Satin-, and Cat-birds, feeding in the same trees and on similar berries and fruits, particularly the small wild fig. I often observed it capturing insects on the wing and flying very high, frequently above the tops of the loftiest trees.

The sexes when fully adult differ so little in colour that they can scarcely be distinguished; the male is however of a more uniform tint about the head, neck, and throat, and has the yellowish olive of the upper surface of a deeper tint than the female.

Head and all the upper surface yellowish olive; wings and tail-feathers dark brown; the outer webs of the coverts and secondaries grey, margined and broadly tipped with white; all but the two centre tail-feathers with a large oval-shaped spot of white on the inner, and the extremity of the outer web white, the white mark gradually increasing in size as the feathers recede from the centre until it becomes an inch long on the external one; under surface white, washed with olive-yellow on the sides of the chest, each feather with an elongated pear-shaped mark of black down the centre; bill dull flesh-red; irides scarlet; feet lead-colour.

The young bird during the first year has the bill blackish brown instead of dull flesh-red; the upper surface olive-brown, each feather strongly streaked down the centre with dark brown; wings brown, under surface of the shoulder and all the wing-feathers except the primaries margined with sandy red; the black streaks on the breast more decided, and the white spot at the tip of the lateral tail-feathers much smaller, than in the adult.

Sp. 29-t.   MIMETA AFFINIS, *Gould.*

Mur-re-a rwoo of the Aborigines.

Oriolus affinis, Gould, Birds of Australia, fol., vol. i. Introd. p. liii.

This species inhabits the neighbourhood of Port Essington, and only differs from the preceding in having a smaller body, a shorter wing, a much larger bill, and in the white spots at the tips of the lateral tail-feathers being much smaller in extent. Although I have not at this moment any specimens wherewith to institute a comparison, I have but little doubt that this bird is quite distinct from its southern representative, *M. viridis.*

Gilbert informed me that it is abundant in every part of the Cobourg Peninsula and the adjacent islands, in every variety of situation. Its note is loud, distinct, and very unlike that of every other bird he had ever heard; the sound usually uttered is a loud clear whistle, terminating in a singular guttural harsh catch; but in the cool of the evening, when perched among the thick foliage of the topmost branches of the *Eucalypti* and other trees, it pours forth a succession of very pleasing notes.

A nest taken on the 4th of December contained two nearly hatched eggs; it was attached by the rim to a drooping branch of the swamp *Melaleuca*, about five feet from the ground, was very deep and large, and formed of very narrow strips of the paper bark mixed with a few small twigs, the bottom of the interior lined with very fine wiry twigs.

The eggs, which are large for the size of the bird, are of a beautiful bluish white, sparingly spotted all over with deep umber brown and bluish grey; the latter appear as if beneath the surface of the shell; their medium length is one inch and three lines long by eleven lines broad.

Sp. 285.   MIMETA FLAVOCINCTA.

CRESCENT-MARKED ORIOLE.

Mimetes flavo-cinctus, King, Survey of Intertropical Coasts of Australia, vol. ii. p. 419.
Mimeta flavo-cincta, Vig. and Horsf. in Linn. Trans., vol. xv. p. 327.

Oriolus flavocinctus, Gould, Birds of Australia, fol., vol. iv. pl. 14.

This species was discovered on the north coast of Australia by Captain Philip Parker King, R.N., and described by him in his "Survey of the Intertropical Coasts of Australia," referred to above; Gilbert procured two specimens at Port Essington, and Commander Ince, R.N., subsequently obtained an additional example in the same locality. All the information that has reached me respecting its habits and economy is contained in a short note sent to me by Gilbert, which merely states that his specimens were obtained in the forests of mangroves bordering the coast.

The *Mimeta flavocincta* is the largest and by far the most gaily coloured species of the genus yet discovered in Australia. In the islands to the northward of that country there are other species of still larger size, but none of them are so richly coloured.

The male has the head, neck, and all the upper surface dull greenish yellow, with a stripe of black, broad at the base and tapering to a point, down the centre of each feather; under surface greenish yellow, passing into pure yellow on the under tail-coverts; wings black, all the feathers margined with greenish yellow and broadly tipped with pale yellow; tail black, washed on the margins with greenish yellow, and largely tipped, except the two middle feathers, with brightyellow, which increases in extent as the feathers recede from the centre; irides reddish orange; bill dull red; feet lead-colour.

The female differs in being of smaller size, in having the under surface striated with black, and the markings of the wings straw-white instead of yellow.

Genus SPHECOTHERES, *Vieillot.*

Australia presents us with two well-defined species of this genus; others inhabit New Guinea and the neighbouring islands; but as yet we have no evidence of the form occurring on the continent of India.

These birds appear to offer an alliance to the members of the genera *Oriolus* and *Mimeta.*

Sp. 260. SPHECOTHERES MAXILLARIS.

Southern Sphecotheres.

Turdus maxillaris, Lath. Ind. Orn., Suppl. p. xliii.
Sphecotheres viridis, Vig. and Horsf. in Linn. Trans., vol. xv. p. 215.
—— *virescens*, Jard. and Selb. Ill. Orn., vol. ii. pl. 79.
—— *australis*, Swains.
—— *canicollis*, Swains. Anim. in Menag., p. 320.
—— *grisea*, Less. Traité d'Orn., p. 351.
—— *maxillaris*, Gray Gen. of Birds, vol. i. p. 231, *Sphecotheres*, sp. 1.

Sphecotheres australis, Gould, Birds of Australia, fol., vol. iv. pl. 15.

I killed a fine specimen of this bird on Mosquito Island, at the mouth of the river Hunter, in September 1839; it was perched on a dead branch which towered above the green foliage of one of the high trees of the forest, and my attention was drawn to it by its loud and singular note: this was the only example that came under my observation; but it is more plentiful in the neighbourhood of the river Clarence, is abundant at Moreton Bay, and that it enjoys a wide range is proved by Mr. Bynoe having procured an adult male on the north coast. It appears to be peculiar to the brushes, and its food doubtless consists of the berries and fruits which abound in those districts. Nothing is known of its nidification.

The sexes differ very widely from each other in colour.

The male has the crown of the head and the cheeks glossy

black; orbits and a narrow space leading to the nostrils naked and of a light buffy yellow; throat, chest, and collar at the back of the neck dark slate-grey; all the upper surface, greater wing-coverts, outer webs of the secondaries, abdomen, and flanks yellowish green; lesser wing-coverts, primaries, and inner webs of the secondaries slaty black, fringed with grey; vent and under tail-coverts white; tail black, the apical half and the outer web of the external feather pure white; the apical half of the second feather on each side white, the next on each side with a large spot of white at the extremity, and the six central feathers slightly fringed with white at the tip; bill black; irides very dark brown in some, red in others; feet flesh-colour.

The female has the upper surface brown, washed with olive, each feather with a darker centre; wings dark brown, the coverts and secondaries conspicuously, and the primaries narrowly, edged with greenish grey; under surface buffy white, each feather with a broad and conspicuous stripe of brown down the centre; flanks washed with yellowish green; under tail-coverts white, with a narrow stripe of brown down the centre; tail brown, each feather narrowly edged on the inner web with white, and all but the two lateral ones on each side washed with yellowish green.

Sp. 297. SPHECOTHERES FLAVIVENTRIS, *Gould*.

NORTHERN SPHECOTHERES.

Sphecotheres flaviventris, Gould in Proc. of Zool. Soc., 1849, p. 111.

Sphecotheres flaviventris, Gould, Birds of Australia, fol., Supplement, pl.

This bird may always be distinguished from its near ally the *S. maxillaris* by the beautiful jonquil-yellow of its under surface. Mr. Macgillivray informed me that it is very common in the neighbourhood of Cape York, where he daily observed it either in pairs or in small parties of three or four individuals, which were generally very shy and difficult of approach. It frequents the open forest land in company with the *Tropidorhynchus argenticeps*, and resorts to the branches for its food, which consists of fruit of various kinds, such as figs, &c. His specimens were procured by keeping himself carefully concealed beneath one of its favourite feeding trees and watching until an opportunity offered of getting a shot. He once saw several nests which he had no doubt belonged to this species; nearly all of them were built among the topmost branches of very large gum-trees, which he could not induce the natives to attempt to climb; a deserted nest was however within reach, being placed on an overhanging branch not more than twenty feet from the ground; it measured about a foot in diameter, and was composed of small sticks lined with finer ones.

As is the case with the other members of the genus, the sexes offer a marked difference in colour.

The male has the crown of the head and cheeks glossy black; orbits, and a narrow space leading to the nostrils naked, and of a light buffy yellow, or flesh-colour; all the upper surface, wing-coverts, outer webs of the secondaries, and a patch on either side of the chest, olive-green; chin, chest, abdomen, and flanks beautiful yellow; vent and under tail-coverts white; primaries and inner webs of secondaries black, edged with grey; tail black, the external web and the apical half of the internal web of the outer feather on each side white; the apical half of the second feather on each side white; the next, or third, on each side with a large spot of white at the tip; bill black; feet flesh-colour.

The female is striated on the head with brown and whitish; has the upper surface olive-brown; the wing-feathers narrowly edged with greenish grey; the under surface white, with a conspicuous stripe of brown down the centre of each feather; and the vent and under tail-coverts white, without striæ.

Total length $10\frac{1}{4}$ inches; bill $1\frac{1}{8}$; wing $5\frac{5}{8}$; tail $4\frac{1}{4}$; tarsi $\frac{7}{8}$.

Family ——— ?

Genus CORCORAX, *Lesson*.

A genus containing only one species, which possesses many singular habits. So far as is yet known, it is confined to Australia.

Sp. 288.   CORCORAX MELANORHAMPHUS.

WHITE-WINGED CORCORAX.

Coracia melanoramphus, Vieill. Nouv. Dict. d'Hist. Nat., tom. viii. p. 2.
Pyrrhocorax leucopterus, Temm. Man. d'Orn., tom. i. p. 121.
Fregilus leucopterus, Vig. and Horsf. in Linn. Trans., vol. xv. p. 265.
Corvus leucopterus, Wagl. Syst. Nat., *Corvus*, sp. 14.
Corcorax australis, Less. Traité d'Orn., p. 325.
—— *leucopterus*, G. R. Gray, List of Gen. of Birds, 2nd edit. p. 52.
—— *melanorhynchus*, Gray, Gen. of Birds, vol. ii. p. 321.
Corcorones melanorhynchus, Cab. Orn. Nat. in Wiegm. Archiv, 1847, p. 325.
—— *melanorhamphus*, Cab. Mus. Hein., Theil i. p. 228.
Waybung, Aborigines of New South Wales.

Corcorax leucopterus, Gould, Birds of Australia, fol., vol. iv. pl. 16.

This bird is distributed over all parts of New South Wales and South Australia; it is very abundant in the whole of the Upper Hunter district, I killed it in the interior of South Australia; and Mr. Elsey met with it at the edge of a dense scrub on the Burdekin in lat. 10° 30′ S. It usually occurs in small troops of from six to ten in number, feeding upon the ground, over which it runs with considerable rapidity. In disposition it is extremely tame, readily admitting of a very close approach, and then merely flying off to the low branch of some neighbouring tree. During flight the white marking of the wing shows very conspicuously, and on alighting the bird displays many curious actions, leaping from branch to branch with surprising quickness, at the same time spreading the tail and moving it up and down in a very singular manner; on being disturbed it peeps and pries down upon the intruder below, and generally utters a harsh, grating, disagreeable and tart note; at other times, while perched among the branches of the trees, it makes the woods ring with its peculiar soft, low, very pleasing but mournful pipe.

During the pairing-season the male becomes very animated, and his manners so remarkable, that it would be necessary for my readers to witness the bird in its native wilds to form a just conception of them : while sitting on the same branch close to the female, he spreads out his wings and tail to the fullest extent, lowers his head, puffs out his feathers and displays himself to the utmost advantage, and when two or more are engaged in these evolutions, the exhibition cannot fail to amuse and delight the spectator. A winged specimen gave me more trouble to catch than any other bird I ever chased; its power of passing over the ground being so great, that it bounded on before me and cleared every obstacle, hillocks and fallen trees, with the utmost facility.

The White-winged Corcorax is a very early breeder, and generally rears more than one brood in a year, the breeding-season extending over the months of August, September, October, and November. The nest is a most conspicuous fabric, composed of mud and straw, resembling a bason, and is usually placed on the horizontal branch of a tree near to or overhanging a brook. The eggs vary from four to seven in number, and are of a yellowish white, boldly blotched all over with olive and purplish brown, the latter tint appearing as if beneath the surface of the shell; they are one inch and a half long by one inch and one line broad.

It has often struck me that more than one female deposited her eggs in the same nest, as four or five females may be frequently seen either on the same or the neighbouring trees, while only one nest is to be found.

The bird generally evinces a preference for open forest land, but during the breeding-season affects the neighbourhood of brooks and lagoons, which may be accounted for by the fact of such situations being necessary to enable it to procure the mud to build its nest, besides which they also afford it an abundance of insect food.

The whole of the plumage black, with glossy green reflexions, with the exception of the inner webs of the primaries, which are white for three parts of their length from the base; irides scarlet; bill and feet black.

Family ——?

Genus STRUTHIDEA, *Gould*.

The only known species of this genus is confined to the stony ridges of the southern and eastern parts of Australia. Probably some peculiarity in the construction of the bill is requisite for the extraction of the seeds in the cones of the *Callitris pyramidalis*, upon which it is mostly seen, and if the short arched bill of this bird be given for this purpose, it is one of the most striking instances of means to an end with which I am acquainted. I consider this to be one of the most anomalous forms comprised in the avi-fauna of Australia.

Sp. 289.   STRUTHIDEA CINEREA, *Gould*.

GREY STRUTHIDEA.

Struthidea cinerea, Gould in Proc. of Zool. Soc., part iv. p. 143.
Brachystoma cinerea, Swains. An. in Menag., p. 297.
Brachyprorus cinereus, Cab. Mus. Hein., Theil i. p. 217.

Struthidea cinerea, Gould, Birds of Australia, fol., vol. iv. pl. 17.

From what I personally observed of this bird, it would seem to be a species peculiar to the interior, and, so far as is yet known, confined to the southern and eastern portions of

Australia. I found it inhabiting the pine ridges, as they are termed by the colonists, bordering the extensive plains of the Upper and Lower Namoi, and giving a decided preference to the *Callitris pyramidalis*, a fine fir-like tree peculiar to the district. It was always seen in small companies of three or four together, on the topmost branches of the trees, was extremely quick and restless, leaping from branch to branch in rapid succession, at the same time throwing up and expanding the tail and wings; these actions being generally accompanied with a harsh unpleasant note; their manners, in fact, closely resemble those of the White-winged *Corcorax* and the *Pomatorhini*.

The following notes on this species I find in Gilbert's journal of the occurrences during his expedition with Leichardt from Moreton Bay to Port Essington. They were written on the sixteenth day after his departure, and will not be devoid of interest:—

"Oct. 10.—Strolled about in search of novelties, and was amply repaid by finding the eggs of *Struthidea cinerea*. I disturbed the bird several times from a rosewood-tree growing in a small patch of scrub, and felt assured it had a nest, but could only find one, which I considered to be that of a *Grallina*; determined, if possible, to solve the difficulty, I lay down at a short distance within full view of the tree, and was not a little surprised at seeing the bird take possession of, as I believed, the *Grallina*'s nest; I immediately climbed the tree and found four eggs, the medium length of which was one inch and a quarter by seven-eighths of an inch in breadth; their colour was white, with blotches, principally at the larger end, of reddish brown, purplish grey and greenish grey; some of the blotches appearing as if they had been laid on with a soft brush. From the appearance of the nest I should say it was an old one of *Grallina*, particularly as it contained a much greater quantity of grass for a lining than I ever observed in the nest of a *Grallina* while that bird had possession of it; if this be not the case, then the nest of *Struthidea* is precisely similar, being like a great basin of mud, and placed in the same kind of situation, on a horizontal branch.

"Oct. 21.—In the evening I again met with the *Struthidea*, which I disturbed from a nest like the one above described, and from the new appearance of the structure I am inclined to believe it to be constructed by the bird itself, although it does so closely resemble that of *Grallina*, especially as in this case the nest was placed in a situation far from water, and there were no *Grallinæ* in the neighbourhood. This nest, like the last, had a very thick lining of fine grass, and appeared as if just finished for the reception of the eggs."

There is no doubt that the nests above described were those of *Struthidea*; those of *Corcorax* and *Grallina* are precisely similar; and we now know that all three birds build the same kind of mud nests.

The food, as ascertained by dissection, is insects; the stomachs of those examined were tolerably hard and muscular, and contained the remains of coleoptera.

The sexes assimilate so closely in size and in the colouring of their plumage, that they are to be distinguished only by dissection.

Head, neck, back, and under surface grey, each feather tipped with lighter grey; wings brown; tail black, the middle feathers glossed with deep rich metallic green; irides pearly white; bill and legs black.

Total length $11\frac{1}{2}$ inches; bill $\frac{3}{4}$; wing $5\frac{1}{4}$; tail 6; tarsi $1\frac{1}{4}$.

Family CORVIDÆ.

Genus CORVUS, *Linnæus*.

It is exceedingly interesting to trace the range of the members of this genus or the true Crows; not so much on account of their wide distribution, as from the circumstance of the form being non-existent in some countries which appear admirably adapted for their well-being; thus, while the species are widely distributed over the whole of Europe, Asia, Africa, North America, the Indian Islands, and Australia, none are to be found in South America or New Zealand

Sp. 290.   CORVUS AUSTRALIS, *Gmelin*.

White-eyed Crow.

Corvus australis, Gmel. Edit. Linn. Syst. Nat., vol. i. p. 365.
—— *coronoides*, Wagl. Isis, 1829, p. 748.
—— *coronoïdes*, Vig. and Horsf. in Linn. Trans., vol. xv. p. 261.—
Schlegel, Not. sur Gen. *Corvus* in Nat. Art. Mag., Achtste Aff. p. 6.
Wir-dang, Aborigines of Western Australia.
Ou-bo-lak, Aborigines of Port Essington.
Crow of the Colonists.

Corvus coronoides, Gould, Birds of Australia, fol., vol. iv. pl. 18.

This species is so intermediate in size, in the development of the feathers of the throat, in its voice, and in many parts of its economy, between the Carrion Crow and Raven of our own island, that it is difficult to say to which of those species it is most nearly allied; I prefer, however, placing it among the true Crows to assigning it to a companionship with the larger members of the family. Every part of Australia yet explored has been found to be inhabited by it; some slight difference, however, is observable between individuals from Port Essington, Swan River, Tasmania, and New South Wales, but these differences appear to me to be too trivial to be regarded as specific; specimens from Western Australia are somewhat less in size than those procured in the other localities mentioned, while Port Essington examples have the basal portion of the feathers on the back of the neck greyish white, which is not the case with those inhabiting the south coast. When the birds are fully adult, the colour of the eye is white, I believe, in the whole of them—a circumstance which tends to strengthen the opinion I entertain of their being one and the same species.

In Western Australia, for the greater part of the year, this bird is met with in pairs or singly; but in May and June it congregates in families of from twenty to fifty, and is then very destructive to the farmer's seed crops, which appear to be its only inducement for assembling together, as it is not known to congregate at any other period. In New South Wales and Tasmania it is also usually seen in pairs, but occasionally congregated in small flocks. At Port Essington, where it is mostly seen in pairs, in quiet secluded places, it is not so abundant as in other parts of Australia.

The stomach is tolerably muscular, and the food consists of insects, carrion of all kinds, berries, seeds, grain, and other vegetable substances.

Its croak very much resembles that of the Carrion Crow, but differs in the last note being lengthened to a great extent.

Its nest, which is formed of sticks and of a large size, is usually placed near the top of the largest gum-trees. The eggs, which are three or four in number, are very long in form, and of a pale dull green colour, blotched, spotted, and freckled all over with umber-brown, the blotches being of a much greater size at the larger end; they are about one inch and three quarters long by one inch and an eighth broad.

The whole of the plumage rich shining purplish black, with the exception of the elongated feathers on the throat, which are slightly glossed with green; bill and feet black; irides in some white, in others brown.

Family STURNIDÆ.

Genus CALORNIS, *G. R. Gray*.

But one species of this form has yet been discovered in Australia; others inhabit Batchian and New Guinea, and, I believe, Java and Sumatra. Of their habits and economy but little is known; the Australian member is perhaps the most beautiful of the whole.

Sp. 201.   CALORNIS METALLICA.

SHINING CALORNIS.

Lamprotornis metallicus, Temm. Pl. Col. 266.
Calornis metallica, Gray, Gen. of Birds, vol. ii. p. 327, *Calornis*, sp. 2.
Mooter, Goodang tribe of Aborigines at Cape York.

Aplonis metallica, Gould, Birds of Australia, fol., Supplement, pl.

This species inhabits the northern portion of Australia, New Guinea, Timor, the Celebes, Amboyna, and New Ireland.

Mr. Macgillivray has obligingly furnished me with the following interesting account of its habits and nidification:—

"During the early part of our last sojourn at Cape York, this bird was often seen passing rapidly over the tops of the trees in small flocks of a dozen or more. In their flight they reminded me of the Starlings, and, like them, made a chattering noise while on the wing. One day a native took me to a breeding-place in the centre of a dense scrub, where I found a gigantic cotton-tree standing alone, with its branches literally hung with the pensile nests of the bird: the nests, averaging two feet in length and one in breadth, are of a somewhat oval form, slightly compressed, rounded below and above, tapering to a neck, by the end of which they are suspended; the opening is situated in the centre of the widest part; they are almost entirely composed of portions of the stem and the long tendrils of a climbing-plant (*Cissus*) matted and woven together, and lined with finer pieces of the same, a few leaves (generally strips of *Pandanus* leaf), the hair-like fibres of a palm (*Caryota cereus*), and similar materials: the eggs, usually two, but often three in number, are an inch long by eight-tenths of an inch broad, and of a bluish grey, speckled with reddish pink, chiefly at the larger end; some have scarcely any markings, others a few minute dots only. The note of the bird is short, sharp, and shrill, and resembles '*twee-twee*,' repeated, as if angrily, several times in quick succession.

"On the tree above mentioned the nests were about fifty in number, often solitary, but usually three or four together in a cluster—sometimes so closely placed as to touch each other.

"The bird appears to enjoy a wide range. During the progress of the expedition two were shot at the Duchateau Isles, in the Louisiade Archipelago, and I saw a specimen on board H.M.S. Meander, which had been procured at Carteret Harbour, in New Ireland.

"The stomachs of those examined contained triturated seeds and other vegetable matter."

When fully adult, the two sexes are so precisely alike that dissection must be resorted to to distinguish them.

The general plumage is a mixture of dark rich bronzy green and purple, the green hue predominating on the lower part of the throat and the upper part of the back; wings and tail bluish black, washed on the margins with bronzy green; bill and feet black; irides vermilion.

The young of both sexes have the upper surface similarly coloured, but not so bright as in the adult; wings brown, narrowly margined with brownish white; all the under surface buffy white, streaked on the breast, flanks, and under tail-coverts with brownish black.

Family CRATEROPODIDÆ?

Genus POMATOSTOMUS, *Cabanis*.

The members of this genus range over all parts of Australia, but do not extend to India, where their place is supplied by numerous species of the allied form *Pomotorhinus*.

Sp. 292. POMATOSTOMUS TEMPORALIS.

TEMPORAL POMATORHINUS.

Pomatorhinus temporalis, Vig. and Horsf. in Linn. Trans., vol. xv. p. 330.
—— *trivirgatus*, Temm. Pl. Col., 443.
—— *frivolus, temporalis, et trivirgatus*, Gray, Gen. of Birds, vol. i. p. 229, *Pomatorhinus*, sp. 8, 9, 10.
Turdus frivolus, Lath. Ind. Orn., Supp. p. xliii?
Pomatostomus temporalis, Cab. Mus. Hein., Theil i. p. 83.

Pomatorhinus temporalis, Gould, Birds of Australia, fol., vol. iv. pl. 20.

This species inhabits New South Wales, particularly those districts where *Angophoræ* and *Eucalypti* abound; it is gregarious in its habits, and is exceedingly noisy and garrulous. Commencing with the branches nearest the ground, it gradually ascends, in a succession of leaps, to the very tops of the trees, whence, with elevated tail, it peers down, and continually utters its peculiar chattering cry; it is frequently to be seen on the ground, but on the slightest alarm it resorts to the trees, and ascends them in the manner described. Its powers of flight are not very great, and appear to be only employed to convey it from the top of one tree to another, the whole troop following one after the other.

The situation of the nest is somewhat varied; on the *Eucalypti* it is mostly built at the extremity of the branch: it is of a large size, and very much resembles that of the

Magpie of Europe, being of a completely domed form, outwardly composed of small long twigs about the size of a thorn, crossing each other, and but very slightly interwoven: the entrance is in the form of a spout, about half the length of a man's arm, and the twigs are placed in such a manner that the points incline towards each other, rendering it apparently impossible for the bird to enter without breaking them, while egress, on the other hand, is very easy; the nest has a thick inner lining of the fine inner bark of trees and fine grasses. In traversing the pasture-lands at Camden, any part of the Upper Hunter district, and some portions of the Liverpool Plains, the attention of the traveller is often attracted by the large nest of this bird; three or four are often to be seen on the same tree.

The eggs, which are four in number, and one inch in length by nine lines in breadth, are buffy brown, clouded with dark brown and purple, and streaked with hair-like lines of black, which generally have a tendency to run round the egg; in some instances, however, they take a diagonal direction, and give the surface a marble-like appearance.

The food consists of insects of various kinds.

The sexes do not differ in outward appearance, and may be thus described:—

Throat, centre of the breast, and a broad stripe over each eye white; lores and ear-coverts dark brown; centre of the crown, back, and sides of the neck greyish brown, gradually deepening into very dark brown on the wing-coverts, back, and scapularies; wings very dark brown, with the exception of the inner webs of the primaries, which are rufous for three-fourths of their length from the base; tail-coverts and tail black, the latter largely tipped with pure white; abdomen and flanks dark brown, stained with rusty red; bill blackish olive brown, except the basal portion of the lower mandible, which is greyish white; irides in the adult straw-yellow, in the young brown; feet blackish brown.

Sp. 293. POMATOSTOMUS RUBECULUS, *Gould*.

RED-BREASTED POMATORHINUS.

Pomatorhinus rubeculus, Gould in Proc. of Zool. Soc., part vii. p. 144.
Pomatostomus rubeculus, Cab. Mus. Hein., Theil i. p. 83, note.

Pomatorhinus rubeculus, Gould, Birds of Australia, fol., vol. iv. pl. 21.

This bird is rather numerously dispersed over the northern parts of Australia, where it takes the place of the *Pomatostomus temporalis* of New South Wales, from which it differs but little either in size or colouring; its slightly smaller dimensions and the red hue of the breast are, however, characteristics by which it may at all times be distinguished from its prototype. On the Cobourg Peninsula it inhabits the open parts of the country, and when disturbed takes to the higher branches of the gums, first mounting upon one of the lower boughs, and then, by a succession of hops and leaps, ascending to the top. In its actions and economy it very closely assimilates to the other species of the genus, being, like them, a noisy and restless bird; and feeding on insects, which are frequently sought for on the ground under the canopy of the larger trees.

Throat and stripe over each eye white; chest and upper part of the abdomen dull brownish red; stripe from the nostrils, through each eye to the occiput, blackish brown; centre of the crown, back, and lower part of the abdomen dark brown, slightly tinged with olive; upper and under tail-coverts and tail black, all the feathers of the latter tipped with white; irides straw-yellow; bill blackish grey, becoming paler at the base; legs and feet greenish grey.

The sexes are alike in plumage.

Total length $9\frac{1}{2}$ inches; bill $1\frac{1}{4}$; wing 4; tail $4\frac{3}{8}$; tarsi $1\frac{1}{4}$.

Sp. 294.   POMATOSTOMUS SUPERCILIOSUS.

WHITE-EYEBROWED POMATORHINUS.

Pomatorhinus superciliosus, Vig. and Horsf. in Linn. Trans., vol. xv. p. 330.

Pomatostomus superciliosus, Cab. Mus. Hein., Theil i. p. 84.

Gnow-un, Aborigines of the mountain districts of Western Australia.

Pomatorhinus superciliosus, Gould, Birds of Australia, fol. vol. iv. pl. 22.

This species ranges over the whole of the southern portion of the continent of Australia, where it must be regarded as a bird peculiar to the interior, rather than as an inhabitant of the districts near the coast. It is common on the Liverpool Plains, and it was particularly noticed by my friend Captain Sturt during his expedition to the Darling. I myself met with it near the bend of the river Murray, and it has also been found in the York district of Western Australia, but I have never heard of its having been seen either in the north or north-western parts of the country. It usually moves about in small troops of from six to ten in number, and is without exception the most restless, noisy, querulous bird I ever observed. Its mode of progression among the branches of the trees is no less singular than is its voice different from that of other birds; it runs up and down the branches of the smaller trees with great rapidity and with the tail very much spread and raised above the level of the back. It usually feeds upon the ground under the Banksias and other low trees, but upon the least intrusion flits on to the lowest branch, and by a running or leaping motion quickly ascends to the highest, when it flies off to the next tree, uttering at the same time a jarring, chattering, and discordant jumble of notes, which are sometimes preceded by a rapidly repeated, shrill, piping whistle.

When a troop are engaged in ascending the branches, which they usually do in line, they have a singular habit of suddenly assembling in a cluster, spreading their tails and wings, and puffing out their plumage until they resemble a great ball of feathers.

The breeding-season commences in September and continues during the three following months. The nest is a large domed structure of dried sticks, with an entrance in the side, which is hidden from view by the sticks of the upper part of the nest being made to project over it for four or five inches like the thatch of a shed; the inside is generally lined with the soft parts of flowers and the dust of rotten wood, but occasionally with feathers. In Western Australia the nest is usually constructed in a dead jam-tree, the branches of which are drawn together at the top like a broom. It often happens that three or four pairs of birds build their nests in the same small clump of trees. The eggs are very like those of $P.$ $temporalis$, the ground-colour being olive-grey clouded with purplish brown, and streaked with similar hair-like lines of black; they are usually four in number, eleven and a half lines long by eight lines broad.

The sexes as well as the young so closely resemble each other, that they can only be distinguished by the aid of dissection.

Lores, space surrounding the eye and the ear-coverts dark silky brown; a broad line of white, bounded above and beneath with a narrow one of dark brown, commences at the base of the upper mandible, passes over the eye and continues to the occiput; crown of the head and all the upper surface, flanks, and under tail-coverts olive-brown, passing into a purer and deeper brown on the primaries; tail dark brown, crossed by very indistinct bars of a darker colour, the five lateral feathers on each side tipped with white; chin, throat, and chest white; bill blackish brown, the lower part of the under mandible greyish white; irides in the adult straw-yellow, in the young brown; feet blackish brown.

Sp. 295. POMATOSTOMUS RUFICEPS, *Hartlaub*.

CHESTNUT-CROWNED POMATORHINUS.

Pomatorhinus ruficeps, Hartl. in Cabanis's Journ. für Orn., vol. i. p. 21.

Pomatorhinus ruficeps, Gould, Birds of Australia, fol., Supplement, pl.

When I visited South Australia in 1838 the colony was in its infancy, and the city of Adelaide a chaotic jumble of sheds and mud huts, with trees growing here and there in the newly marked-out streets and squares. Among these trees Parrakeets of various kinds, and Honey-eaters still more numerous, were busily occupied in search of food or otherwise engaged; here and there also might be seen groups of newly-arrived emigrants, both English and Irish, who had chosen this distant country for their future home; groups of Germans, too, whose fatherland no longer offered opportunities for enterprise, were dotted about the country busily engaged in constructing their little villages and getting their gardens under cultivation. It was one of these German emigrants who, inspired by the works of nature with which he was so profusely surrounded, employed some of his leisure hours in collecting the birds which came under his notice and in transmitting them to the Museum at Bremen. Among the birds so collected and transmitted was the present new and very beautiful *Pomatostomus*, which Dr. Hartlaub has the merit of first describing. Since that period the bird has been discovered in other parts of Australia, and I am indebted to Professor M'Coy for fine examples procured in the interior of Victoria.

"Of this fine and typical species," says Dr. Hartlaub, "the Bremen Collection received two examples, scarcely differing in colour, in a collection of South Australian birds sent from Adelaide. It is remarkable that the bird escaped the researches of Mr. Gould and his collectors, and one cannot help imagining that it must have recently arrived from some part of the interior of the country, and accompanied other stragglers towards the coast.

"In size and colour *P. ruficeps* is more nearly allied to *P. superciliosus* than to any other, but it differs from that species in the brown-red colour of the head, in the white bars on the wings, and in the black mark which separates the reddish brown of the flanks from the white of the breast. In our two specimens the sexes have not been ascertained; one of them is rather less brilliantly coloured than the other."

Crown of the head and nape chestnut- or brown-red, bounded below by a conspicuous line of white; lores blackish brown; behind the eye and ear-coverts brown; upper part of the back and wing-coverts grey, each feather with a dark brown centre, giving those parts a mottled appearance; lower part of the back and rump pure dark grey; greater and lesser wing-coverts and secondaries tipped with white; throat, breast, and centre of the abdomen white; flanks reddish brown, separated from the white of the abdomen by a stripe of black; under tail-coverts brown, spotted with greyish white; four central tail-feathers dark brown, indistinctly rayed with black; the three outer feathers on each side brown, largely tipped with pure white; bill and feet blackish horn-colour, the base of the mandibles lighter.

Family MELIPHAGIDÆ.

The Honey-eaters, or that group of birds forming the family *Meliphagidæ*, are unquestionably the peculiar and most striking feature in Australian ornithology. They are in fact to the fauna what the *Eucalypti*, *Banksiæ*, and *Melaleucæ* are to the flora of Australia. The economy of these birds is so strictly adapted to those trees that the one appears essential to the other; for what can be more plain than that the brush-like tongue is especially formed for gathering the honey from the flower-cups of the *Eucalypti*, or that their diminutive stomachs are especially formed for this kind of food, and the peculiar insects which constitute a portion of it? When I say that there are at least fifty species of Meliphagous birds in Australia, my readers will naturally expect that they are divisible into many genera, and this is really the case, as will be seen as we proceed.

Genus MELIORNIS, *G. R. Gray.*

No example of this genus has yet been discovered in the northern or intertropical regions of Australia, all the species known being confined to the southern parts of the continent, the islands in Bass's Straits, and Tasmania. They feed principally upon the pollen and honey of the flowers, but occasionally upon insects; in disposition they are tame and familiar; and they frequent the *Banksia* in preference to other trees.

The sexes are generally alike in plumage, and the young assume the adult livery at an early period of their existence.

Sp. 296. MELIORNIS NOVÆ-HOLLANDIÆ.

New Holland Honey-eater.

Certhia novæ-hollandiæ, Lath. Ind. Orn., p. 290.
New Holland Creeper, White's Journ., pl. in p. 186.
L'Héorotaire tacheté, Vieill. Ois. Dor., tom. ii. p. 91, pl. 57.
Meliphaga novæ-hollandiæ, Vig. and Horsf. in Linn. Trans., vol. xv. p. 311.
Melitreptus novæ-hollandiæ, Vieill. 2nde édit. du Nouv. Dict. d'Hist. Nat., tom. xiv. p. 328.
Meliphaga balgonera, Steph. Cont. of Shaw's Gen. Zool., vol. xiv. p. 261.
—— *barbata*, Swains. Class. of Birds, vol. ii. p. 326.
Meliornis novæ-hollandiæ, G. R. Gray, List of Gen. of Birds, 2nd edit. p. 19.

Maliphaga novæ-hollandiæ, Gould, Birds of Australia, fol., vol. iv. pl. 23.

The *Meliornis novæ-hollandiæ* is one of the most abundant

and familiar birds inhabiting the colonies of New South
Wales, Tasmania, and South Australia: all the gardens of
the settlers are visited by it, and among their shrubs and
flowering plants it annually breeds. The belts of Banksias,
growing on sterile, sandy soils, also afford it so congenial an
asylum, that I am certainly not wrong in saying they are
never deserted by it, or that the one is a certain accom-
paniment of the other. The range enjoyed by this species
appears to be confined to the south-eastern portions of Australia: it is abundant on the sandy districts of South Australia
wherever the Banksias abound. In Tasmania it is much more
numerous on the northern than on the southern portion of
the island. It evinces a more decided preference for shrubs
and low trees than for those of a larger growth; consequently
it is a species particularly subject to the notice of man; nor
is it the least attractive of the Australian avi-fauna; the
strikingly-contrasted markings of its plumage, and the beautiful appearance of its golden-edged wings, when passing with
its quick jumping flight from shrub to shrub, rendering it a
most conspicuous and pleasing object.

It has a loud, shrill, liquid, although monotonous note. Its
food, which consists of the pollen and juices of flowers, is procured while clinging and creeping among them in every
variety of position: it also feeds on fruits and insects.

It usually rears two or three broods during the course of
the season, which lasts from August to January: the nest is
very easily found, being placed in any low open bush. One
of those in my collection was taken from a row of peas in
the kitchen-garden of the Government House at Sydney. It
is a somewhat compact structure, composed of small wiry
sticks, coarse grasses, and broad and narrow strips of bark;
the inside lined with the soft woolly portion of the blossoms
of small ground plants: the eggs, which are two or three in
number, are of a pale buff, thinly spotted and freckled with
deep chestnut-brown, particularly at the larger end, where they

not unfrequently assume the form of a zone; their medium length is nine lines and a half, and breadth nearly seven lines.

The sexes are alike in colour, and may be thus described:—

Crown of the head and cheeks black, with minute white feathers on the forehead round the base of the upper mandible; a superciliary stripe, a moustache at the base of the upper mandible, and a small tuft of feathers immediately behind the ear-coverts white; feathers on the throat white and bristle-like; upper surface brownish black, becoming browner on the rump; wings brownish black, the outer edges of the quills margined at the base with beautiful wax-yellow, and faintly margined with white towards the extremities; tail brownish black, margined externally at the base with wax-yellow, and all but the two centre feathers with a large oval spot of white on the inner web at the tip; under surface white, broadly striped longitudinally with black, the black predominating on the breast and the white on the abdomen; irides white; bill and feet black.

Sp. 297. MELIORNIS LONGIROSTRIS, *Gould*.

LONG-BILLED HONEY-EATER.

Meliphaga longirostris, Gould in Proc. of Zool. Soc., part xiv. p. 83.
Meliornis longirostris, Cab. Mus. Hein., Theil i. p. 117.
Ban-dene, Aborigines of the lowland districts of Western Australia.
Yellow-winged Honey-eater of the Colonists of Swan River.

Meliphaga longirostris, Gould, Birds of Australia, fol., vol. iv. pl. 24.

Although the *Meliornis longirostris* and *M. novæ-hollandiæ* are very similar, they will on comparison prove to be specifically distinct; they are, in fact, beautiful representatives of each other on the opposite sides of the great Australian continent, the *M. longirostris* inhabiting the western, and the *M. novæ-hollandiæ* being spread over the eastern portion of the country, and it would be a matter of some interest to know at what degree of longitude the two species inosculate. Several points of difference are found to exist in the two species, the most material of which are in the shape and length of the bill, and in the size of the white mark on the fore-part of the cheeks; the *M. longirostris*, as its name implies, has the bill much more lengthened and comparatively stouter than that of its near ally, and it moreover has the white patch on the face much less defined, and blended to a greater extent with the neighbouring black colouring; in the size of the body the two species are very much alike.

The *M. longirostris*, like the other species of the group, is very pugnacious, and when fighting utters a rapidly repeated chirrup, very much resembling that of the European Sparrow.

It is a very early breeder, commencing in the first days of July and continuing as late as the last week in November. The nest consists of small sticks and fibrous roots, lined with Zamia wool or the buds of flowers, and is built in a variety of situations, sometimes in small thinly-branched trees, at about twelve feet from the ground, at others in small clumps of grass, only a few inches above it; the eggs are ordinarily two in number, but towards the latter end of the breeding-season three are often found; their ground-colour is a delicate buff, with the larger end clouded with reddish buff, and thickly spotted and blotched with chestnut-brown and chestnut-red arranged in the form of a zone; their medium length is nine lines, and breadth seven lines.

The sexes are alike in colouring, but the female is about one-fifth smaller than her mate in all her admeasurements.

Crown of the head and cheeks black, with minute white feathers on the forehead round the base of the upper mandible; a superciliary stripe, a moustache at the base of the lower mandible, and a small tuft of feathers immediately behind the ear-coverts white; feathers on the throat white and bristle-like; upper surface brownish black, becoming browner on the rump; wings brownish black, the outer edges of the quills margined at the base with beautiful wax-yellow, and faintly margined with white towards the extremities; tail brownish black, margined externally at the base with wax-yellow, and all but the two centre feathers with a large oval spot of white on the inner web at the tip; surface white, broadly striped with black, the black predominating on the breast and the white on the abdomen; irides white; bill and feet black.

Total length 7 inches; bill 1; wing ⅔; tail 3½; tarsi ¾.

Sp. 208. MELIORNIS SERICEA, *Gould.*

WHITE-CHEEKED HONEY-EATER.

New Holland Creeper, female, White's Voy., pl. in p. 297?
L'Héorotaire noir, Vieill. Ois. Dor., tom. ii. p. 106, pl. 71.
Meliphaga sericea, Gould in Proc. of Zool. Soc., part iv. p. 144.
—— *sericeola*, Gould in Proc. of Zool. Soc., part v. p. 152, female.
Meliornis sericea, Cab. Mus. Hein., Theil i. p. 117.

Maliphaga sericea, Gould, Birds of Australia, fol., vol. iv. pl. 25.

The White-cheeked Honey-eater is an inhabitant of New South Wales, and certainly proceeds as far to the eastward as Moreton Bay; but the birds inhabiting the country to the northward of this being comparatively unknown, it is impossible to say how far its range may extend in that direction. It has not been discovered in Tasmania or South Australia. It differs materially in its habits and diposition from the *Meliphaga novæ-hollandiæ*, being less exclusively confined to the brushes, and affecting localities of a more open character. I observed it to be tolerably abundant in the Illawarra district, particularly among the shrubs surrounding the open glades of the forest; it is also common at Botany Bay, and on most parts of the sea-coast between that place and the river Clarence; but I never met with it during any of my excursions into the interior of the country.

Unlike its near ally, it is a remarkably shy species; so much so, that I had much difficulty in getting within gun-shot of it. When perched on the trees it is a most showy bird, its white cheek-feathers and contrasted tints of colouring rendering it very conspicuous.

I did not succeed in finding its nest.

The sexes are alike in colour, but the female is somewhat smaller than the male. The white cheeks and the absence of white tips to the tail-feathers will at all times distinguish it from the *M. novæ-hollandiæ*.

Crown of the head, throat, and space round the eye black; an obscure band of white crosses the forehead and passes over each eye; a beautiful plume of hair-like white feathers spreads over the cheeks and ear-coverts; back dusky brown, striped longitudinally with black; under surface white, each feather having a central longitudinal mark of black; wings dark brown, the outer edge of all the primaries and secondaries wax-yellow; tail dark brown, the external edges margined with yellow; irides dark brown; feet and bill black.

Total length 6¼ inches; bill ⅞; wing 2¾; tail 2½; tarsi ⅞.

Sp. 299. MELIORNIS MYSTACALIS, *Gould*.

MOUSTACHED HONEY-EATER.

Meliphaga mystacalis, Gould in Proc. of Zool. Soc., part viii. p. 161.
Meliornis mystacalis, Cab. Mus. Hein., Theil i. p. 117.
Ben-dene, Aborigines of Swan River.

Meliphaga mystacalis, Gould, Birds of Australia, fol., vol. iv. pl. 26.

At the time I described this new species of *Meliornis* in the 'Proceedings of the Zoological Society,' I was not aware that Temminck had applied the term *mystacalis* to another species of Honey-eater, or I should have selected a different appellation; as, however, Temminck's bird belongs to a distinct section of this great family, any alteration would rather tend to produce confusion than otherwise.

The *Meliornis mystacalis* is a native of Western Australia, in which country it beautifully represents the *M. sericea* of New South Wales. It is abundant in the vicinity of Perth and Fremantle, and is sparingly dispersed over many other districts of the Swan River colony; according to Gilbert it is remarkably shy, and only found in the most secluded places in the bush, or on the summits of the limestone hills running parallel with the beach; it generally feeds on the topmost branches of the *Banksiæ*, and is very pugnacious, defending its young from intruders with the most determined courage.

Its flight is very varied, and is occasionally characterized by a great degree of rapidity: during the season of incubation it frequently rises above its nest in a perpendicular direction, and having attained a considerable height, suddenly closes its wings, and descends abruptly until it reaches the top of the scrub, when the wings are again expanded, and it flies horizontally for a few yards, perches, and then utters its peculiar sharp chirping note; it also occasionally hovers over small trees, and captures insects after the manner of the Flycatchers.

It is a very early breeder, young birds ready to leave the nest having been found on the 8th of August; it has also been met with breeding as late as November; it doubtless, therefore, produces more than one brood in the course of the season. The nest is generally built near the top of a small, weak, thinly-branched bush, of about two or three feet in height, situated in a plantation of seedling mahogany or other *Eucalypti*; it is formed of small dried sticks, grass, and narrow strips of soft bark, and is usually lined with Zamia wool; but in those parts of the country where that plant is not found, the soft buds of flowers, or the hairy flowering part of grasses, form the lining material, and in the neighbourhood of sheep-walks, wool collected from the scrub. The eggs are usually two in number. They are nine lines long by seven lines broad, and are usually of a dull reddish buff, spotted very distinctly with chestnut and reddish brown, interspersed with obscure dashes of purplish grey.

Head, chin, and throat black; over the eye a narrow line of white; ears covered by a conspicuous tuft of white feathers, which are closely set, and terminate in a point towards the back; upper surface brownish black, the feathers edged with white; under surface white, with a broad stripe of black down the centre of each feather; wings and tail blackish brown, conspicuously margined with bright yellow; irides brown; bill black; feet blackish brown.

Total length 6¼ inches; bill 1; wing 3; tail 2¾.

Genus LICHMERA, *Cabanis*.

Of this form I consider there is only one species known, the *L. australasiana*, for I cannot agree with M. Cabanis in associating with it the *Glycephala* (*Stigmatops*) *ocularis*.

Lichmera differs from *Meliornis* in presenting a considerable variation in the colouring of the sexes, in other respects it is very similar.

Sp. 300. LICHMERA AUSTRALASIANA.

Tasmanian Honey-eater.

L'Héoroteire noir et blanc, Vieill. Ois. Dor., t. ii. pl. 55. p. 89.
Certhia australasiana, Shaw, Gen. Zool., vol. viii. p. 226.
—— *pyrrhoptera*, Lath. Ind. Orn., Supp. p. xxxviii?
Meliphaga australasiana, Vig. and Horsf. in Linn. Trans., vol. xv. p. 313.
Meliphaga inornata, Gould in Proc. of Zool. Soc., part v. 1837, p. 152.
Melithreptus melanoleucus, Vieill., 2ᵉ édit. du Nouv. Dict. d'Hist. Nat., tom. xiv. p. 326.
Lichmera australasiana, Cab. Mus. Hein., Theil i. p. 118.

Meliphaga australasiana, Gould, Birds of Australia, fol. vol. iv. pl. 27.

This little Honey-eater is abundantly dispersed over every part of Tasmania, South Australia, and New South Wales.

It is one of the few species which enliven with their presence the almost impenetrable forests that cover a great portion of Tasmania, giving preference to such parts as are clothed with a thick brush of dwarf shrubby trees growing beneath the more lofty gums. It also resorts to the thick beds of the *Epacris impressa*, whose red and white heath-like flowers bespangle the sides of the more open hills; the blossoms of this beautiful plant afford it an abundant supply of food, which it seeks so intently as to admit of a sufficiently close approach to enable one to observe its actions without disturbing it; while thus occupied it may be seen clinging to the stems in every possible attitude, and inserting its slender brush-like tongue up the tube of every floret with amazing rapidity. Independently of honey it feeds on insects of various kinds, particularly those of the orders *Diptera* and *Hymenoptera*. When disturbed it flits off with a quick darting flight, settling again at the distance of a few yards among the thickest tufts of the *Epacris*, or shrouds itself from observation among the foliage of the sapling gums. It breeds in September and the four following months.

The nest, which is always placed on a low shrub, is of a round, open form, outwardly constructed of the inner rind of the stringy bark gum-tree, and generally lined with fine grasses.

The male has a black stripe passing from the base of the bill through the eye, and a lunar-shaped mark down each side the breast, nearly meeting in the centre, black; a narrow stripe above the eye and one behind the lunar marks on the breast white; all the upper surface dusky black; wings blackish brown, the primaries and secondaries margined externally, particularly at their base, with golden yellow; tail-feathers brownish black, fringed with golden yellow at the base, the two lateral feathers having a long oval spot of white on their inner webs at the tip; throat and chest white, with a streak of brown down the middle of each feather; centre of the abdomen white; flanks and under tail-coverts sooty grey; irides red; bill and feet black.

The female is of a nearly uniform dusky brown; is destitute of the white stripe over the eye and the white spots on the lateral tail-feathers; has only a faint tinge of the golden yellow on the wings and tail.

Genus GLYCIPHILA, *Swainson*.

The members of this genus resort to higher trees than the *Meliornes*, are more shy in disposition, possess considerable power of flight, and partake more exclusively of insect-food.

The young differ considerably from the adult in their markings.

Sp. 301.   GLYCIPHILA FULVIFRONS.

FULVOUS-FRONTED HONEY-EATER.

Certhia fulvifrons, Lewin, Birds of New Holl, pl. 22.
Meliphaga fulvifrons, Vig. and Horsf. in Linn. Trans., vol. xv. p. 317.
Glyciphila fulvifrons, Swains. Class. of Birds, vol. ii. p. 326.
—— *melanops*, Gray, Gen. of Birds, vol. i. p. 119, *Glyciphila*, sp. 1.
Philedon rubrifrons, Less. Voy. de la Coq.
Wy-ro-dju-dong, Aborigines of the lowland districts of Western Australia.
White-throated Honey-sucker, Colonists of Swan River.
Certhia melanops, Lath. Ind. Orn., Supp. p. xxxvi?
—— *mellivora*, Shaw, Gen. Zool., vol. viii. part 1, p. 245.
Meliphaga albiventris, Steph. Cont. of Shaw's Gen. Zool., vol. xiv. p. 201.

Glyciphila fulvifrons, Gould, Birds of Australia, fol., vol. iv. pl. 28.

This species would appear to be distributed over the whole of the southern portion of the Australian continent, since it is to be found in New South Wales, South Australia, and at Swan River, where it is particularly abundant on the limestone hills near the beach around Fremantle; it is also an inhabitant of the northern parts of Tasmania, and all the islands in Bass's Straits.

Its flight is very rapid, and it frequently mounts high in the air, and flies off to a distance with an horizontal and even motion. It is an exceedingly active bird among the branches, clinging about and around the flowers of the *Eucalypti* in search of food in every variety of position.

The site generally chosen for its nest, as observed at Swan River, is some low bush or scrubby plant, in which it is often placed near the ground; it is of a deep cup-shaped and compact form, constructed of dried grasses, and frequently lined with Zamia wool, or buds of the Banksia cones; sometimes, however, sheep's wool is employed to impart warmth and softness; the materials, in fact, depend entirely upon the nature of those that the locality may furnish, while in the form of the nest little or no variation occurs. The eggs are large for the size of the bird, and are often of a lengthened form; and sometimes quite white, without the least trace of spots, but they are generally blotched with marks of chestnut-red; occasionally this colour is very faint, and spread over the surface of the shell as if stained with it; in other instances the marks are very bold and decided, forming a strong contrast to the whiteness of the other part of the surface: the medium length of the eggs is ten lines and a half, and breadth seven lines; they are usually two in number, but the bird very frequently lays only one. The breeding-season lasts from August to February.

The song is rather remarkable, commencing with a single note slowly drawn out, and followed by a quick repetition of a double note, repeated six or eight times in succession; it is mostly uttered when the bird is perched on the topmost branch of a tree.

The sexes present the usual difference in size, the female being somewhat less than her mate; but in the colour and disposition of the markings they are alike.

Forehead and under surface of the wing fulvous or tawny; over each eye a narrow line of white; a line of brownish black commences at the base of the bill, surrounds the eye, passes down the sides of the neck and chest, and nearly meets on the breast; behind the ear-coverts a narrow stripe of buffy white, separated from the line over the eye by a small patch of black; centre of the back dark brown, with a stripe of ashy brown down the centre of each feather; the remainder of the upper surface and flanks ashy brown; throat and abdomen white; wings and tail dark brown, the wing-coverts and primaries margined with olive; irides brown; bill blackish brown; legs and feet greenish grey.

The young has all the upper surface dark brown, streaked with buffy white, and is entirely destitute of the fulvous covering of the forehead and the lunulate markings on the sides of the chest; the throat, moreover, is of a dull wax-yellow, the chest mottled dark brown and buffy white, and the primaries edged with a dull wax-yellow.

Sp. 302.   GLYCIPHILA ALBIFRONS, *Gould*.

WHITE-FRONTED HONEY-EATER.

Glyciphila albifrons, Gould in Proc. of Zool. Soc., part viii. p. 160.
Gool-be-gool-burn, Aborigines of the mountain districts of Western Australia.
White-throated Honey-eater, of the Colonists.

Glyciphila albifrons, Gould, Birds of Australia, fol., vol. iv. pl. 29.

I first observed this fine species of *Glyciphila* in the great Murray scrub of South Australia, where I succeeded in killing several specimens of both sexes; it is an inhabitant of the York and other inland districts of Western Australia, and it is also found in the interior of Victoria and New South Wales.

In its disposition the present bird is remarkably shy, a trait common, it would seem, to all the members of the genus. All those I observed were busily engaged in collecting their insect and saccharine food from the flowers of a species of dwarf *Eucalyptus*.

Its note is rapidly repeated, and much resembles the double call of the *Pardalotus striatus*, but is much louder and more distinct.

The breeding-season lasts from August to February. The nests observed were constructed in the fork of a small dead branch in an exposed situation; they were very similar to that of *Meliornis longirostris*, but more shallow and less neatly formed. The eggs also closely resembled those of that bird, the ground-colour being delicate buff, clouded with a reddish tint at the larger end, and distinctly spotted with chestnut and purplish grey, thickly disposed at the larger end, but very sparingly over the rest of the surface; the eggs are nine and a half lines long by seven lines broad.

The sexes present no difference in colour or markings, but, as usual, the female is much less in size.

Forehead, lores, a narrow ring round the eye, and a narrow line running from the angle of the lower mandible white; crown of the head black, each feather slightly margined with white; ear-coverts blackish grey, behind which an irregular line of white; all the upper surface brown, irregularly margined with white, producing a mottled appearance; wings and tail brown, the primaries margined externally with yellowish green; chin and throat brownish black, the former minutely speckled with white; under surface of the wing buff; chest and abdomen white, striped with blackish brown on the flanks; irides dark brown; naked skin round the eyes dark brownish black in front, arterial blood-red behind; bill black; legs and feet greenish grey.

Total length $5\frac{3}{4}$ inches; bill $\frac{3}{4}$; wing $3\frac{1}{2}$; tail $2\frac{3}{4}$; tarsi $\frac{7}{8}$.

Sp. 303.　GLYCIPHILA FASCIATA, *Gould.*
FASCIATED HONEY-EATER.

Glyciphila fasciata, Gould in Proc. of Zool. Soc., part x. p. 187.

Glyciphila fasciata, Gould, Birds of Australia, fol., vol. iv. pl. 30.

All the specimens hitherto collected of this species have been obtained from the Cobourg Peninsula, where, according to Gilbert, it is far from being common, for in his notes he says, "I only once observed it near the settlement, and once again met with it on the neck of the peninsula near the mainland. Its favourite haunts appeared to be the upper branches of the *Melaleucæ*, from the blossoms of which it collects its food. In both instances I observed small families of about twelve in number. Its note is a sharp shrill piping call, very rapidly repeated."

The fasciated markings of the under surface, by which this species is at once distinguished from every other member of the genus *Glyciphila*, would seem to indicate the propriety of its being separated therefrom; as, however, it is precisely of the same structure, and agrees with them in the colouring of the upper surface, I have preferred retaining it where it was originally placed.

Its food consists of insects generally, the pollen and occasionally the buds of flowers.

Crown of the head brownish black, with a small crescent of white at the extremity of each feather; feathers of the back very dark brown, margined with buffy brown; rump tinged with rufous; wings and tail dark brown, fringed with light brown; sides of the face, throat, and under surface white; from the angle of the mouth, down the side of the neck, a narrow stripe of brownish black; chest crossed by a number of semicircular brownish-black fasciæ; flanks and under tail-coverts buff, the former with a stripe of brownish black down the centre; irides reddish brown; bill greenish grey; feet aurora-red.

Total length $4\frac{3}{4}$ inches; bill $\frac{5}{8}$; wing $2\frac{7}{8}$; tail $2\frac{1}{2}$; tarsi $\frac{5}{8}$.

Genus STIGMATOPS, *Gould.*

Of this form, I believe two, if not three species inhabit Australia, and as many more the islands to the northward.

Sp. 304.   STIGMATOPS OCULARIS, *Gould.*

BROWN HONEY-EATER.

Glyciphila ? ocularis, Gould in Proc. of Zool. Soc., part v. p. 154.
Lichmera ocularis, Cab. Mus. Hein., Theil i. p. 118.
Jin-jo-gow, Aborigines of the mountain districts of Western Australia.
Brown Honey-sucker of the Colonists.

Glyciphila ocularis, Gould, Birds of Australia, fol., vol. iv. pl. 31.

I met with the Brown Honey-eater in abundance on Baker's Island at the mouth of the Hunter, and on the banks of the Namoi in the interior of New South Wales; and Gilbert records that he found it equally numerous at Swan River.

In its actions and manners it displays the usual activity of the Honey-eaters generally, creeping and clinging among the branches with the greatest ease, and particularly affecting those most laden with blossoms, into which it inserts its brush-like tongue to procure the sweet pollen: like other species of the group, it also feeds with avidity upon all kinds of small insects.

Its powers of song are considerable: the most frequently repeated note being remarkably shrill, rich, clear, and distinct in tone. While the female is sitting upon her eggs, the male sings all day long with scarcely any intermission.

The situations chosen for the site of the nest are various; the most favourite position appears to be the side of a tea-tree, the bark of which is hanging down in tatters; it is also sometimes suspended from the drooping branches of the stinkwood; and in one instance Gilbert found it attached to two slender fibrous roots, hanging from beneath a bank over a pool of water. The nest is generally formed of soft strips of paper bark or dried grasses, matted together with small spiders' cocoons or vegetable fibres, and so closely resembles the branch upon which it is placed, as to render it very difficult of detection; it is usually lined with fine grasses, zamia wool, the soft part of the cones of the *Banksiæ*, delicate white buds of flowers, or sheep's wool collected from the bushes of the sheep-runs.

September, October, and November constitute the breeding-season. The eggs, which are two in number, vary considerably in their colouring, some being pure white without a trace of spots or markings, others having a zone round the larger end formed of freckled markings of light reddish brown; others again are thinly sprinkled with this colour over the whole of their surface, and one or two procured at Swan River were bespeckled with numerous fine freckles of bluish grey; the average length of a number of eggs was eight lines by six lines in breadth.

Crown of the head, all the upper surface, wings, and tail dark olive-brown, passing into yellowish brown on the rump and bases of the tail-feathers; primaries and secondaries margined with wax-yellow; immediately behind the eye a very small patch of glossy brownish-yellow feathers, the anterior portion of which is silvery; throat and chest greyish brown; abdomen and under tail-coverts olive-grey; irides light red; bill dark brown; legs and feet bluish grey; tarsi tinged with green.

Total length 5¼ inches; bill ⅞; wing 2¾; tail 2¼; tarsi ¾.

Sp. 305.   STIGMATOPS SUBOCULARIS, *Gould.*

LEAST HONEY-EATER.

Glyciphila ? subocularis, Gould in Proc. of Zool. Soc., part v. p. 154.

In the folio edition of the 'Birds of Australia' I united this bird with *S. ocularis*; but upon further examination and comparison I have come to the conclusion that it is different.

I believe that another species of this form exists on the north-west coast.

The *S. subocularis* is a smaller bird than *S. ocularis*, and consequently one of the most diminutive of the *Meliphagidæ*; besides differing in size, a yellower tint pervades the entire plumage, and the little spangle-like feathers behind the eye are scarcely observable; in all other respects the two birds are very similar.

The *S. subocularis* was shot on the north-west coast, and the skin kindly sent to me by Lieut. Emery of H.M.S. Beagle.

Total length 4¾ inches; bill ⅝; wing 2⅜; tail 2¼; tarsi ⅝.

Genus PTILOTIS, *Swainson*.

The species of this group are not only more numerous than those of any other division of the *Meliphagidæ*, but they also comprise some of the most beautiful and gaily-coloured members of the family. Nearly all the species are either prettily marked about the face, or have the ear-coverts largely developed and characterized by a colouring different from that of the other parts of the plumage. The species with olive-green backs, such as *P. flavigula* and *P. leucotis*, frequent the dwarf and thickly-leaved *Eucalypti*; the more gaily-attired species with bright yellow cheeks and ear-coverts, such as *P. ornata* and *P. plumula*, are most frequently found among the flowering *Acaciæ*; some species, particularly *P. penicillata*, descend from the trees and seek for insects on the ground; while the *P. chrysotis*, *P. chrysops*, and *P. fusca* are almost entirely confined to the brushes and seek honey and insects from among the hanging festoons of *Tecoma* and other beautiful creepers. The members of this group are principally Australian, but some inhabit New Guinea and the adjacent islands; they are generally alike in plumage, but the females are smaller than the males, and the young assume the adult livery from the nest.

Sp. 306.   PTILOTIS LEWINII, *Swainson.*

LEWIN'S HONEY-EATER.

Meliphaga chrysotis, Lewin, Birds of New Holl., pl. 5.
Ptilotis lewinii, Swains. Class. of Birds, vol. ii. p. 326.
Spot-eared Creeper, Shaw, Gen. Zool., vol. viii. p. 244.

Ptilotis chrysotis, Gould, Birds of Australia, fol., vol. iv. pl. 32.

This bird is certainly the *Meliphaga chrysotis* of Lewin's "Birds of New Holland," where it is beautifully figured, but it is equally certain that it does not correspond with Latham's description of his *Certhia chrysotis* as given in his "General History;" neither is it figured by Vieillot in his "Oiseaux Dorés," to which Latham refers. I shall, therefore, adopt the specific name *Lewinii* proposed for it by Swainson.

The Yellow-eared Honey-eater is very common in New South Wales, where it inhabits the thick brushes. I found it especially abundant in all parts of the river Hunter, as well as on the Liverpool and other ranges. No examples came under my notice in South Australia, and I do not believe that it extends so far to the westward. In its habits and disposition it assimilates very closely to the *Ptilotis flavigula* of Tasmania. It prefers low shrubby trees to those of a larger growth. I have often been permitted to approach within a few yards of it while threading the dense brushes without causing it the least alarm. Like the rest of its genus, this species feeds on insects, the pollen of flowers, and occasionally fruits and berries. It is not celebrated for the richness of its notes or for the volubility of its song, but its presence, when not visible among the foliage, is always to be detected by the loud ringing whistle note, which it continually pours forth during the months of spring and summer.

The sexes are alike in colour, but the female presents the same disparity of size that is observable between the sexes of the other species of the genus; the young at an early age assume the plumage of the adults, but the colour is not so rich or decided.

I found a nest of this species in a gully under the Liverpool range; it was placed in the thickest part of one of the creeping plants which overhung a small pool of water; like that of the rest of the genus, it was cup-shaped in form, suspended by the brim, and very neatly made of sticks and lined with very fine twigs; the eggs are two in number, of a pearly white spotted with purplish brown, the spots forming a zone at the larger end; they are eleven and a half lines long by eight lines broad.

Upper surface olive-green; under surface the same colour but paler; behind the ears an oval spot of fine yellow; region of the eyes blackish; below the eye a narrow stripe of yellow; bill black at the tip, yellow at the base; legs purplish flesh-colour; irides dark lead-colour; gape white.

Sp. 307. PTILOTIS SONORA, *Gould.*

SINGING HONEY-EATER.

Ptilotis sonorus, Gould in Proc. of Zool. Soc., part viii. p. 160.
Meliphaga sonora, Gray and Mitch. Gen. of Birds, vol. i. p. 122, *Meliphaga*, sp. 12.
Doo-run-doo-run, Aborigines of the lowlands, and
Gool-bo-ort, Aborigines of the mountain districts of Western Australia.
Larger Honey-sucker, Colonists of Swan River.

Ptilotis sonorus, Gould, Birds of Australia, fol., vol. iv. pl. 33.

I have abundant evidence that the range of this species extends across the entire continent of Australia from east to west; I found it very numerous on the Namoi and other portions of the interior of New South Wales, and equally plentiful in South Australia; it is one of the commonest birds of the colony of Swan River, and we know that it extends very far north, for examples were procured by Gilbert during Dr. Leichardt's expedition from Moreton Bay to Port Essington.

Moderately-sized trees, particularly *Casuarina* and *Banksia*, thinly scattered over grassy plains and the crowns and sides of low hills, are its usual places of resort. In Western Australia it enters the gardens and commits considerable havoc among the fruit-trees, particularly figs, of the seeds of which it appears to be very fond. It also feeds upon insects, which are principally sought for among the branches; but it frequently seeks for them and small seeds on the ground, when it hops around the boles and beneath the branches of the trees in a most lively manner.

Its natural notes are full, clear, and loud, and may be heard at a considerable distance. In South Australia I heard it in full song in the midst of winter, when it was one of the shiest birds of the country.

It is exceedingly pugnacious in disposition, often fighting with the Wattle Birds (*Anthochæræ*), and other species even larger than those.

The breeding-season commences in August and terminates in December. The nest is a frail, round, cup-shaped structure, the materials of which vary in different situations; those observed by me in New South Wales being composed of fine dried stalks of annuals thinly lined with fibrous roots woven together with spiders' webs, and suspended by the rim to two or three fine twigs near the centre of the tree; on the other hand, those observed by Gilbert in Western Australia were formed of green grasses, which become white and wiry when dry, matted together with the hair of kangaroos or opossums, lined with fine grasses and the down of flowers, and placed in a thick scrubby bush at about three feet from the ground.

The eggs are usually two, but occasionally three in number, of a light yellowish buff, thickly freckled with small indistinct reddish brown marks; or of a nearly uniform fleshy buff without spots or markings, but of a deeper tint at the larger end; their medium length is eleven lines, and breadth eight lines.

Crown of the head and all the upper surface greyish olive; wings and tail brown, margined on their external webs with greenish yellow; lores, space round the eye and broad line down the sides of the neck black; ear-coverts pale yellow, behind which is an obscure spot of greyish white; throat and under surface pale yellowish grey striated with light brown; irides dark brown; bill black; legs and feet greenish grey.

Total length 7¼ inches; bill 1; wing 3¾; tail 3¼; tarsi 1.

The female is like the male in colour, but smaller in all her dimensions.

Sp. 308.   PTILOTIS VERSICOLOR, *Gould.*

Varied Honey-eater.

Ptilotis versicolor, Gould in Proc. of Zool. Soc., part x. p. 136.
Meliphaga versicolor, Gray, Gen. of Birds, vol. i. p. 122; *Meliphaga*, sp. 17.

Ptilotis versicolor, Gould, Birds of Australia, fol., vol. iv. pl. 34.

This fine species, which is a native of the northern portion of Australia, is only known to me from a specimen contained in a collection from that part of the country. That its whole habits and economy will hereafter be found to assimilate most closely to those of the *Ptilotis sonora* is certain, as it is most intimately allied to that species, but may be readily distintinguished from it by its larger size, its much longer and stouter bill, by the more contrasted character of its markings, and the sulphur or wax-yellow colour which pervades the breast and upper surface. It is one of the finest species yet discovered of the genus to which it belongs, and is at present so rare, that my own specimen is probably the only one that has been brought to Europe.

All the upper surface brownish olive, tinged with yellowish olive on the margins of the feathers; outer webs of the primaries and tail wax-yellow; inner webs brown; under surface of the wing and tail yellowish buff; stripe over the eye to the back of the neck black; ear-coverts dark grey; below the ear-coverts a stripe of bright yellow; throat and under surface yellow, becoming paler as it approaches the vent, each feather with a stripe of brown down the centre.

Total length 8 inches; bill 1; wing 4; tail 3¼; tarsi 1.

Sp. 309. PTILOTIS FASCIOGULARIS, *Gould.*
FASCIATED HONEY-EATER.

Ptilotis fasciogularis, Gould in Proc. of Zool. Soc., part xix. p. 285.

Ptilotis fasciogularis, Gould, Birds of Australia, fol., Supplement, pl.

It is pleasing to record for the first time a species so well marked as the present, and which differs from the other members of its genus, in the distinct bars of pale yellow and brown which occupy the throat and fore part of the neck. All the specimens that have yet come under my notice were sent to me a few years since by Strange, who collected them on the low swampy islands lying off the eastern coast of Australia, northward of Moreton Bay; they comprise examples of both sexes, ascertained by dissection, and the only difference between them consists, as is usual with the other members of the genus, in the smaller size of the female.

For a *Ptilotis* this is a large and robust species, equalling in size the *P. sonora,* to which it has a close affinity.

All the upper surface, wings, and tail olive-brown, the feathers of the head and back with darker centres, and the primaries and tail-feathers narrowly margined externally with wax-yellow; lores and a streak down the side of the head from the posterior angle of the eye blackish-brown; ear-coverts pale yellow; on each side of the neck a patch of yellowish-white; feathers of the throat brownish-black, each bordered with pale yellow, presenting a fasciated appearance; breast blackish-brown; under surface striated with brown and buff, becoming paler towards the vent; irides lead-colour; bill bluish-black, with a yellow gape; feet black.

Total length 7¼ inches; bill ⅞; wing 3¾; tail 3½; tarsi 1⅛.

Sp. 310. PTILOTIS FLAVIGULA, *Gould.*

YELLOW-THROATED HONEY-EATER.

Ptilotis flavigula, Gould in Proc. of Zool. Soc., part vi. p. 24.
Meliphaga flavigula, Gray and Mitch. Gen. of Birds, vol. i. p. 122, *Meliphaga*, sp. 15.
Melithreptus flavicollis, Vieill., 2ᵉ édit. du Nouv. Dict. d'Hist. Nat., tom. xiv. p. 325 ?

Ptilotis flavigula, Gould, Birds of Australia, fol., vol. iv. pl. 35.

This fine and conspicuous species of *Ptilotis* is abundant in all the ravines round Hobart Town, and is very generally dispersed over the whole of Tasmania. If I mistake not, I have also seen specimens from Victoria. Its colouring assimilating in a remarkable degree with that of the leaves of the trees it frequents, it is somewhat difficult of detection. When engaged in searching for food it frequently expands its wings and tail, creeps and clings among the branches in a variety of beautiful attitudes, and often suspends itself to the extreme ends of the outermost twigs. It flies in an undulating manner like a Woodpecker, but this power is rarely exercised.

Its note is a full, loud, powerful, and melodious call.

The stomach is muscular, but of a very small size, and the food consists of bees, wasps, and other Hymenoptera, to which are added Coleoptera of various kinds, and the pollen of flowers.

It is a very early breeder, as proved by my finding a nest containing two young birds covered with black down, and about two days old, on the 29th of September.

The nest of this species, which is generally placed in a low bush, differs very considerably from those of all the other

Honey-eaters with which I am acquainted, particularly in the character of the material forming the lining; it is the largest and warmest of the whole, and is usually formed of ribbons of stringy bark, mixed with grass and the cocoons of spiders; towards the cavity it is more neatly built, and is lined internally with opossum or kangaroo fur; in some instances the hair-like material at the base of the large leaf-stalks of the tree-fern is employed for the lining, and in others there is merely a flooring of wiry grasses and fine twigs. The eggs, which are either two or three in number, are of the most delicate fleshy buff, rather strongly but thinly spotted with small, roundish, prominent dots of chestnut-red, intermingled with which are a few indistinct dots of purplish grey; their average length is eleven lines, and breadth eight lines.

The only external difference in the sexes is the smaller size of the female, which is nearly a third less than that of the male.

Lores and cheeks black; crown of the head, ear-coverts, breast, and under surface dark grey, with silvery reflexions; a few of the ear-coverts tipped with yellow; chin and upper part of the throat rich gamboge-yellow; all the upper surface, wings, and tail rich yellowish olive, brightest on the margins of the quill- and tail-feathers; inner webs of the primaries and secondaries dark brown; under surface of the shoulder and wing gamboge-yellow; abdomen and flanks washed with olive; bill black; interior of the bill, throat, and tongue rich orange; irides wood-brown; legs and feet brownish lead-colour.

Total length 8 inches; bill 1; wing $4\frac{1}{4}$; tail $4\frac{1}{4}$; tarsi 1.

The young birds assume the adult colouring from the time they leave the nest.

Sp. 311.   PTILOTIS LEUCOTIS.
WHITE-EARED HONEY-EATER.

Turdus leucotis, Lath. Ind. Orn., p. xliv.
White-eared Honey-eater, Lewin, Birds of New Holl., pl. 20.
—— *Thrush*, Lath. Gen. Syn. Supp., vol. ii. p. 373.
Meliphaga leucotis, Vig. and Horsf. in Linn. Trans., vol. xv. p. 314.

Ptilotis leucotis, Gould, Birds of Australia, fol., vol. iv. pl. 36.

The White-eared Honey-eater enjoys a very wide range of habitat; I found it in abundance in the belts of the Murray and other parts of South Australia, and in the brushes near the coast as well as in the open forests of *Eucalypti* in New South Wales; it is very common in the Bargo brush on the road to Argyle, and Gilbert mentions that he shot a specimen near York in the interior of Western Australia, but it is there so rare that he believed the individual he procured was the only one that had been seen. It is as much an inhabitant of the mountainous as of the lowland parts of the country, and is always engaged in creeping and clinging about among the leafy branches of the *Eucalypti*, particularly those of a low or stunted growth.

Its note is loud, and very much resembles that of the *Ptilotis penicillata*. The stomach is small and membranous, and the food consists of insects of various kinds.

The sexes are alike in their markings, but they differ considerably in size, the male being much larger than the female.

Upper surface and abdomen yellowish olive; crown of the head grey, streaked longitudinally with black; throat and chest black; ear-feathers pure silvery white; tips of the tail-feathers yellowish white; bill black; irides greenish grey, with a narrow ring of pale wood-brown; legs and feet leaden greenish grey.

Sp. 312.   PTILOTIS AURICOMIS.

YELLOW-TUFTED HONEY-EATER.

Muscicapa auricomis, Lath. Ind. Orn., Supp. p. xlix.
—— *mystacea*, Lath. Ind. Orn., Supp. p. li.
—— *novæ-hollandiæ*, Lath. Ind. Orn., vol. ii. p. 478.
Mustachoe Honey-eater, Lath. Gen. Syn., Supp. vol. ii. p. 221 ?
Yellow-tufted Flycatcher, Lath. Gen. Syn., Supp. vol. ii. p. 215.
Certhia auriculata, Shaw, Gen. Zool., vol. viii. p. 236.
L'Héorotaire à oreilles jaunes, Vieill. Ois. dor., tom. ii. p. 123. pl. 85.
Tufted-eared Honey-eater, Lath. Gen. Hist., vol. iv. p. 197.
Meliphaga auricomis, Swains. Zool. Ill., vol. i. pl. 43.
Philemon erythrotis, Vieill. 2ᵉ édit. du Nouv. Dict. d'Hist. Nat., tom. xxvii. p. 429.

Ptilotis auricomis, Gould, Birds of Australia, fol. vol. iv. pl. 37.

The Yellow-tufted Honey-eater is abundant in New South Wales, inhabiting at one season or other every portion of the country; the brushes near the coast, the flowering trees of the plains, and those of the sides and crowns of the hills towards the interior being alike tenanted by it. It is an active, animated species, flitting with a darting flight from tree to tree; and threading the most thickly-leaved branches with a variety of sprightly actions.

I never succeeded in finding the nest of this species, but E. P. Ramsay, Esq., has contributed an interesting account of its nidification to the 'Ibis' for 1864, from which I extract the following passages:—

"The Yellow-tufted Honey-eater is perhaps one of the most beautiful birds of New South Wales; nor are its eggs less beautiful than the bird itself. It evinces a preference for the more open underwood of young *Eucalyptus* and Wattle-trees (*Acacia decurrens*), which are plentiful near Dobroyde, Enfield, and Parramatta, rather than for the dense scrub-land near the coast. I have met with it as far as Manar, between Braidwood and Goulburn. Like most of its tribe, the Yellow-tufted Honey-eater is very partial to fruit, and during the latter end of February, and throughout the month of March, the pear-trees swarm with this and many other species. During the orange season also they visit us in great numbers, and many may be seen fighting over the half-decayed fruit with which the ground at that time is literally strewed.

The *Ptilotis auricomis* "remains with us throughout the whole year, and breeds earlier than the generality of Honey-eaters. Eggs have been taken early in June, and as late as the end of October, during which month they sometimes have a third brood; but August and September appear to be their principal breeding months. The nest is a neat, but somewhat bulky structure, open above, and composed of strips of the Stringy-bark-tree. The eggs, which are usually two in number, are of a pale flesh-pink, darkest at the larger end, where they are also spotted and blotched with markings of a much deeper hue, inclining to salmon colour; in some these markings form a zone, in others one irregular patch, with a few dots upon the rest of the surface. When taken they have a beautiful blush of pink, but it generally disappears a few days after they are blown. Their length is from ten to eleven lines by seven to eight in breadth. Some have a few obsolete dots of faint lilac; others are without markings, save one patch at the larger extremity. The site selected is usually some low bushy scrub among the rich clusters of *Tecoma australis*, or amidst the thick tufts of *Blechnum cartilagineum*, which often covers a space of many square yards in these clumps, where it clings to the stems of the ferns. I have several times found two or three pairs breeding at the same time within a few yards of each other. The ferns and *Tecomæ* seem to be their favourite places for breeding; but the nests are often found suspended between forks in the small bushy oaks (*Casuarinæ*).

"In the nest of this Honey-eater I have several times found

the egg of a Cuckoo; this egg is of a very pale flesh-colour, eleven and a half lines long by eight and a half broad, and usually without any markings; but one specimen had a few dots of black and dark reddish brown upon a pale flesh-coloured ground. I have also taken similar eggs from the nests of a species of *Ptilotis* and of *Melithreptus lunulatus*. I have not been able to determine to which species of Cuckoo they belong—most probably to the *Cuculus cinereus*," i. e. *Cacomantis flabelliformis*.

The female of this species, as is the case with others of the genus, is smaller than the male, but exhibits no difference whatever in the colouring of her plumage.

Crown of the head olive-yellow; throat bright yellow; a black line commences at the base of the bill, surrounds the eye, and extends over the ear-coverts; behind the ear springs a lengthened tuft of rich yellow feathers; upper surface, wings, and tail dark brown, with a tinge of olive; primaries and tail-feathers margined with olive-yellow; chest and under surface brownish yellow; bill black; irides reddish brown; feet blackish brown.

Sp. 313.   PTILOTIS CRATITIA, *Gould*.

WATTLE-CHEEKED HONEY-EATER.

Ptilotis cratitia, Gould in Proc. of Zool. Soc., part viii. p. 160.
Meliphaga cratitia, Gray, Gen. of Birds, vol. i. p. 122, *Meliphaga*, sp. 18.
Lichenostomus cratitius, Cab., Mus. Hein., Theil. i. p. 119, note.
—— *occidentalis* Cab. Mus. Hein., Theil i. p. 110 ?

Ptilotis cratitius, Gould, Birds of Australia, fol., vol. iv. pl. 38.

I first met with this new species of Honey-eater on the 26th of June 1839, on the ranges near the Upper Torrens in South Australia: it appeared to be a most pugnacious bird, driving every other species from the tree upon which it was feeding. I afterwards met with it on Kangaroo Island and in the Belts of the Murray. In all these situations it evinced a decided preference for the *Eucalypti*, among the smaller branches and flowers of which it was busily engaged in extracting pollen and honey from the flower-cups. The trees in the Belts of the Murray and on Kangaroo Island are of a dwarf character, while those of the Upper Torrens are very lofty; yet each appeared to be equally resorted to.

I have never seen this bird from any other parts of Australia than those I have mentioned; further research may, however, enable us to assign to it a much greater range of habitat. It is very closely allied to the *Ptilotis auricomis*, but may at all times be distinguished from that, as well as from every other known species by the stripe of beautiful lilac-coloured bare skin, which stretches from the corner of the mouth and extends down the sides of the cheeks; after death, this skin becomes dry and discoloured.

The sexes are nearly alike in plumage, and both have the fleshy appendage on the cheeks, but the female is somewhat smaller than the male.

Crown of the head grey; all the upper surface olive-green; wings and tail brown, margined with greenish yellow; lores, a large space surrounding the eye and the ear-coverts black, below which is a narrow line of bright yellow; from the gape, down each side of the throat for five-eighths of an inch, a naked fleshy appendage, free at the lower end, of a beautiful lilac-colour and very conspicuous in the living bird; anterior to this is a tuft of bright yellow feathers; throat and under surface olive-yellow; irides and eyelash black; bill black; feet blackish brown tinged with olive.

Total length 7 inches; bill $\frac{7}{8}$; wing $2\frac{1}{8}$; tail $3\frac{1}{4}$; tarsi $\frac{7}{8}$.

Although I have placed M. Cabanis's *Lichenostomus occidentalis* as a synonym of this species, with which, after carefully reading his description, I believe it to be identical, I shall restore his bird to the rank of a species, whenever I may obtain evidence that it is really different.

Sp. 314.   PTILOTIS ORNATA, *Gould*.
         GRACEFUL PTILOTIS.

Ptilotis ornatus, Gould in Proc. of Zool. Soc., part vi. p. 24.
Meliphaga ornata, Gray, Gen. of Birds, vol. i. p. 122, *Meliphaga*, sp. 10.

Ptilotis ornatus, Gould, Birds of Australia, fol., vol. iv. pl. 39.

It was a source of much gratification to myself to have unexpectedly found this elegant little bird in that rich arboretum, the Belts of the Murray, which had already supplied me with so many novelties. It was there confined to trees of a dwarf growth, while in the country in the neighbourhood of Swan River I am informed it is seen on the topmost branches of the gum- and mahogany-trees, clinging and flitting about the blossoms, not unfrequently descending to the ground and hopping about beneath the branches and near the boles of the larger trees, doubtless in search of insects.

It has rather a loud ringing and not unpleasing song, which is constantly poured forth.

The nest is generally suspended from a horizontal forked branch, frequently in an exposed situation, and is of a neat, small, open, cup-shaped form, composed of fine vegetable fibres and grasses matted together with spiders' webs, and sometimes wool. The eggs are either two or three in number, of a deep salmon-colour, becoming paler at the smaller end, and minutely freckled with reddish brown, particularly at the larger end; they are nine lines long by seven broad.

The female differs from the male in being somewhat less in size, and those I collected had the nostrils, eyelash, and basal portion of the bill orange instead of black, as in the male; still I am not fully satisfied that this orange colouring may not indicate immaturity, and that the fully adult female may not have these, as in her mate.

Crown of the head, external edge of the wings, rump, and

tail-feathers olive; back olive-brown; all the under surface greyish white, each feather having a longitudinal mark of brown down the centre; under tail-coverts lighter; on each side of the neck a lengthened tuft of rich yellow feathers; eye black, surrounded in the male by a narrow black eyelash except for a third of the space, behind which is yellow; feet purplish brown; bill black.

Total length 6½ inches; bill ½; wing 3½; tail 3¼; tarsi ¾.

Sp. 315. PTILOTIS PLUMULA, *Gould*.

PLUMED PTILOTIS.

Ptilotis plumulus, Gould in Proc. of Zool. Soc., part viii. p. 150.
Meliphaga plumula, Gray, Gen. of Birds, vol. i. p. 122, *Meliphaga*, sp. 11.

Ptilotis plumulus, Gould, Birds of Australia, fol., vol. iv. pl. 40.

In size this species is rather less than *Ptilotis ornata*, and, independently of the accessory black tuft on the sides of the neck, the breast is of more delicate and paler colour, with the feathers much more faintly marked with brown down the centre. All the specimens I have seen were collected in the district of York, about 60 miles eastward of Swan River, where it inhabits the white-gum forests, resorting to the tops of the highest trees, and is seldom to be seen on the ground. Its note is much varied, consisting of a loud shrill shake, somewhat resembling the sportsman's pea-whistle, continued without intermission for a great length of time. When disturbed it flits among the branches with a quick darting flight; while at other times it soars from tree to tree with the most graceful and easy movement.

Its small, elegant, cup-shaped nest is suspended from a slender horizontal branch, frequently so close to the ground as to be reached by the hand; it is formed of dried grasses lined with soft cotton-like buds of flowers. The breeding-season continues from October to January; the eggs being two in number, ten lines long by seven lines broad, of a pale salmon colour, with a zone of a deeper tint at the larger end, and the whole freckled with minute spots of a still darker hue. The stomach is diminutive and slightly muscular, the food consisting of insects and honey.

The sexes appear to present no difference in the colour of their plumage; but the female, as is the case with the other members of the genus, is considerably smaller than her mate.

Crown of the head and all the upper surface bright olive-yellow, approaching to grey on the back; lores black; ear-coverts, throat, and under surface pale yellowish grey, faintly striated with a darker tint; behind the ear two tufts, the upper of which is narrow and black; the lower, which is more spread over the sides of the neck, of a beatiful yellow; primaries and tail-feathers brown, margined with bright olive-yellow; irides very dark reddish brown; bill black; legs and feet apple-green.

Total length 4¾ inches; bill ½; wing 3¼; tail 2¾; tarsi ½.

Sp. 310.   PTILOTIS FLAVESCENS, *Gould*.

YELLOW-TINTED HONEY-EATER.

Ptilotis flavescens, Gould in Proc. of Zool. Soc., part vii. p. 142.
Meliphaga flavescens, Gray, Gen. of Birds, vol. i. p. 122, *Meliphaga*, sp. 16.

Ptilotis flavescens, Gould, Birds of Australia, fol. vol. iv. pl. 41.

The only example of this new species that I have seen is from the north coast of Australia, where it was procured and subsequently presented to me by my friend Benjamin Bynoe, Esq., late of Her Majesty's Surveying Ship the 'Beagle.' It differs from all the other members of its genus in the uniform yellow colouring of its plumage, for which reason I have assigned to it the specific appellation of *flavescens*.

I regret to say that nothing whatever is at present known of its habits or economy.

Head and all the under surface delicate citron-yellow, the yellow prevailing over the head; immediately under the ear-coverts a spot of blackish brown, posterior to which is a spot of bright yellow; the remainder of the plumage olive-grey.

Total length 4¼ inches; bill ½; wing 2⅞; tail 2½; tarsi ¾.

Sp. 317.   PTILOTIS FLAVA, *Gould.*

YELLOW HONEY-EATER.

Ptilotis flava, Gould in Proc. of Zool. Soc., part x. p. 130.
Meliphaga flava, Gray, Gen. of Birds, vol. i. p. 122, *Meliphaga,* sp. 7.

Ptilotis flava, Gould, Birds of Australia, fol. vol. iv. pl. 42.

This new species may be distinguised from all its congeners by the uniform colouring of its plumage; it is in fact a most remarkable bird, inasmuch as I scarcely recollect one similarly coloured in any genus that has come under my notice. I regret that, as regards the history of this Honey-eater, its range over the Australian continent, its habits and economy, all is a perfect blank; a single specimen is all I have at present seen; this was procured by one of the officers of Her Majesty's Ship the 'Beagle,' while employed on the north coast. The names of Captain Stokes, Lieutenant Emery, and Mr. Bynoe have been repeatedly mentioned in this work, with feelings of personal gratification that their labours have been useful to science. It now only remains for me to describe the colours of this bird; having I trust thrown out a sufficient hint to those who may visit its native country, and may have opportunities of observing it, that contributions to its history are very desirable.

Head and all the under surface delicate citron-yellow, the yellow prevailing over the head; immediately under the ear-coverts is a spot of blackish brown, posterior to which is a patch of bright yellow, the remainder of the plumage olive-grey.

Total length 6¼ inches; bill ⅞; wing 3½; tail 3½; tarsi ⅞.

Sp. 318.   PTILOTIS PENICILLATA, *Gould.*

WHITE-PLUMED HONEY-EATER.

Meliphaga penicillata, Gould in Proc. of Zool. Soc., part iv. p. 143.

Ptilotis penicillatus, Gould, Birds of Australia, fol., vol. iv. pl. 43.

This species, which is rarely met with in New South Wales, is very abundant in South Australia; I met with it even in the streets and gardens of Adelaide; and it doubtless enjoys a wide range over the interior of the country. From what I observed of its habits, it appears to differ from the generality of Honey-eaters in the partiality it evinces for the ground; for although most of its time is spent among the leafy branches of the gums and wattles, it is often to be seen hopping about under the trees in search of insects and seeds, which with the pollen of the flowers of the *Eucalypti* and *Acaciæ* constitute its food.

Its silvery white neck-plumes present a character by which it is at once distinguished from all other known species. The smaller size of the female is the only external difference between the sexes, for when fully adult their markings are precisely alike. Some of the specimens killed had the bill entirely black, while others had the base yellowish white, which is doubtless indicative of immaturity.

Its slightly-constructed nest, formed of grasses and wool, is cup-shaped, and is suspended by the rim, like those of the other Honey-eaters. "The *Ptilotis penicillata,*" says Mr. Angas, "builds in the Acacias close to my house at Collingrove, near Angaston. I can sit at dinner and watch the young ones being fed. One female sat hatching close to the window with the strong light of a moderator lamp shining on her at night. The eggs are three in number."

Sides of the face and ear-coverts pale yellow; behind the ear-coverts a small tuft of white silky feathers; upper surface rich yellowish grey, the outer edges of the quill- and tail-feathers tinged with a richer colour; under surface light yellowish brown; bill black; legs purplish flesh-colour; irides very dark brown.

Total length 6¼ inches; bill ⅝; wing 3; tail 3; tarsi ¾.

Sp. 310.   PTILOTIS FUSCA, *Gould.*

FUSCOUS HONEY-EATER.

Meliphaga fusca, Gould in Syn. Birds of Australia, part ii.
Ptilotis fusca, Gould, id., part iv.
Certhia chrysotis, Lath. Ind. Orn., Supp. p. xxxviii ?

Ptilotis fusca, Gould, Birds of Australia, fol., vol. iv. pl. 44.

This species of Honey-eater, which is not distinguished by any brilliancy in its plumage, is abundantly dispersed over the thick brushes of New South Wales; and in the months of August and September, when the beautiful *Tecoma* is in blossom, it may be seen flitting about among the thick clusters of the pendent flowers in search of insects, which are sometimes captured while on the wing, but more generally extracted from the tubular florets.

I observed nothing remarkable in its economy, or in which it differed from the other members of the group. Like them it is generally found among the flowers and the most leafy branches of the trees. I have never seen it on the plains, nor have I received specimens from any other part of Australia than New South Wales, where it is to be met with both in winter and summer.

The sexes are very nearly alike in colouring; in fact, with the exception of the female being a trifle smaller than the male, no outward distinction is visible.

The whole of the upper surface greyish brown with a tinge of olive; a ring of black feathers surrounds the eye; ear-coverts blackish brown; behind the ear a small patch of yellow; throat, chest, and under surface light greyish brown; irides light yellow; eyelash light yellow; gape and corners of the mouth yellow; bill dull yellow at the base and black at the tip; feet fleshy brown.

Total length 6¼ inches; bill ½; wing 3½; tail 2⅔; tarsi ¾.

Sp. 320.   PTILOTIS CHRYSOPS.
    Yellow-faced Honey-eater.

Sylvia chrysops, Lath. Ind. Orn., Supp. p. liv.
Black-cheeked Honey-eater, Lath. Gen. Syn., Supp. vol. ii. p. 248.
Meliphaga chrysops, Vig. and Horsf. in Linn. Trans., vol. xv. p. 316.
Yellow-eared Flycatcher, White's Voy., pl. in p. 161 ?

Ptilotis chrysops, Gould, Birds of Australia, fol., vol. iv. pl. 45.

The *Ptilotis chrysops* may be regarded as one of the commonest species of Honey-eaters inhabiting the colonies of New South Wales and South Australia, its distribution over those countries being almost universal. On reference to my journal I find that it was equally abundant in the gardens of Sydney, in the brushes near the coast, in the district of the Upper Hunter, and on the Liverpool range; and that in South Australia it was quite as numerous in the mangrove thickets on the coast, as in the interior of the country. It is very animated and sprightly in its actions, and during the months of spring and summer is constantly engaged in singing; its melodious song being poured forth while the bird is perched on the topmost branches of the trees.

A nest found near the Liverpool range in October was very neatly constructed, rather small in size, round and open in form, and so thin that I could see through it; it was suspended to the fine twigs of a *Casuarina* at some height from the ground, while another suspended to the lower branches of a sapling gum was within reach of the hand. They were outwardly composed of the inner bark of trees, moss, &c., lined with fine vegetable fibres and grasses. The eggs, which are two and sometimes three in number, are of a lengthened form, and of a deep reddish buff, strongly marked at the larger end with deep chestnut-red and purplish grey; the remainder of the surface ornamented with large spots and blotches of the same colour, somewhat thinly dispersed; their medium length is ten lines and a half by seven lines in breadth.

The sexes are so much alike that no visible difference is perceptible, except in the smaller size of the female.

Crown of the head, back of the neck, all the upper surface, wings, and tail dark brown with a slight tinge of olive; throat and under surface dark greyish brown, the latter colour predominating on the chest; a fine line of black runs from the nostrils through the eye; this black line is bounded below by a stripe of yellow which runs under the eye and over the ear-covert, and below this runs another parallel line of black, which commences at the base of the lower mandible and extends beyond the line of the ear-coverts; immediately above the eye behind is a small spot of yellow, and behind the ear-coverts a like spot of white; bill blackish brown; irides and eyelash dark brown; legs leaden brown.

Sp. 321. PTILOTIS FILIGERA, *Gould.*

STREAKED HONEY-EATER.

Ptilotis filigera, Gould in Proc. of Zool. Soc., part xviii. p. 278.

Ptilotis filigera, Gould, Birds of Australia, fol., Supplement, pl.

The *P. filigera* is one of the novelties which rewarded the researches of Mr. Wilcox, who obtained two examples among some mangroves at Cape York, where he observed it in company with another species of the same genus. Although a dull-coloured species, it is rendered interestingly different from all its congeners by the thread-like streak beneath the ear-coverts, and by the small striæ which decorate the back of the neck and the upper part of the mantle.

Upper surface, wings, and tail rich olive-brown, with numerous small marks of greyish white on the apical portion of the nuchal feathers; the wing-coverts broadly, and the remainder of the feathers narrowly edged with brownish buff; from the gape beneath the eye a streak of white; ear-coverts blackish grey; from the centre of the lower angle of the ear-coverts a very narrow streak of silky yellow, which, proceeding backwards, joins the line of white from beneath the eye; throat brownish grey; under surface sandy buff, the feathers of the breast and the middle of the abdomen with lighter centres; bill olive-black; naked space beneath the eye yellow; legs and feet slate-colour.

Total length 7¼ inches; bill 1; wing 4; tail 3; tarsi ¾.

The young are destitute of the white marks on the nape, and have the under surface more rufous and without the lighter centres.

Genus STOMIOPERA, *Reichenbach.*

Dr. Reichenbach considers the following species sufficiently different from the true *Ptilotes*, to warrant its separation into a distinct genus, and if the difference in its singular habits be taken into consideration, the separation is justifiable.

Sp. 322. STOMIOPERA UNICOLOR, *Gould.*

UNIFORM-COLOURED HONEY-EATER.

Ptilotis unicolor, Gould in Proc. of Zool. Soc., part x. p. 136.
Meliphaga unicolor, Gray, Gen. of Birds, vol. i. p. 122, *Meliphaga,* sp. 8.
Stomiopera unicolor, Reich. Handb. der Spec. Orn., p. 109.

Ptilotis unicolor, Gould, Birds of Australia, fol., vol. iv. pl. 46.

This bird, which differs from the true *Ptilotes* in some parts of its structure, in the uniform colouring of its plumage, and in its habits and manners, is one of the many species that rewarded Gilbert's researches at Port Essington; where he states it was seldom met with in the immediate vicinity of the harbour, but that it gradually increased in number as he approached the narrow neck of the peninsula and the mainland about Mountnorris Bay. The situations in which it was usually observed were those adjacent to swampy thickets, and here it was generally seen in pairs: it appears to be of a most lively disposition, being always in motion; its actions much resemble those of the *Tropidorhynchus argenticeps*, with which bird it often fights severe battles. When among the trees its movements are very amusing, and its agility in running upon and creeping round the branches in search of insects is fully equal to that of the *Sittella*. Its flight is very short, feeble, and peculiar, rarely extending to a greater distance than from branch to branch, or from tree to tree, and is performed with a very rapid motion of the wings; the tail being at the same time much retroverted over the back, gives the bird a most ludicrous appearance. It emits a great variety of notes and calls; frequently giving utterance to a loud chattering cry, much resembling that of the *Mayzantha*, but more often a note so similar to the well-known chirrup of the common English Sparrow, that it might be easily mistaken for the note of that bird.

The stomach is diminutive but muscular, and the food consists of honey, insects of various kinds, seeds, and berries.

Lores and orbits deep brown; all the plumage brownish olive; the under surface paler than the upper; primaries margined with brighter olive than the other parts of the body; under surface of the shoulder pale buff; irides obscure red; bill dark olive-brown; naked gape fleshy white, passing into yellow at the corner of the mouth; legs and feet light ash grey.

Total length 7 inches; bill 1; wing $3\frac{2}{3}$; tail $3\frac{1}{4}$; tarsi 1.

Genus PLECTORHYNCHA, *Gould*.

Of this singular form only one species has yet been discovered. It inhabits the plains of the eastern portion of Australia, where it dwells among the *Eucalypti* and *Acaciæ*, and is a very noisy garrulous bird.

The sexes are alike in plumage, and the young assume the adult plumage at a very early age.

Sp. 323. PLECTORHYNCHA LANCEOLATA, *Gould*.

LANCEOLATE HONEY-EATER.

Plectorhyncha lanceolata, Gould in Proc. of Zool. Soc., part v. p. 153.
Melithreptus lanceolatus, Gray, Gen. of Birds, vol. i. p. 128, *Melithreptus*, sp. 11.

Plectorhyncha lanceolata, Gould, Birds of Australia, fol., vol. iv. pl. 47.

The Liverpool Plains and the country immediately to the northward are, I believe, the only portions of the Australian continent in which this bird has been seen. I found it rather sparingly dispersed over the forests bordering the rivers Mokai and Namoi, and it appeared to increase in number as I descended the latter stream towards the interior. It was generally observed alone, or in pairs, keeping almost exclusively to the *Acaciæ* and *Eucalypti*. Its chief food is the pollen of flowers and insects, for procuring which and for constructing its beautiful nest its pointed spine-like bill is admirably adapted. I find it stated, in my notes taken on the spot, that this bird possesses the peculiar habit of sitting motionless among the thickest foliage of the topmost branches of the highest trees, where it cannot be seen without the closest observation, although its immediate locality is indicated by its powerful whistling note. Upon one occasion only did I discover the nest; it was suspended from the extreme tip of a branch of a *Casuarina* overhanging a stream, was outwardly composed of grasses, interwoven with wool and the cotton-like texture of flowers, and contained two eggs rather lengthened in shape, being eleven and a half lines long by eight lines broad, and of a flesh-white, very minutely sprinkled with reddish buff, forming an indistinct zone at the larger end.

So closely do the sexes resemble each other in colour, that by dissection alone can they be distinguished; the male, however, rather exceeds the female in size.

Crown of the head, ear-coverts, and back of the neck mottled with black and white, a longitudinal mark of black running down the centre of each feather; throat and under surface greyish white, the stem of each feather, which ends lanceolate, pure white; back, wings, and tail light brown; irides brown; bill dark bluish horn-colour; legs and feet light blue.

Total length 9 inches; bill 1; wings $4\frac{1}{4}$; tail $4\frac{1}{2}$; tarsi 1.

The young, of which I killed several specimens in the month of January, had even at that early age assumed the general markings of the adult; and the circumstance of there being fully-fledged young and eggs at the same time, proves that these birds rear at least two broods in the season.

Genus MELIPHAGA, *Lewin*.

The only species of this form known appears to be confined to the south-eastern portions of Australia. It is extremely bold and pugnacious, and generally frequents the highest branches of the lofty *Eucalypti*, both of the brushes and of the plains, but is most abundant in the districts near the coast.

The sexes are alike in plumage, and but little difference is observable between nestling and adult birds.

Sp. 324.　　MELIPHAGA PHRYGIA.

Warty-faced Honey-eater.

Merops phrygius, Lath. Ind. Orn., Supp. p. xxxiv.
Warty-faced Honey-eater, Lewin, Birds of New Holl., pl. 14.
Black and Yellow Bee-eater, Lath. Gen. Syn., Supp. vol. ii. p. 154.
Black and Yellow Honey-eater, Lath. Gen. Hist., vol. iv. p. 165.
Philemon phrygius, Vieill. Ency. Méth., part ii. p. 617.
Meliphaga phrygia, Lewin, Birds of New Holl., p. 13. pl. 4.
Anthochera phrygia, Vig. and Horsf. in Linn. Trans., vol. xv. p. 322.
Zanthomiza phrygia, Swains. Class. of Birds, vol. ii. p. 326.
Xanthomyza phrygia, Gould, Birds of Australia, fol., vol. iv. pl. 48.
Mock Regent Bird, Colonists of New South Wales.

Zanthomyza phrygia, Gould, Birds of Australia, fol., vol. iv., text to pl. 48.

This is not only one of the handsomest of the Honey-eaters, but is also one of the most beautiful birds inhabiting Australia, the strongly contrasted tints of its black and yellow plumage rendering it a most conspicuous and pleasing object, particularly during flight. It is a stationary species, and enjoys a range extending from South Australia to New South Wales; I also met with it in the interior nearly as far north as the latitude of Moreton Bay. Although it is very generally distributed, its presence appears to be dependent upon the state of the *Eucalypti*, upon whose blossoms it mainly depends for subsistence; it is consequently only to be found in any particular locality during the season that those trees are in blossom. It generally resorts to the loftiest and most fully-flowered tree, where it frequently reigns supreme, buffeting and driving every other bird away from its immediate neighbourhood; it is, in fact, the most pugnacious bird I ever saw, evincing particular hostility to the smaller *Meliphagidæ*, and even to others of its own species that may venture to approach the trees upon which two or three have taken their station. While at Adelaide, in South Australia, I observed two pairs that had possessed themselves of one of the high trees that had been left standing in the middle of the city, which tree, during the whole period of my stay, they kept sole possession of, sallying forth and beating off every bird that came near. I met with it in great abundance among the brushes of New South Wales, and also found it breeding in the low apple-tree flats of the Upper Hunter. I have occasionally seen flocks of from fifty to a hundred in number, passing from tree to tree as if engaged in a partial migration from one part of the country to another, or in search of a more abundant supply of food.

The nest, which is usually constructed on the overhanging branch of a *Eucalyptus*, is round, cup-shaped, about five inches in diameter, composed of fine grasses, and lined with a little wool and hair. The eggs are two in number, of a deep yellowish buff, marked all over with indistinct spots and irregular blotches of chestnut-red and dull purplish grey, particularly at the larger end, where they frequently form a zone; they are eleven lines long by eight lines and a half broad.

The stomachs of the specimens I killed and dissected on the Hunter were entirely filled with liquid honey; insects, however, doubtless form a considerable portion of their diet.

The sexes are nearly alike in colouring, but the female is much smaller than the male, and the young are destitute of the warty excrescences on the face, that part being partially clothed with feathers.

Head, neck, upper part of the back, chin, and chest black; scapularies black, broadly margined with pale yellow; lower part of the back black, margined with yellowish white; upper tail-coverts like the scapularies; wings black, the coverts margined with yellow; spurious wing yellow; primaries black, with an oblong stripe of yellow occupying the margin of the outer and a portion of the inner web next the quill, which is black; secondaries black, broadly margined on the outer web with yellow; under surface black, with an arrow-shaped mark of yellowish white near the extremity of each feather; two centre tail-feathers black, slightly tipped with yellow; the remainder black at the base, and yellow for the remainder of their length, the black decreasing and the yellow increasing as the feathers recede from the two central ones; irides reddish brown; bill black; feet blackish brown; warty excrescences covering the face dirty yellowish white.

Genus LICHNOTENTHA, *Cabanis*.

The generic term of *Melicophila* proposed by me for this form, having been previously employed, I adopt that substituted by M. Cabanis.

The single species known is, I believe, confined to Southern and Western Australia. It possesses many singular habits, and differs from most other species of the *Meliphagidæ* in the totally different colouring of the sexes.

Sp. 325. LICHNOTENTHA PICATA, *Gould*.

PIED HONEY-EATER.

Entomophila picata, Gray Gen. of Birds, vol. i. p. 118, *Entomophila*, sp. 4.
Lichnotentha picata, Cab. Mus. Hein., Theil i. p. 116.

Melicophila picata, Gould, Birds of Australia, fol., vol. iv. pl. 49.

The actions of this bird when on the wing are extremely varied, and some of them very graceful; it frequently ascends in a perpendicular direction to a considerable height above the trees, when the contrast presented by its black and white plumage renders it a conspicuous and pleasing object. It is at all times exceedingly shy, and invariably perches on the top of an isolated bush or dead branch. It usually utters a peculiar plaintive note, slowly repeated several times in succession; it also emits a single note, which so closely resembles that of the *Myzomela nigra*, as to be easily mistaken for it. It is at all times extremely difficult of approach, and the female is even more shy and wary than the male.

Gilbert states that this species assembles in vast flocks, which continue soaring about during the greater portion of the day. It is a periodical visitant to Western Australia, where it arrives in the latter part of October. It also inhabits the plains round Adelaide in the neighbouring colony.

The male has the head, throat, sides of the chest, back, wings, inner webs of the upper tail-coverts, two centre and the tips of the remaining tail-feathers black; the wing-coverts, the base and the margins of both webs of the secondaries, the rump, outer webs of the upper tail-coverts, the under surface, and the lateral tail-feathers for three-fourths of their length pure white; irides reddish brown; bill bluish grey, becoming black on the culmen near the tip; naked skin and a small fleshy appendage beneath the eye ash grey; legs and feet greenish grey.

The female is light brown, each feather being darkest in the centre; wings and tail dark brown, the former margined with buffy white; under surface buffy white, with a small streak of black near the tip of each feather.

Total length 6¼ inches; bill ⅞; wing 3½; tail 2⅞; tarsi ⅞.

Genus ENTOMOPHILA, *Gould.*

But one species of this form has yet been discovered. It is strictly Australian, and appears to be confined to the interior of the country.

Sp. 326.   ENTOMOPHILA PICTA, *Gould.*

PAINTED HONEY-EATER.

Entomophila picta, Gould in Proc. of Zool. Soc., part v. p. 154.
——
Entomophila picta, Gould, Birds of Australia, fol, vol. iv. pl. 50.

This beautiful little Honey-eater is an inhabitant of the

interior of New South Wales, where it frequents the myalls (*Acacia pendula*) and other trees bordering the extensive plains of that part of Australia. On a comparison of skins of this species with those of the other *Meliphagidæ*, prior to my visit to the country, I had been led to suspect that its actions and economy would be found to differ materially from those of the other members of its family, and such proved to be the case, for it is much more active among the branches, captures insects on the wing, and darts forth and returns to the same spot much after the manner of the Flycatchers. Its song is a loud but not very harmonious strain, which is frequently uttered when on the wing. During flight it repeatedly spreads its tail, when the white portion of the feathers shows very conspicuously; the yellow colouring of the wing also contributes to the beauty of its appearance.

I found the nest of this bird with two nearly fledged young on the 5th of September; the nest was the frailest structure possible, most ingeniously suspended by the rim to the twigs and thick drooping leaves of the *Acacia pendula*, and entirely composed of very fine fibrous roots.

Head, cheeks, and all the upper surface black, the posterior edges of the ear-coverts tipped with white; wings black, the outer edges of the primaries and secondaries rich yellow at their base, forming a conspicuous broad mark on the wing; tail black, margined externally with rich yellow, each feather except the two centre ones more or less largely tipped on the internal web with white; throat and all the under surface white, the flanks having a few longitudinal faint spots of brown; bill soft and pulpy, and of a deep pink red; irides hazel; eyelash darker hazel; feet purplish lead-colour.

Total length 5¼ inches; bill ½; wing 3⅛; tail 2⅔; tarsi ¾.

The female is much less brilliant than the male, but does not differ in the distribution of the markings.

Total length 5⅛ inches; bill ¾; wing 3⅛; tail 2¾; tarsi ¾.

Genus CONOPOPHILA, *Reichenbach*.

The members of this genus are two in number; both are confined to Australia.

Sp. 327. CONOPOPHILA ALBIGULARIS, *Gould*.
WHITE-THROATED HONEY-EATER.

Entomophila? albogularis, Gould in Proc. of Zool. Soc., part x. p. 137.
Conopophila albigularis, Reich. Handb. der Spec. Orn., p. 119.
Me-lud-be-re, Aborigines of Port Essington.

Entomophila albogularis, Gould, Birds of Australia, fol., vol. iv. pl. 51.

This species is a native of the northern portion of Australia. "I first met with it," says Gilbert, "on Mayday Island in Van Diemen's Gulf, where it appeared to be tolerably abundant; I afterwards found it to be equally numerous in a large inland mangrove swamp near Point Smith. It is an extremely active little bird, constantly flitting from branch to branch and taking irregular flights, during which it utters its pretty song; it also pours forth its agreeable melody for a length of time without intermission while sitting on the topmost branches of the trees. I never observed it in any other than swampy situations, or among the mangroves bordering the deep bays and creeks of the harbours. Its small pensile nest is suspended from the extremity of a weak projecting branch in such a manner that it hangs over the water, the bird always selecting a branch bearing a sufficient number of leaves to protect the entrance from the rays of the sun; in form the nest is deep and cup-like, and is composed of narrow strips of the soft paper-like bark of the *Melaleucæ*, matted together with small vegetable fibres, and slightly lined with soft grass. I found a nest in the latter part of November and another in the early part of December which contained three eggs in each, while a third procured towards the end of January had only two; the eggs are rather lengthened in form, and not very unlike those of *Malurus cyaneus* in the colour and disposition of their markings; their ground-colour being white, thinly freckled all over with bright chestnut-red, particularly at the larger end; they are nine lines long and six lines broad. During the breeding-season it exhibits considerable pugnacity of disposition, and instead of its usual pretty note, utters a chattering and vociferous squeaking.

"The stomach was very small, but tolerably muscular, and its food consisted of insects generally."

Head dark grey; all the upper surface brown; wings and tail darker brown; primaries, secondaries, and basal half of the tail-feathers margined with wax-yellow; throat pure white; chest and flanks reddish buff; centre of the abdomen and under tail-coverts white; irides bright reddish brown bill blackish grey; feet bluish grey.

Total length 4¼ inches; bill ½; wing 2⅜; tail 2; tarsi ¾.

Sp. 328. CONOPOPHILA RUFIGULARIS, *Gould.*

RED-THROATED HONEY-EATER.

Entomophila rufogularis, Gould in Proc. of Zool. Soc., part x. p. 137.
Conopophila rufigularis, Reich. Handb. der Spec. Orn., p. 120.

Entomophila rufogularis, Gould, Birds of Australia, fol., vol. iv. pl. 52.

This is another of the novelties that has rewarded the researches of the Officers of H.M.S. 'Beagle' on the northern coast of Australia. It is the least of the genus yet discovered, and is nearly allied to *C. albigularis*, but from which it may at once be distinguished by the red colouring of its throat. The sexes, judging from the specimens sent me by Bynoe, are very similar in their markings.

Nothing whatever is known of its habits and economy.

Head and all the upper surface brown; wings and tail

darker brown; primaries, secondaries, and tail-feathers margined externally with wax-yellow; throat rust-red; sides of the head and all the under surface very pale brown; bill and feet dark purplish brown.

Total length 4¾ inches; bill ½; wing 2½; tail 2½; tarsi ⅝.

Genus ACANTHOGENYS, *Gould.*

The genus *Acanthogenys*, of which only one species is known, presents us with a form intermediate in size and in structure between the smaller Honey-eaters (*Meliornes, Ptilotes,* &c.) on the one hand, and the larger kinds (*Anthochæra*) on the other.

The sexes are alike in plumage, and the young are very similar, but are destitute of the spines on the cheek, which are scarcely assumed during the first year. The *Banksiæ* are the trees most frequented by it.

Sp. 329. ACANTHOGENYS RUFIGULARIS, *Gould.*

SPINY-CHEEKED HONEY-EATER.

Acanthogenys rufogularis, Gould in Proc. of Zool. Soc., part v. p. 153.
Meliphaga rufogularis, Gray and Mitch. Gen. of Birds, vol. i. p. 122.
 Meliphaga, sp. 28.

Acanthogenys rufogularis, Gould, Birds of Australia, fol., vol. iv. pl. 53.

Numerous and diversified as are the forms of the great family of the *Meliphagidæ,* the present species has always appeared to me more than usually interesting, because in the first place few are more elegantly formed, and in the second it differs widely from all others in plumage, and in the singular spiny processes which adorn its cheeks and ear-coverts. In its habits and general economy it bears a close alliance to the Wattle-birds (*Anthochæra*), but still presents in these respects sufficient differences to warrant its separation into a distinct genus or subgenus, as ornithologists may think fit to designate the division.

The Spiny-cheeked Honey-eater ranges very widely over the interior of Australia. I observe it to be very numerous on the Lower Namoi to the northward of the Liverpool Plains in New South Wales. It was the commonest species of the *Meliphagidæ* I met with in the interior of South Australia; and I have also received a pair of this or a closely allied species from the interior of Western Australia; as, however, some difference exists between these latter and the birds from New South Wales, I refrain, until I have seen other examples, from stating that it goes so far to the westward as the Swan River Settlement. Like the Brush Wattle-bird it is rather a shy species, but its presence may at all times be detected by the loud hollow whistling note which it frequently utters while on the wing, or while passing with a diving flight from tree to tree. It appears to give a decided preference to the Banksia and other trees growing upon sandy soil; its presence therefore is a certain indication of the poverty of the land. It is very active among the branches, clinging and creeping about with the greatest ease and elegance of position.

The nest, which is a round, rather deep, cup-shaped structure, is suspended from a fine branch of a low tree, and is composed of long wiry grasses, and now that the sheep is a denizen of the country, matted together both internally and externally with wool. The eggs are three in number, of a dull olive-buff, strongly dotted with deep chestnut-brown and bluish grey, the markings being most numerous at the larger end. The average length is one inch, and breadth nine lines.

The sexes are so much alike, that, with the exception of the female being slightly inferior to the male in size, no difference is perceptible.

Crown of the head, back, and wings dusky brown, each feather margined with pale brown; upper tail-coverts with each feather dusky brown in the centre; stripe behind the eye and on the sides of the neck black, above which on the side of the neck another line of whitish mingled with dusky; hairs on the cheeks white; below the lower mandible a line of feathers, which are white crossed by black lines; throat and fore part of the chest pale rufous; under surface dirty white, each feather striated with dusky brown; tail blackish brown, tipped with white; bare part of the face and base of the bill soft, pulpy, and of a pinky flesh-colour; irides bluish lead-colour; feet olive.

Total length $9\frac{3}{4}$ inches; bill $1\frac{1}{2}$; wing $4\frac{1}{4}$; tail $4\frac{1}{4}$; tarsi 1.

Genus ANTHOCHÆRA, *Vigors and Horsfield*.

The two species of this peculiarly Australian genus are exclusively confined to the southern or extra-tropical parts of the country; one to Tasmania, the other to the continent.

Sp. 330.   ANTHOCHÆRA INAURIS.
Wattled Honey-eater.

Anthochæra carunculata, Vig. and Horsf. in Linn. Trans., vol. xv. p. 321.
Creadion carunculatus, Vieill. Gal. des Ois., tom. i. pl. 94.
Wattle Bird of the Colonists of Tasmania.

Anthochæra inauris, Gould, Birds of Australia, fol., vol. iv. pl. 54.

The vast primæval forests of *Eucalypti* clothing the greater portion of Tasmania are the habitual resort of this bird; from these retreats however it frequently emerges, and visits the flowering *Eucalypti* of the more open parts, where forty or fifty individuals may be frequently seen on a single tree, even in the vicinity of Hobart Town and the islands of South Arm and Bruni. The neighbourhood of the Macquarrie Plains is also a locality particularly favourable to it; from this district hundreds are annually sent to the markets of Hobart Town for the purposes of the table. It is highly prized as an article of food, and in winter becomes excessively fat, the entire body and neck, both internally and externally, being completely enveloped. This bird feeds almost exclusively on honey and the pollen of the *Eucalypti*; the only other food detected in its very diminutive stomach being the remains of coleopterous insects. Its whole structure is admirably adapted for procuring this kind of food; its long tongue, with its brush-like tip, being protruded into the honey-cups of the newly opened flowers, a succession of which appears with every rising sun throughout the year, upon one or other of the numerous species of the *Eucalypti*.

The same restless disposition seems to be common to all the tribe of Honey-eaters, and this bird is as active and quick in its movements as the smallest of the genus, hanging and clinging to the branches in every possible variety of position; and when thirty or forty are seen on a single tree, they present a very animated appearance. Its flight, which seldom extends farther than from tree to tree, is very similar to that of the Magpie of Europe. Its note is a harsh and disagreeable scream, resembling in loudness and somewhat in tone the call of the Pheasant. Both sexes have the wattled appendages beneath the ear, but they are less developed in the female, which moreover is smaller than the male.

The nest is a moderately large cup-shaped structure, formed of fine twigs and grasses intermingled with wool, and is usually built on some low tree, such as the *Casuarina* or *Acacia*. I failed in procuring the eggs, but my son Charles, now engaged in a Geological Survey of Tasmania, having transmitted some to me I am enabled to state that they are of a pale salmon-colour sprinkled all over, but particularly at the larger end, with small specks and blotches of yellowish red and here and there with grey; they are one inch and three-eighths long by seven-eighths broad. They are very like those of *A. carunculata*, but are more thickly blotched with yellowish red.

In size this bird nearly equals the Magpie (*Pica caudata*) of the British Islands.

Crown of the head and back of the neck striped with black and grey, the centre of each feather being black, and its external edges grey; back and shoulders dusky brown, the shaft of each feather buffy white; wings deep blackish brown, the external margins of the primaries slightly, and the secondaries broadly fringed with grey; tips of all the primaries white; tail much graduated; the upper tail-coverts and two middle tail-feathers grey, the remainder blackish brown, and the whole tipped with white; chin and under tail-coverts white; throat, breast, and flanks grey, each feather having a central mark of blackish brown, which is much enlarged on the lower part of the breast; centre of the abdomen rich yellow; bill black; corner of the mouth yellow; irides very dark brownish black; feet light flesh-colour; claws black; bare skin round the ear, and the upper part of the long pendulous wattle which hangs from below the ear white, gradually deepening into rich orange at its extremity.

Sp. 331. ANTHOCHÆRA CARUNCULATA.

Wattled Honey-eater.

Merops carunculatus, Lath. Ind. Orn., vol. i. p. 276.
Corvus paradoxus, Lath. Ind. Orn., Supp. p. 26.
—— *carunculatus*, Shaw, Gen. Zool., vol. vii. p. 378.
Pie à pendeloques, Daud. Orn., tom. ii. p. 246. pl. 16.
Wattled Crow, Lath. Gen. Syn., Supp. vol. ii. p. 119.
Wattled Bee-eater, Lath. Ib., vol. ii. p. 150.
Anthochæra lewinii, Vig. and Horsf. Linn. Trans., vol. xv. p. 322, note.
Djung-gung, Aborigines of Western Australia.
Wattle Bird of the Colonists.

Anthochæra carunculata, Gould, Birds of Australia, fol., vol. iv. pl. 55.

This long known species, the *Merops carunculatus* of Latham, enjoys a wide range of habitat, extending as it does over the whole of the southern portion of the continent, the bird being equally as abundant in Southern and Western Australia as in New South Wales; how far it may proceed to the northward has not yet been ascertained; it does not inhabit Tasmania. I observed it to be very numerous in all the high gum-trees around Adelaide, in most parts of the interior, and in all the *Angophora* flats and forests of *Eucalypti* of New South Wales. It is a showy active bird, constantly engaged in flying from tree to tree and searching among the flowers for its food, which consists of honey, insects, and occasionally berries. In disposition it is generally shy and wary, but at times is confident and bold: it is usually seen in pairs, and the males are very pugnacious. Its habits and manners, in fact, closely resemble those of the *A. inauris*, and, like that bird, it utters with distended throat a harsh disagreeable note.

It breeds in September and October. The nests observed by myself in the Upper Hunter district were placed on the horizontal branches of the *Angophora*, and were of a large rounded form, composed of small sticks, and lined with fine grasses; those found by Gilbert in Western Australia were formed of dried sticks, without any kind of lining, and were placed in the open bushes. The eggs are two or three in number, one inch and three lines long by ten lines and a half broad; their ground-colour is reddish buff, very thickly dotted with distinct markings of deep chestnut, umber, and reddish brown, interspersed with a number of indistinct marks of blackish grey, which appear as if beneath the surface of the shell: eggs taken in New South Wales are somewhat larger than those from Western Australia, and have markings of a blotched rather than of a dotted form, and principally at the larger end.

In size this bird is about equal to *A. inauris*, and the sexes are only distinguished by the smaller size of the female.

Crown of the head, a line running from the base of the bill beneath the eye, and the ear-coverts blackish brown; space under the eye silvery white, bounded behind by an oblong naked flesh-coloured spot, below which is a short pendulous wattle of a pinky blood-red colour; back of the neck and all the upper surface greyish brown, each feather having a stripe of white down the centre; upper tail-coverts greyish brown, broadly margined with grey; primaries and secondaries deep blackish brown, the former slightly, and the latter broadly edged with grey; all the primaries tipped with white; two middle tail-feathers greyish brown, the remainder deep blackish brown, the whole largely tipped with white; throat, breast, and flanks grey, the centre of each feather being lighter; middle of the abdomen yellow; irides bright hazel-red; legs brownish flesh-colour; inside of the mouth yellow.

Genus ANELLOBIA, *Cabanis*.

Besides the two species of this form known to inhabit Australia, others, I believe, exist in the islands adjacent to and in New Guinea. These birds are nearly allied to the *Anthochærœ*, but differ in their plumage and in the absence of auricular appendages. They frequent low swampy places, and are particularly fond of the Leptospermum or tea-tree, in the midst of which they often secrete themselves. They are alike in plumage; and the egg, or eggs to the number of two or three are deposited in a round, cup-shaped nest.

It will be seen by the synonyms of *A. mellivora* that both Cabanis and Reichenbach have proposed generic names for this form; that of the former author, having the priority of a year, is necessarily the one adopted.

Sp. 332.   ANELLOBIA MELLIVORA.

Brush Wattle-Bird.

Certhia mellivora, Lath. Ind. Orn., Supp. p. xxxvii.
Goruck Creeper, Shaw, Gen. Zool., vol. viii. p. 243.
Mellivorous Creeper, Lath. Gen. Syn., Supp. vol. ii. p. 160.
Merops chrysopterus, Lath. Ind. Orn., Supp. p. xxxiii.
Mellivorous Honey-eater, Lath. Gen. Hist., vol. iv. p. 161.
Anthochæra mellivora, Vig. and Horsf. in Linn. Trans., vol. xv. p. 321.
Anellobia mellivora, Cab. Mus. Hein., Theil i. p. 120.
Melichæra mellivora, Reich. Hand. der Spec. Orn., p. 180.
Goo-gwar-ruck, Aborigines of the coast of New South Wales.

Anthochæra mellivora, Gould, Birds of Australia, fol., vol. iv. pl. 56.

This bird is a native of Tasmania, New South Wales, and South Australia; and in all these countries may be found in such situations as are favourable to the growth of Leptospermums. In the former country it is especially abundant on the banks of the Tamar, and in the belts of Banksias that stretch along the northern shores of that island. Among the places in which it is most numerous on the continent, are near the Port of Adelaide, in South Australia; and Illawarra, Newcastle, and Sydney, in New South Wales. The Botanic Garden at the latter place, although in the midst of a populous city, is visited by great numbers of this bird, and I may mention that two of their nests with eggs, forming part of my collection, were taken from the shrubs growing on the borders of this place of public resort. It is but sparingly dispersed in the interior of New South Wales and South Australia: how far its range may extend to the westward of Spencer's Gulf I have had no means of ascertaining: I have never yet received it from Swan River or any part of the western coast, its place being there supplied by an allied species, the *A. lunulata*.

The Brush Wattle-bird is a bold and spirited species,

evincing a considerable degree of pugnacity, fearlessly attacking and driving away all other birds from the part of the tree on which it is feeding. During the months of spring and summer the male perches on some elevated branch and screams forth its harsh and peculiar notes, which have not unaptly been said to resemble a person in the act of vomiting, whence the native name of *Goo-gwar-ruck*. While thus employed it frequently jerks up its tail, throws back its head, and distends its throat, as if great exertion was required to force out these harsh and guttural sounds.

The breeding-season commences in September and continues during the three following months. The nest, which is round, open, and rather small in size, is generally placed in the fork of a small branch often within a few feet of the ground, and is formed of fine twigs and lined with fibrous roots.

The eggs are two and sometimes three in number, of a beautiful salmon-colour, strongly blotched at the larger end, and here and there over the remainder of the surface with deep chestnut-brown; thirteen lines long by nine lines broad.

In size this bird is about equal to the following, *A. lunulata*.

The sexes are only to be distinguished from each other by the smaller size of the female; and the young from the nest have all the marks of the adult, but much less apparent.

All the upper surface dark brown, each feather marked down the centre with a minute line of white; primaries chestnut-brown on the inner webs for three parts of their length from the base; outer webs and remainder of the inner brown, tipped with white; secondaries, wings, and tail-coverts greyish brown, tipped with white; tail brown, tinged with olive, and all the feathers tipped with white; feathers of the throat and chest blackish brown at the base and white at the tip; feathers of the under surface the same as the upper, but with the white mark broader and more conspicuous; bill black; irides grey; feet vinous brown.

Sp. 333.   ANELLOBIA LUNULATA, *Gould.*

LUNULATED WATTLE-BIRD.

Anthochæra lunulata, Gould in Proc. of Zool. Soc., part v. p. 153.
Anellobia lunulata, Cab. Mus. Hein., Theil i. p. 120.
Melichæra lunulata, Reich. Haub. der Spec. Orn., p. 132.
Djong-gong, Aborigines of the lowland, and
Tor-dal-l, Aborigines of the mountain districts of Western Australia.
Little Wattle-bird, Colonists of Swan River.

Anthochæra lunulata, Gould, Birds of Australia, fol., vol. iv. pl. 57.

This species is very nearly allied to the *Anthochæra mellivora*, but differs from that bird in the greater length of its bill, in the entire absence of the striæ down the head and the back of the neck, and in the possession of a lunulate mark of white on either side of the neck. Its natural habitat is Western Australia, where it generally frequents the Banksias bordering rivers and lakes, and in fact all situations similar to those resorted to by its near ally: it is to be found in every part of the colony, but appears to be more abundant in the neighbourhood of Swan River than elsewhere. In its habits it is very solitary and shy, and is moreover very pugnacious, attacking every bird, both large and small, that approaches its domicile.

Its flight is rapid and uneven, and its general note is a discordant cackling sound of the most disagreeable description.

"A remarkable circumstance," says Gilbert, "connected with the incubation of this bird is, that it appears to lay but a single egg, and to have no regular time of breeding, its nest being found in abundance from August to November. It is rather small in size, and is deposited in the fork of a perpendicular growing branch: the tree most generally chosen is that called by the colonists of Swan River the stink-wood, but it has been found in the parasitic clump of a Banksia, and also in a small scrubby brush two or three feet from the ground; but it is more frequently constructed at a height of at least eight or twelve. It is formed of dried sticks, and lined with Zamia wool, soft grasses or flowers, and sometimes with sheep's wool. The egg is rather lengthened in form, being one inch and two lines long by nine and a half lines broad; its ground-colour is a full reddish buff, thinly spotted and marked with deep chestnut-brown and chestnut-red, some of the spots and markings appearing as if beneath the surface of the shell, and being most thickly disposed near the larger end."

The stomach, which is slightly muscular, is diminutive in size, and the food consists of honey and insects with which the young are also fed.

The female is considerably smaller than the male, but does not differ in the colouring of her plumage.

Crown of the head, back of the neck, and upper part of the back olive-brown, the feathers being darkest in the middle; lower part of the back and rump olive-brown, each feather having a line of white down the stem, dilated into a spot at the extremity; upper tail-coverts olive-brown, with a crescent-shaped mark of white at the tip; primaries brown, the inner webs for nearly their whole length deep chestnut; secondaries and tertiaries brown margined with grey; two middle tail-feathers greyish brown, very slightly tipped with white, the remainder dark brown largely tipped with white; feathers of the sides of the neck long, narrow, pointed, and of a silvery grey; throat and fore part of the neck greyish brown, with a round silvery grey spot at the extremity of each feather; feathers of the chest and under surface greyish brown, with a fine line of white down the centre, dilated into an oblong spot at the extremity, the white predominating on the hinder part of the abdomen and under tail-coverts; on each side of the chest an oblique mark of pure white; irides bright hazel; bill blackish brown; feet and legs yellowish grey, the former the darkest with a tinge of olive.

Total length 12 inches; bill 1⅜; wing 6¼; tail 5¼; tarsi 1⅜.

Genus TROPIDORHYNCHUS, *Vigors and Horsf.*

The law of representation in Australia appears to be chiefly confined to the species inhabiting the eastern and western coasts, but in this case it takes the opposite direction, for more singular and perfect representations cannot be found than the *T. corniculatus* and *T. citreogularis* of the south-eastern parts of the country, are of the *T. argenticeps* and *T. sordidus* of the north-western; another species, *T. buceroides*, inhabits the north-east coast, and others are found in New Guinea and the neighbouring islands.

Sp. 334. TROPIDORHYNCHUS CORNICULATUS,
Vig. and Horsf.

Friar Bird.

Merops corniculatus, Lath. Ind. Orn., vol. i. p. 270.
Corbi calau, Le Vaill. Ois. d'Am. et des Indes., tom. i. p. 69, pl. 24.
Knob-fronted Honey-eater, Lath. Gen. Hist., vol. iv. p. 161.
—— *Bee-eater*, Lath. Gen. Syn. Supp., vol. ii. p. 151.
Tropidorhynchus corniculatus, Vig. and Horsf. in Linn. Trans., vol. xv. p. 324.
Merops monachus, Lath. Ind. Orn. Supp., p. xxxiv, young.
Cowled Bee-eater, Lath. Gen. Syn. Supp., vol. ii. p. 155.
—— *Honey-eater*, Lath. Gen. Hist., vol. iv. p. 162, young.
Knob-fronted Bee-eater, White's Voy., pl. in p. 190, young.
Coldong, Aborigines of New South Wales.
Friar, Poor Soldier, &c., of the Colonists.
Buceros corniculatus, Temm.

Tropidorhynchus corniculatus, Gould, Birds of Australia, fol., vol. iv. pl. 58.

There are few birds more familiarly known in the colony of New South Wales than this remarkable species of Honey-eater: it is generally dispersed over the face of the country, both in the thick brushes near the coast and in the more open forests of the interior. My own observations induce me to consider it as a summer visitant only to New South Wales; but as a lengthened residence in the country would be necessary to determine this point, my limited stay may have led me into error. It does not visit Tasmania, neither have I traced it so far to the westward as South Australia.

The Friar Bird, selecting the topmost dead branch of the most lofty trees whereon to perch and pour forth its garrulous and singular notes, attracts attention more by its loud and extraordinary call than by its appearance. From the fancied resemblance of its notes to those words, it has obtained from the Colonists the various names of "Poor Soldier," "Pimlico," "Four o'clock," &c. Its bare head and neck have also suggested the names of "Friar Bird," "Monk," "Leather Head," &c.

Its flight is undulating and powerful, and it may frequently be seen passing over the tops of the trees from one part of the forest to another. While among the branches it displays a more than ordinary number of singular positions; its curved and powerful claws enabling it to cling in every variety of attitude, frequently hanging by one foot with its head downwards, &c. If seized when only wounded, it inflicts with its sharp claws severe and deep wounds on the hands of its captor.

Its food consists of the pollen of the *Eucalypti* and insects, to which are added wild figs and berries.

It commences breeding in November, when it becomes animated and fierce, readily attacking hawks, crows, magpies (*Gymnorhina*), or other large birds that may venture within the precincts of its nest, never desisting from the attack until they are driven to a considerable distance. The nest, which is rather rudely constructed, and of a large size for a Honey-eater, is cup-shaped, and outwardly composed of the inner rind of the stringy bark and wool, to which succeeds a layer of fine twigs lined with grasses and fibrous roots, the whole being suspended to the horizontal branch of an apple- (*Angophora*) or gum-tree without the least regard to secresy, frequently within a few feet of the ground. So numerous were they breeding in the Apple-tree Flats near Aberdeen and Yarrundi, on the Upper Hunter, that they might almost be termed gregarious. The eggs are generally three in number, of a pale salmon-colour with minute spots of a darker tint, one inch and five lines long by eleven lines broad.

There is no observable difference in the plumage of the sexes, but the female is somewhat smaller in size.

The adults have the bill and head dull ink-black; all the upper surface, wings, and tail greyish brown, the feathers of the latter tipped with white; chin and lanceolate feathers on the chest silvery white, with a fine line of brown down the centre; remainder of the under surface brownish grey; eye red, fading immediately after death to brown and sometimes to greyish hazel; feet lead-colour.

The young, although having the same general colouring as the adult, have the head less denuded of feathers, and a mere rudiment of the knob on the bill; the feathers on the breast are also less lanceolate in form, and those on the sides of the chest are margined with yellow; eye dark brown, surrounded with short brown feathers lengthening into a tuft at the back of the head; feet much more blue than in adults.

Sp. 335. TROPIDORHYNCHUS BUCEROIDES.

HELMETED FRIAR BIRD.

Philedon buceroides, Swains. Anim. in Menag., p. 325.
Tropidorhynchus buceroides, Gray and Mitch. Gen. of Birds, vol. i. p. 125, *Tropidorhynchus*, sp. 2.

Tropidorhynchus buceroides, Gould, Birds of Australia, fol., Supplement, pl.

This bird may be regarded as the representative on the north coast of *Tropidorynchus corniculatus* of the southern part of the country, for it was in the Cape York Peninsula that it was obtained; not, however, by Mr. Macgillivray, who, I believe, mistook it for the common species, and did not procure examples; which is much to be regretted, since the bird is so extremely rare in our collections.

The *Tropidorhynchus buceroïdes* differs very considerably from the *T. corniculatus* and every other Australian species in its much larger size, in the great elevation of the culmen, and in the crown of the head being clothed with feathers.

Feathers of the crown and nape brown, with pale greyish or silvery edges; all the upper surface, wings, and tail light brown; feathers of the under surface lighter brown with a silky lustre, those of the throat with darker centres; face leaden-black; bill black; feet blackish brown.

Total length 11 inches; bill 1¼; wing 5; tail 4¾, tarsi 1⅛.

Sp. 330. TROPIDORHYNCHUS ARGENTICEPS, *Gould.*
SILVERY-CROWNED FRIAR BIRD.

Tropidorhynchus argenticeps, Gould in Proc. of Zool. Soc., part vii. p. 144.
—— *monachus*, Vig. and Horsf. in Linn. Trans., vol. xv. p. 324, young female? (nec *Merops monachus*, Lath.)

Tropidorynchus argenticeps, Gould, Birds of Australia, fol., vol. iv. pl. 59.

For the first knowledge of this species of *Tropidorhynchus*, science is indebted to the late Mr. Bynoe, Surgeon of Her Majesty's Surveying ship 'Beagle,' who, on my visiting Sydney, placed his specimens at my disposal; after my return, other examples were sent to me by Sir George Grey.

Bynoe's specimens were all obtained during the survey of the north-west coast, a portion of Australia the natural productions of which are but little known; and Sir George Greys' during his expedition into the interior, from the same coast.

In size the Silvery-crowned Friar-bird is somewhat inferior

to the common species (*Tropidorhynchus corniculatus*), from which it may also be readily distinguished by the crown of the head being clothed with well-defined, small, lanceolate feathers. Of its habits and economy nothing is known; but as it is very nearly allied to the last-mentioned species, we may reasonably conclude that they are very similar.

Crown of the head silvery grey; the remainder of the head naked, and of a blackish brown; throat and all the under surface white; back, wings, and tail brown; bill and feet blackish brown.

Total length 10½ inches; bill 1⅜; wing 5½; tail 4½; tarsi 1¼.

337. TROPIDORHYNCHUS CITREOGULARIS, *Gould*.

YELLOW-THROATED FRIAR BIRD.

Tropidorynchus citreogularis, Gould in Proc. of Zool. Soc., part iv. p. 143.

Yellow-throated Friar, Colonists of New South Wales.

Tropidorhynchus citreogularis, Gould, Birds of Australia, fol., vol. iv. pl. 60.

This is strictly a bird of the interior of the south-eastern portion of Australia, and is never, so far as I am aware, found on the sea-side of the mountain ranges. I observed it in tolerable abundance during my tour to the Namoi; first meeting with it in the neighbourhood of Brezi, whence as I descended the river to the northward it gradually became more numerous. I killed both adult and young birds in December, the latter of which had just left the nest, consequently the breeding-season must have been about a month previous. The yellow colouring of the throat peculiar to the period of immaturity is entirely wanting in the adult, and the bird is one of the plainest-coloured species of the Australian Fauna.

Its habits and manners are very similar to those of the *Tropidorhynchus corniculatus*; like that bird it feeds on insects, berries, fruits, and the flowers of the *Eucalypti*, among the smaller branches of which it may constantly be seen hanging and clinging in every possible variety of attitude.

The adult has the whole of the upper surface, wings, and tail light brown; all the under surface pale greyish brown; bill and legs leaden olive; irides and eyelash nearly black; naked part of the face mealy bluish lead-colour.

Total length 10¼ inches; bill 1½; wing 5¾; tail 4½; tarsi 1⅛.

The young are similar to the adult, but have the feathers of the upper surface fringed with grey, and those of the wings slightly margined with greenish yellow; the throat and sides of the chest lemon-yellow; face blackish, and not so mealy as in the adult.

Sp. 338. TROPIDORHYNCHUS SORDIDUS, *Gould*.

SORDID FRIAR BIRD.

Ar-dulk and *Wul-be-rat*? Aborigines of Port Essington.
Leather-neck of the Colonists of Port Essington.

Tropidorhyncus sordidus, Gould, Birds of Australia, fol., vol. I. Introd., p. lviii.

This species inhabits the Cobourg Peninsula, and is very similar to *T. citreogularis*, but is smaller in all its admeasurements except in the bill, which is more developed.

Gilbert informed me that its habits and manners are precisely similar to those of *T. argenticeps*, but that it is less abundant, less active, and has not so deep a voice as that species.

The young has the yellow colouring of the throat still more extensive than in immature examples of *T. citreogularis*.

Genus ACANTHORHYNCHUS, *Gould*.

This genus, like many others of the family, may be regarded as strictly Australian: it comprises two, if not three, well-marked species, each of which is confined to a particular part of the country; the *A. tenuirostris* dwelling on the eastern coast, and the *A. superciliosus* on the western; both inhabit countries precisely in the same degree of latitude, and form beautiful representatives of each other.

Sp. 339.- ACANTHORHYNCHUS TENUIROSTRIS.

SPINE-BILL.

Certhia tenuirostris, Lath. Ind. Orn., Suppl. p. xxxvi.
Le Cap noir, Vieill. Ois. Dor., tom. ii. p. 94. pl. 60.
Slender-billed Honey-eater, Lath. Gen. Hist., vol. iv. p. 194. pl. 62.
Flapping Honey-eater, Lath. Ib., vol. iv. p. 195.
Slender-billed Creeper, Lath. Gen. Syn., Supp. vol. ii. p. 165. pl. 129.
Meliphaga tenuirostris, Vig. and Horsf. in Linn. Trans., vol. xv. p. 317.
Acanthorhynchus tenuirostris, Gould, Syn. Birds of Australia, part ii.
—— *dubius?*, Gould in Proc. of Zool. Soc., part v. p. 25.
Leptoglossus cucullatus, Swains. Class of Birds, ii. p. 327.
Cobbler's Awl, Colonists of Tasmania.
Spine-bill, Colonists of New South Wales.

Acanthorhynchus tenuirostris, Gould, Birds of Australia, fol. vol. iv. pl. 61.

On referring to the above list of synonyms, it will be seen that I formerly entertained an opinion that there were two species of this genus very nearly allied to each other, the one a native of Tasmania, and the other of the continent of Australia; the former being distinguished from the latter by its smaller size in all its admeasurements, by the crescent-shaped markings of the neck, and by the brown of the abdomen being much deeper in colour; I am now, however, inclined to believe they are identical; but should the Tasmanian bird to which I have given the name of *dubius* prove to be merely a local variety, this species will be found to range over Tasmania and all the south-eastern portions of Australia.

There is no member of the large family of Honey-eaters to which it belongs that enjoys a structure more especially

adapted for the purposes of its existence than the present
species, whose fine and extremely delicate bill is peculiarly
suited for the extraction of insects and honey from the inmost
recesses of the tubular flowers which abound in many parts of
Australia, particularly of the various species of *Epacris*, a tribe
of plants closely allied to the Heaths (*Erica*) of Africa and
Europe, which when in bloom are always frequented by
numbers of these birds; so much so, indeed, that it would
seem as if the one was expressly designed for the other.
Those who have traversed the immense beds of *Epacris
impressa*, so abundantly dispersed over Tasmania, must
have often observed the bird darting out from beneath
his feet, flitting off to a very short distance, and descending again to the thickest parts of the beds. It also
frequents the wattles and gums during their flowering-season,
and appears to be attracted to their blossoms quite as much
for the insects as for the nectar, the stomachs of all those
dissected containing the remains of coleoptera and other
insects. It is rather shy in disposition except when closely
engaged in procuring food, when it may be approached
within a few yards or so.

Its flight is extremely quick and darting, and performed
with a zigzag motion; and its note, which is a monotonous
shriek, is somewhat loud for so small a bird.

The nest is a small cup-shaped and rather neat structure, although not so compact or nicely formed as that
of many other birds; those I found, both in Tasmania
and New South Wales, were built on some low shrubs a few
feet from the ground, mostly in a species of *Leptospermum*;
it is outwardly constructed of moss and grasses, and lined
with feathers; the eggs are two in number, of a delicate
buffy white, increasing in depth of colour towards the larger
end; in some instances I have found them marked with a
zone of reddish chestnut spots shaded with indistinct markings
of grey, intermingled with very minute ink-like dots; in form

the eggs are much lengthened and pointed; their medium length is nine lines, and breadth six lines.

Crown of the head shining greenish black; space between the bill and the eye, ear-coverts, lunated band on the sides of the chest, primaries, and six middle tail-feathers black; the remainder of the tail-feathers black, largely tipped with white, and slightly margined on the external web with brown; back of the neck rufous chestnut, passing into chestnut-brown on the upper part of the back; secondaries, greater wing-coverts, rump, and upper tail-coverts grey; throat, checks, and chest white, the first with a patch of chestnut-brown in the centre, deepening into black on its lower edge; abdomen, flanks, and under tail-coverts light chestnut-brown; irides scarlet; bill black; feet reddish brown.

Specimens from Tasmania have the patch in the centre of the throat and the lunated marks on the sides of the neck much deeper, and the whole of the under surface richer chestnut.

340. ACANTHORHYNCHUS SUPERCILIOSUS, *Gould*.

White-eyebrowed Spine-bill.

Acanthorhynchus superciliosus, Gould in Proc. of Zool. Soc., part v. p. 24.

Bool-jeet, Aborigines of the lowland districts of Western Australia.

Acanthorhynchus superciliosus, Gould, Birds of Australia, fol., vol. iv. pl. 62.

Hitherto I have only received this fine and well-marked species of Spine-billed Honey-eater from Western Australia, but hereafter it will doubtless be found to range over a much greater extent of country; although a very local bird, it is tolerably abundant both at Swan River and King George's Sound, and is found to give a decided preference to the forests of Banksias, upon the blossoms of which trees it almost solely subsists. Its food consists of insects and honey, for obtaining which its delicately organised bill is peculiarly adapted. Like its congeners, this species occasionally frequents the low shrub-like trees, and sometimes is even to be observed upon the ground in search of food. In its actions it displays great activity, darting about from branch to branch with a rapid zigzag motion; its flight is irregular and uneven, but it often rises perpendicularly in the air, uttering at the same time a rather pretty song.

The nest, which is constructed among the large-leaved Banksias, is of a round compact form, and is composed of dried fine grasses, tendrils of flowers, narrow threads of bark, and fine wiry fibrous roots matted together with zamia wool, forming a thick body, which is warmly lined with feathers and zamia wool mingled together; the external diameter of the nest is three inches, and that of the cavity about one inch and a quarter. The eggs are two in number, nine lines long by six and a half broad; their ground-colour in some instances is a delicate buff, in others a very delicate bluish white, with a few specks of reddish brown distributed over the surface, these specks being most numerous at the larger end, where they frequently assume the form of a zone. The breeding-season is in October.

The sexes present little or no difference in external appearance, but the female may generally be distinguished from the male by her more diminutive size and the more slender contour of her body.

Crown of the head, all the upper surface, wings, and six middle tail-feathers greyish brown, the remainder of the tail-feathers black, largely tipped with white, and narrowly margined on their external edges with brown; space between the bill and eye, and the ear-coverts blackish brown; stripe over the eye, chin, and a broader stripe beneath the eye white; back part of the neck light chestnut-brown; centre of the throat rich chestnut, bounded below by a crescent of white, which is succeeded by another of black; abdomen and under tail-coverts light greyish brown, in some specimens deepening into buff; irides reddish brown; bill black; legs dark brown.

Total length 5¼ inches; bill 1⅛; wing 2½; tail 2¼; tarsi ¾.

Genus MYZOMELA, *Vigors and Horsfield.*

Five well-marked species of this genus are distributed over Australia; numerous others are found in New Guinea and the neighbouring islands; the form also occurs in the Polynesian Islands.

Sp. 341.   MYZOMELA SANGUINOLENTA.

SANGUINEOUS HONEY-EATER.

Certhia sanguinolenta, Lath. Ind. Orn., Supp. p. xxxvii.
―― *dibapha*, Lath. ib., p. xxxvii.
―― *erythropygia*, Lath. ib., p. 88.
―― *australasia*, Leach, Zool. Misc., vol. i. pl. 11.
Meliphaga cardinalis, Vig. and Horsf. in Linn. Trans., vol. xv. p. 316.
Cinnyris rubrater, Less. Kittl. Küpf., t. 8. f. 1.
Blood-bird of the Colonists of New South Wales.

Myzomela sanguinolenta, Gould, Birds of Australia, fol., vol. iv. pl. 63.

This beautiful little bird is an inhabitant of the thick brushes of New South Wales, particularly those near the coast and those clothing the hilly portions of the interior, and I have reason to believe that it is rarely, if ever, found among the trees of the open parts of the country. I have not yet seen specimens from the western, and only a single example from the northern coasts, whence I infer that the south-eastern part of the continent is its natural habitat. It gives a decided preference to those parts of the forest that abound with flowering plants, whose fragrant blossoms attract large numbers of insects, upon which and the pollen of the flower-cups it chiefly subsists.

The sexes are very dissimilar in colour, the female being of a uniform pale brown above and lighter beneath, while the male is dressed in a gorgeous livery of scarlet and black; the young, as is usually the case where the sexes differ in colour, resemble the female until after the first moult, when they gradually assume the colouring of the male.

The male has the head, neck, breast, back, and upper tail-coverts rich shining scarlet; lores, wings, and tail black, the wing-coverts margined with buffy white, and the primaries with greyish olive; under surface of the wing white; abdomen and under tail-coverts buff; bill and feet black; irides dark brown.

The female is uniform light brown above, becoming much lighter beneath.

Sp. 342. MYZOMELA ERYTHROCEPHALA, *Gould.*

Red-headed Honey-eater.

Myzomela erythrocephala, Gould in Proc. of Zool. Soc., part vii. p. 144.

Myzomela erythrocephala, Gould, Birds of Australia, fol., vol. iv. pl. 64.

The Red-headed Honey-eater is so distinctly marked as almost to preclude the possibility of its being confounded with any known Australian species of the genus.

The northern portion of the country appears to be its true habitat, all the specimens that have come under my notice having been procured at Port Essington, where it is exclusively confined to the extensive beds of mangroves bordering the inlets of the sea. From the flowers of these trees it collects its favourite food, which, like that of the other species of the group, consists of insects and honey. It is a most active little creature, flitting from one cluster of flowers to another, and from branch to branch with the greatest rapidity, uttering at the same time its rather sharp and harsh chirrup. Gilbert states that it is far from being abundant, and is so seldom seen near the settlement that no examples had been procured prior to his visit.

The sexes present the usual difference in the smaller size and sombre colouring of the female.

The male has the head and rump scarlet, the remainder of the plumage deep chocolate-brown; irides reddish brown; bill olive-brown, becoming much lighter on the lower mandible; legs and feet olive-grey.

Total length 4½ inches; bill ¾; wing 2¼; tail 1¾; tarsi ½.

The female is uniform brown above, lighter beneath.

Sp. 343. MYZOMELA PECTORALIS, *Gould*.

BANDED HONEY-EATER.

Myzomela pectoralis, Gould in Proc. of Zool. Soc., part viii. p. 170.

Myzomela pectoralis, Gould, Birds of Australia, fol., vol. iv. pl. 65.

It will be seen from the number of novelties received from the northern portion of Australia that that part of the country possesses a fauna almost peculiar to itself, few species, of the smaller birds at least, being similar to those of the southern districts.

The present interesting bird was forwarded to me by Bynoe as having been shot by him on the north coast, but to my regret it was unaccompanied by any information whatever respecting its habits.

Some of the specimens sent me had the centre of the back of a ferruginous hue, while in others the same part was jet-black; I am inclined to regard the former to be the plumage of the young birds, and it is just possible it may also be characteristic of the adult female.

Forehead, crown of the head, upper surface, wings, tail, and a narrow band across the chest black; throat, upper tail-coverts and all the under surface white; bill and feet black.

Total length 4½ inches; bill ⅔; wing 2⅔; tail 1¾; tarsi ½.

Sp. 344.   MYZOMELA NIGRA, *Gould.*

BLACK HONEY-EATER.

Myzomela nigra, Gould in Birds of Australia, part ii. cancelled.
Dwer-da-ngok-ngan-nin, Aborigines of the mountain districts of Western Australia.

Myzomela nigra, Gould, Birds of Australia, fol., vol. iv. pl. 66.

This most active little bird is peculiar to the interior of Australia, over which it has an extensive range. Gilbert found it at Swan River, and I met with it on the plains near the Namoi; here it was always on the Myalls (*Acacia pendula*), while in Western Australia it generally evinced a preference for the sapling gums. Although it has the feathered tongue and sometimes partakes of the sweets of the flowers, it feeds almost exclusively on insects, which it procures both on the blossoms and among the thickly-foliaged branches. The male frequently pours forth a feeble plaintive note, perched upon some elevated dead branch, where he sits with his neck stretched out and without any apparent motion, except the swelling of the throat and the movement of the bill.

The flight of this bird is remarkably quick, and performed with sudden zigzag starts.

The female differs remarkably from the male in the colouring of the plumage, and, as is the case with many other birds, is much more difficult to detect than the male, which is always more animated, and frequently betrays his presence by his song.

Gilbert was more fortunate than myself in finding the nest of this little bird, and has furnished the following notes respecting its incubation :—

"This species constructs a neat cup-shaped nest, formed of dried grasses. I found two, both of which were built in the most conspicuous situations; one in a fork at the top of a small scrubby bush, unsheltered by even a bough or a leaf; the other was on the dead branch of a fallen tree, in a similar exposed situation, and quite unprotected from wet or heat.

It breeds during the months of October and November, and lays two eggs," which are of a light brownish buff, encircled at the centre with a band of brown, produced by numerous small blotches of that colour, which appear as if beneath the surface of the shell; they are seven lines long by five and a half lines broad.

The male has the head, throat, stripe down the centre of the abdomen, all the upper surface, wings, and tail sooty black; the remainder of the plumage pure white; irides blackish brown; bill and feet black.

Total length 5 inches; bill $\frac{2}{3}$; wing $2\frac{1}{2}$; tail $1\frac{3}{4}$; tarsi $\frac{1}{4}$.

The female differs in having the head, all the upper surface, wings, and tail brown; throat and all the under surface brownish white, the centre of each feather being the darkest; bill brown; legs brownish black.

Sp. 345.   MYZOMELA OBSCURA, *Gould*.

Obscure Honey-eater.

Myzomela obscura, Gould in Proc. of Zool. Soc., part x. p. 136.

Myzomela obscura, Gould, Birds of Australia, fol., vol. iv. pl. 67.

This species is a native of the northern parts of Australia. At Port Essington, where my specimens were procured, it is only to be met with in quiet, secluded and thickly-wooded districts, adjacent to small streams of water; its favourite tree appears to be the *Grevillia*, from the blossoms of which it obtains great quantities of honey and insects. The shy and retiring disposition of this species renders the acquisition of specimens very difficult: "at no time during my stay," remarks Gilbert, "did I succeed in getting sight of more than a solitary individual at a time, and I believe it to be a rare bird in all parts of the Cobourg Peninsula."

This bird differs so much in colour from all the other species yet discovered, that it is readily distinguished from all of them."

The sexes present no external marks of distinction, except that the female is somewhat smaller than the male.

The whole of the plumage is dull brown, with a vinous tinge on the head; under surface paler than the upper; irides bright red; bill dark greenish black; feet dark bluish grey; tarsi tinged with yellow.

Total length 5 inches; bill $\frac{3}{4}$; wing $2\frac{3}{4}$; tail $2\frac{1}{4}$; tarsi $\frac{3}{4}$.

Genus ENTOMYZA, *Swainson*.

Two species of this well-defined genus are comprised in the Australian fauna, one of which inhabits the south-eastern parts of the country, or New South Wales; the other, which so far as we yet know is strictly confined to the north-eastern coast, is very plentiful at Port Essington and in the neighbouring districts.

The form appears to be confined to Australia, for I have never seen it from any other country.

Sp. 346. ENTOMYZA CYANOTIS, *Swains*.

BLUE-FACED HONEY-EATER.

Gracula cyanotis, Lath. Ind. Orn. Supp., p. xxix.
Meliphaga cyanops, Lewin, Birds of New Holland, pl. 4.
Turdus cyaneus, Lath. Ind. Orn. Supp., p. xlii.
Merops cyanops, Lath. Ind. Orn. Supp., p. xxxiv., young.
Tropidorhynchus cyanotis, Vig. and Horsf. in Linn. Trans., vol. xv. p. 325.
Entomyza cyanotis, Swains. Class. of Birds, vol. ii. p. 328.
Batikin, Aborigines of the coast of New South Wales.
Blue-eye of the Colonists.

Entomyza cyanotis, Gould, Birds of Australia, fol., vol. iv. pl. 69.

This attractive and beautiful Honey-eater, one of the finest of the *Meliphagidæ*, is strictly indigenous to New South Wales, where it is abundant and very generally dispersed: I observed it in nearly every part of the colony I visited, both in winter and summer. I also shot a single specimen on the Namoi, but as this was almost the only one I saw beyond the mountain ranges, I believe its most natural habitat to be between the great dividing chain of mountains and the sea. In all probability it may be found far to the northward on the eastern coast, but it has not yet been observed in South Australia, neither is it an inhabitant of Tasmania.

In habits and actions the Blue-faced Honey-eater bears a striking resemblance to the members of the genera *Ptilotis* and *Melithreptus*; like them, it is found almost exclusively on the *Eucalypti*, searching among the blossoms and smaller leafy branches for its food, which is of a mixed character, consisting partly of insects and partly of honey, and probably, berries and fruits, but this latter supposition I was not able to verify. Mr. Cayley states, that he once saw "several of them frequenting a tree, where they were very busy in obtaining something that appeared to have exuded from a wounded part. I do not know what the substance could be otherwise than a kind of gum of a bitter and astringent taste." As I have never detected them in feeding on this or any similar substance, I should rather suppose they were in search of the insects attracted by this exudation.

I have frequently seen eight or ten of these bold and spirited birds on a single tree, displaying the most elegant and easy movements, clinging and hanging in every variety of position, frequently at the extreme ends of the small, thickly-flowered branches, bending them down with their weight; they may be easily distinguished from other birds with which they are frequently in company by their superior size, the brilliancy of their blue face, and the contrasted colours of their plumage; they are rendered equally conspicuous by the pugnacity with which they chase and drive about the other species resorting to the same tree.

It frequently utters a rather loud and monotonous cry, not worthy the name of a song.

I observed a most curious fact respecting the nidification of this bird: in every instance that I found its eggs, they were deposited on the deserted, dome-shaped, large nest of the *Pomatostomus temporalis*, so numerous in the Apple-tree Flats in the district of the Upper Hunter; never within the dome, but in a neat round depression on the top. I had many opportunities of driving the female off the nest, and I can therefore speak with confidence as to this fact. Whether the bird always resorts to these nests, or if, under other circumstances, it constructs a nest for itself, are points to which I would call the attention of those who are favourably situated for investigating them. It is probable that, in places where no suitable substitute is to be found, it makes a nest, like other species of its tribe. It commences breeding early, and rears at least two broods in the year: on reference to my note-book, I find I saw fully-fledged young on the 19th of November, and that I took many of their eggs in December: they were generally two in number, of a rich salmon-colour irregularly spotted with rustbrown, one inch and a quarter long by ten and a half lines broad.

The sexes differ in no respect from each other either in the colouring of the plumage or in the blended richness and delicacy of the blue surrounding the eye, to which it is almost impossible for the artist to do justice.

The young assume the plumage of the adult from the nest, but differ from them in having the naked face and the base of the bill of a pale yellowish olive, which gradually changes to blue after the first season; this has doubtless occasioned the great number of synonyms quoted above.

The adults have the crown of the head and back of the neck black; lower part of the face, chin, and centre of the chest slaty black; a crescent-shaped mark at the occiput, a line from the lower mandible passing down each side of the neck, and all the under surface pure white; the upper surface, wings, and tail golden olive; the inner webs of the primaries and all but the two centre tail-feathers brown; the tail-feathers tipped with white; basal portion of the bill pale bluish grey, passing into blackish horn-colour at the tip; bare space surrounding the eye rich deep blue, becoming of a lighter and greenish hue above the eye; irides yellowish white; eyelash jet-black; feet bluish grey.

The young of the first autumn have the eye dark olive with a black lash, and the denuded parts surrounding it, the base of the under mandible and the gape greenish brimstone-yellow; nostrils and culmen near the head yellowish horn-colour, passing into blackish brown at the tip; feet very similar to those of the adult.

Sp. 346.   ENTOMYZA ALBIPENNIS, *Gould.*

WHITE-QUILLED HONEY-EATER.

Entomyza albipennis, Gould in Proc. of Zool. Soc., part viii. p. 169.
Wur̄-ra-luh, Aborigines of Port Essington.

Entomyza albipennis, Gould, Birds of Australia, fol., vol. iv. pl. 68.

The *Entomyza albipennis* exhibits so many specific differences from the *E. cyanotis,* that it is almost impossible for one to be mistaken for the other: in the first place it is smaller in size, and in the next the tints of the plumage are more strongly contrasted; besides which, the white at the basal portion of the quills is a character which will at all times distinguish it from its near ally. So far as is yet known, its habitat is confined to the northern coast of Australia, where it is said to be rather abundant, particularly in the neighbourhood of the settlement at Port Essington. Gilbert states that it "is one of the first birds heard in the morning, and often utters its plaintive *peet* half-an-hour before daylight; but as soon as the sun is fairly above the horizon, its note is changed to a harsh squeaking tone, which is frequently uttered while the bird is on the wing, and repeated at intervals throughout the day; it often takes long flights, mounts high above the trees, and then progresses steadily and horizontally. It is mostly met with in small families of from six to ten in number, frequenting the topmost branches of the loftiest trees, and is seldom seen on or near the ground."

The sexes present little or no difference in the colouring of the plumage, or, when fully adult, in the colouring of the soft parts, such as the naked skin round the eyes, &c.; immature birds, on the contrary, vary very much in the colouring of the face and bill; in the youthful those parts are saffron-yellow, which changes to rich ultramarine blue in the adult.

The adults have the crown of the head and back of the neck black; lower part of the face, chin, and centre of the chest slaty black; a crescent-shaped mark at the occiput, a line from the lower mandible passing down each side of the neck, and all the under surface pure white; upper surface and wings greenish golden olive; primaries brown, the basal half of their inner webs snow-white; tail-feathers brown, tinged with golden olive, all but the two centre ones tipped with white; point and cutting edges of the upper mandible blackish grey; basal half of the culmen horn-colour; remainder of the bill sulphur-yellow; orbits brilliant blue; legs and feet leek-green.

Total length 12 inches; bill 1½; wing 0; tail 4¾; tarsi 1⅜.

Genus MELITHREPTUS.

No one group of birds is more universally distributed over Australia than the members of this genus, for, like the *Eucalypti*, the trees upon which they are almost exclusively found, their range extends from Tasmania on the extreme south to the most northern part of the continent, and in an equal degree from east to west, each part of country being inhabited by a species peculiarly its own. I believe the form is unknown out of Australia.

Sp. 347. MELITHREPTUS VALIDIROSTRIS, *Gould.*

Strong-billed Honey-eater.

Hæmatops validirostris, Gould in Proc. of Zool. Soc., part iv. p. 144.
Eidopsaris bicinctus, Swains. An. in Menag., p. 341.
Sturnus virescens, Wagl. Syst. Av. Sturnus, sp. 5?
Melithreptus virescens, Gray Gen. of Birds, vol. i. p. 128.
Cherry-picker, Colonists of Tasmania.

Melithreptus validirostris, Gould, Birds of Australia, fol., vol. iv. pl. 70.

This bird, the largest species of the genus yet discovered, is a native of Tasmania, and so universally is it distributed over that island that scarcely any part is without its presence. The crowns of the highest mountains as well as the lowlands, if clothed with *Eucalypti,* are equally enlivened by it. Like all the other members of the genus, it frequents the small leafy and flowering branches; it differs, however, from its congeners in one remarkable character, that of alighting upon and clinging to the surface of the boles of the trees in search of insects. I never saw it run up and down the trunk, but merely fly to such parts as instinct led it to select as the probable abode of insects.

I am indebted to the Rev. Thomas J. Ewing, D.D., for the nest and eggs of this bird, which I failed in procuring during my stay in Tasmania. Like those of the other members of the genus the nest is round and cup-shaped, suspended by the rim and formed of coarse wiry grasses, with a few blossoms of grasses for a lining; the eggs are three in number, eleven lines long by eight lines broad, and of a dull olive-buff, thickly spotted and blotched with markings of purplish brown and bluish grey, the latter appearing as if beneath the surface of the shell.

The song consists of a couple of notes, and is not remarkable for its melody.

The sexes assimilate so closely in size and plumage, that by dissection alone can they be distinguished; the young, on the contrary, during the first autumn differ considerably.

Crown of the head jet-black, with an occipital band of white terminating at each eye; ear-coverts, chin, and back of the neck black; all the upper surface greyish olive, becoming brighter on the rump and external edges of the tail-feathers; wings brown, with a slight tinge of olive; throat pure white; under surface brownish grey; bill black; feet brownish horn-colour; eyes reddish brown; bare skin over the eye white, tinged with bright green.

Total length 6¾ inches; bill ¾; wing 3½; tail 3; tarsi ⅞.

The young have the bill and feet yellow, but the latter paler than the former, and a circle of the same colour round the eye; the band at the occiput is also pale yellow instead of white.

Sp. 348. MELITHREPTUS GULARIS, *Gould*.

BLACK-THROATED HONEY-EATER.

Hæmatops gularis, Gould in Proc. of Zool. Soc., part iv. p. 144.

Melithreptus gularis, Gould, Birds of Australia, fol., vol. iv. pl. 71.

This species is very abundant in all parts of South Australia. It frequents the large *Eucalypti*, and during my stay in Adelaide I frequently saw it on some of the high trees that had been allowed to remain by the sides of the streets in the middle of the city. From this locality it extends its range eastward to Victoria and New South Wales. I killed several specimens in the Upper Hunter district, and observed it to be tolerably numerous on the plains in the neighbourhood of the river Namoi; and that it breeds in these countries is proved by my having shot the young in different stages of growth in all of them. It is a very noisy bird, constantly uttering a loud harsh grating call while perched on the topmost dead or bare branch of a high tree; the call being as frequently uttered by the female as by the male. Like the *Melithreptus lunulatus*, it frequents the leafy branches, which it threads and creeps among with the greatest ease and dexterity, assuming in its progress a variety of graceful attitudes. Insects and the pollen of flowers being almost its sole food, those trees abounding with blossoms are visited by it in preference to others.

With the nest and eggs of this species I am unacquainted; they are therefore desiderata to my cabinet, and would be thankfully received from any person resident in the colonies. That the nest when discovered will be cup-shaped in form, and suspended by the rim to the smaller branches of the *Eucalypti*, and that the eggs will be two or three in number, there can be little doubt.

Crown of the head black, an occipital band of white terminating at each eye; ear-coverts and back of the neck black; back and rump golden olive; wings and tail brown; throat greyish white, with a central stripe of black; under surface greyish brown; bill black; feet and tarsi brownish orange; irides hazel; bare skin above the eye beautiful bluish green.

Total length 6 inches; bill $\frac{3}{4}$; wing $3\frac{1}{8}$; tail $2\frac{3}{4}$; tarsi $\frac{7}{8}$.

There is no variation in the colouring of the sexes, but a very considerable difference between the young and old birds, particularly in the colouring of the soft parts, the young having the gape, lower mandible, and feet yellowish orange.

This bird, the *M. validirostris* and the doubtful *M. brevirostris*, spoken of on page 569, differ from the other members of the genus in having brown wings and a sordid brown under surface, which feature in the colouring is in favour of Dr. Bennett's and Mr. Angas's views of the latter being distinct and not the young of *M. lunulatus*.

Sp. 349. MELITHREPTUS LUNULATUS.

LUNULATED HONEY-EATER.

Certhia lunulata, Shaw, Gen. Zool., vol. viii. p. 224.
Le Fuscalbin, Vieill. Ois. dor., tom. ii. p. 95. pl. 61.
Red-eyed Honey-eater, Lath. Gen. Hist., vol. iv. p. 203. no. 65.
Meliphaga lunulata, Vig. and Horsf. in Linn. Trans., vol. xv. p. 315.
Black-crowned Honey-sucker, Lewin, Birds of New Holl., pl. 24.
Meliphaga atricapilla, Temm. Pl. Col., 335. fig. 1.
———— *torquata*, Swains. Zool. Ill., 1st ser. pl. 110.
Hæmatops lunulatus, Gould in Syn. Birds of Australia, part i.
Gymnophrys torquatus, Swains. Class. of Birds, vol. ii. p. 327.
Melithreptus lunulatus, G. R. Gray, List of Gen. of Birds, 2nd edit. p. 21.
Meliphaga brevirostris, Vig. and Horsf. in Linn. Trans., vol. xv. p. 315 ?

Malithreptus lunulatus, Gould, Birds of Australia, fol., vol. iv. pl. 72.

The Lunulated Honey-eater is very abundantly dispersed over New South Wales and South Australia, where it inhabits almost every variety of situation, but gives a decided preference to the *Eucalypti* and *Angophoræ* trees, among the smaller branches of which it may be constantly seen actively engaged in searching for insects, which, with the pollen and honey of the flower-cups, constitute its food. It is a stationary species, and breeds during the months of August and September; its beautiful, round, cup-shaped, open nest is composed of the inner rind of the stringy bark or other allied gum-trees, intermingled with wool and hair, warmly lined with opossum's fur, and is suspended by the rim to the small leafy twigs of the topmost branches of the *Eucalypti*. The eggs are two or three in number, of a pale buff, dotted all over, but particularly at the larger end, with distinct markings of rich reddish brown and chestnut-red, among which are a few clouded markings of bluish grey; their medium length is nine lines, and breadth six and a half lines.

Like the young of *M. chloropsis*, the young birds of this

species breed some time before they have attained their green livery; at all events I have found examples breeding in a state of plumage, which I believe to be characteristic of youth.

The sexes are alike in plumage, but the female is somewhat smaller than the male.

Upper surface greenish olive; head and chin black; crescent-shaped mark at the occiput and all the under surface white; wings and tail brown, the apical half of the external webs of the primaries narrowly edged with grey; basal half of the external webs of the primaries, the outer webs of the secondaries, and the tail-feathers washed with greenish olive; naked space above the eye scarlet; feet olive; irides very dark brown; bill blackish brown.

Dr. Bennett, of Sydney, and Mr. George French Angas have called my attention to a *Melithreptus* inhabiting New South Wales, which they consider to differ from all those figured by me in the folio edition, and which they state had been found breeding, proving, in their opinion, that it must have attained maturity. The remarks of those gentlemen were accompanied by two very fine skins, which, with two others that had been in my collection for some time, are now before me. At a first glance almost any ornithologist would imagine these birds to be the young of *M. lunulatus*, and I must admit that this was my own impression; but, upon a more minute examination and comparison, I perceive characters which render me somewhat doubtful of this being the case. In the first place, I find all the specimens larger and stouter than any of *M. lunulatus* to which I have access; in the second, I have been informed that the bare space above the eye is greenish blue, and not red; all the under surface of the body is sandy brown in lieu of pure white; the axillary feathers are buff instead of white; the wings are brown, and not wax-yellow; the crown of the head is brownish black instead of pure black; and the lunate band on the occiput is greyish buff, and not white.

In comparing it with another species, *M. gularis*, I find that the latter has cinnamon-brown wings, brown axillaries, and a vinous-brown under surface; consequently the bird is as nearly allied to *M. gularis* as to *M. lunulatus*, except in size. I must therefore leave this matter to the investigation of residents in New South Wales or South Australia, from which latter country one of the specimens was sent.

Total length 4¼ inches; bill ½; wing 2⅞; tail 2¾; tarsi ¾.

Should it ultimately prove to be distinct, then it must bear the inappropriate name of *Melithreptus brevirostris*, as I find it is strictly identical with the type-specimen of the bird so called by Vigors and Horsfield, formerly in the collection of the Linnean Society, and now in the British Museum.

Sp. 350. MELITHREPTUS CHLOROPSIS, *Gould*.

Swan River Honey-eater.

Melithreptus chloropsis, Gould in Proc. of Zool. Soc., part xv. p. 220.
Jill-gee, Aborigines of the lowland, and
Bun-geen, of the mountain districts of Western Australia.
Bell-ril-bell-ril, Aborigines of Swan River.

Melithreptus chloropsis, Gould, Birds of Australia, fol, vol. iv. pl. 73.

This species differs from the *Melithreptus lunulatus* in being of a larger size, and in having the bare space above the eye of a pale green instead of red; in other respects the two birds so closely assimilate, that they are scarcely distinguishable from each other. Individuals in a browner and more dull style of plumage, presenting in fact all the appearances of young birds of the first year, have occasionally been found breeding, a circumstance which has induced many persons to believe them to be distinct; as, however, if I mistake not, I found in New South Wales individuals breeding in a similar style of plumage in company with adults of *M. lunulatus*, I am induced to regard these dull-coloured birds as merely precocious examples of the respective species, affording additional evidence of the extreme fecundity of the Australian birds.

The *Melithreptus chloropsis* is a native of Western Australia, where it is almost always found on the upper branches of the different species of *Eucalypti*, feeding upon the honey of the flowers and insects. Its usual note is a rapidly uttered *twit*, but it occasionally emits a harsh, grating, and lengthened cry.

The nest is usually suspended from the small branches near the top of the gum-trees, where the foliage is thickest, which renders it extremely difficult to detect. A nest found by Gilbert in October was formed of sheep's wool and small twigs; another found by him in November was attached to a small myrtle-like tree, in a thick gum forest, not more than three feet from the ground; both these nests contained three eggs, nine and a half lines long by six and a half lines broad, of a deep reddish buff, thinly spotted all over, but particularly at the larger end, with dark reddish brown, some of the spots being indistinct, while others were very conspicuous.

Upper surface greenish olive; head and chin black; crescent-shaped mark at the occiput and under surface white; wings and tail brown, margined with greenish olive; apical half of the external webs of the primaries narrowly edged with white; irides dull red; bill blackish brown; naked space above the eye greenish white in some, in others pale wine-yellow; tarsi and outer part of the feet light greenish olive; inside of the feet bright yellow.

Total length $5\frac{1}{4}$ inches; bill $\frac{11}{16}$; wing $3\frac{1}{4}$; tail $2\frac{3}{8}$; tarsi $\frac{3}{4}$.

Sp. 351. MELITHREPTUS ALBOGULARIS, *Gould*.

WHITE-THROATED HONEY-EATER.

Melithreptus albogularis, Gould in Proc. of Zool. Soc., part xv. p. 220.

Melithreptus albogularis, Gould, Birds of Australia, fol., vol. iv. pl. 74.

This species, which inhabits the northern and eastern parts

of Australia, is very abundant on the Cobourg Peninsula, and I have received specimens from the east coast. The total absence of any black mark beneath the lower mandible and the pure whiteness of the throat serve to distinguish it from every other known species; the colouring of the back, which inclines to rich wax-yellow, is also a character peculiar to it. It is very numerous around the settlement at Port Essington, where it occurs in families of from ten to fifteen in number; it is of a very pugnacious disposition, often fighting with other birds much larger than itself. While among the leafy branches of the *Eucalypti*, which are its favourite trees, it frequently pours forth a loud ringing whistling note, a correct idea of which is not easily conveyed. Like its near allies the sexes present no other external difference than the smaller size of the female; and the young at the same age present a similar style of colouring to that observable in the *M. lunulatus* and *M. chloropsis*, the head and sides of the neck being brown instead of black, and the naked skin above the eye scarcely perceptible.

The food consists entirely of insects and the pollen of flowers, in searching for which it displays a great variety of positions, sometimes threading the leaves on the smaller branches, and at others clinging to the very extremities of the bunches of flowers.

The nest, which is always suspended to a drooping branch, and which swings about with every gust of wind, is formed of dried narrow strips of the soft bark of the *Melaleuca*. The eggs, which are generally two in number, are of a light salmon-colour, blotched and freckled with reddish brown, and are about nine lines long by six lines broad.

Upper surface greenish wax-yellow; head black; crescent-shaped mark at the occiput, chin and all the under surface white; wings and tail brown margined with greenish wax-yellow; irides dull red; bill brownish black; legs and feet greenish grey, with a tinge of blue on the front of the tarsi.

Total length $4\frac{1}{2}$ inches; bill $\frac{3}{8}$; wing $2\frac{3}{4}$; tail $2\frac{1}{4}$; tarsi $\frac{11}{16}$.

Sp. 352. MELITHREPTUS MELANOCEPHALUS, *Gould.*

BLACK-HEADED HONEY-EATER.

Melithreptus melanocephalus, Gould in Proc. of Zool. Soc., part xiii. p. 62.
Meliphaga atricapilla, Jar. and Selb. Ill. Orn., pl. 134. fig. 1.
—— *affinis,* Less. Rev. Zool., 1839, p. 167 ?

Melithreptus melanocephalus, Gould, Birds of Australia, fol., vol. iv. pl. 75.

This bird I believe to be peculiar to Tasmania, over the whole of which island it is very abundant. The *Eucalypti* are the trees for which it evinces a preference, and it may constantly be seen among their foliage and flowers searching for its food, which, like that of the other members of the *Meliphagidæ,* consists principally of insects, particularly small coleoptera; like the other species of the family also, it creeps and clings about the branches after the manner of the Tits of Europe. It is a lively, animated bird, and generally goes in companies of from ten to twenty in number, according as the supply of food may be more or less plentiful. During the fruit-season it frequents the gardens of the settlers and commits considerable havoc among the fruit, of which it is exceedingly fond.

The sexes are precisely alike in external appearance, but the young differ considerably from the adults, having the throat yellowish white instead of black, and the basal portion of the bill flesh-colour or yellow; their feet also are much lighter than the adults.

This bird is one of the numerous foster-parents of *Cacomantis flabelliformis,* which I have seen it feeding while perched on a bare branch at the edge of the forest.

The whole of the head and throat, and a semilunar mark on either side of the chest deep glossy black; all the upper surface yellowish olive, becoming brighter on the rump; wings and tail brownish grey with lighter margins; breast white; remainder of the upper surface greyish white; bill black; irides reddish brown; feet brown; bare skin over the eye pearly white, slightly tinged with green.

Total length 5¼ inches; bill ⅞; wing 3; tail 2½; tarsi ¾.

Mr. G. R. Gray is of opinion that this is the bird characterized by Latham as *Certhia agilis*; but independently of the difficulty of identifying his description, I may remark that Latham states his bird to be an inhabitant of New South Wales, where I believe the *Melithreptus melanocephalus* is never found.

Genus MYZANTHA, *Vigors and Horsfield.*

A very well defined form among the Honey-eaters, all the species of which are confined to Australia. They are noisy, familiar birds, attractive in their manners, though not in their plumage. The sexes are similarly clothed, and the young assume the adult colouring from the nest.

Sp. 353. MYZANTHA GARRULA, *Vig. and Horsf.*

GARRULOUS HONEY-EATER.

Merops garrulus, Lath. Ind. Orn., Supp. p. xxiv.
Chattering Honey-eater, Lath. Gen. Hist., vol. iv. p. 164.
Philemon garrulus, Vieill. 2nd édit. du Nouv. Dict. d'Hist. Nat., tom. xxvii. p. 427.
Myzantha garrula, Vig. and Horsf. in Linn. Trans., vol. xv. p. 319.
Gracula melanocephala, Lath. Ind. Orn., Supp. p. xxviii.
Manorhina melanocephalus, Wagl. Syst. Av., *Manorhina*, sp. 1.
—— *garrula*, Gray Gen. of Birds, vol. i. p. 127, *Manorhina*, sp. 2.
Cobaygin, Aborigines of New South Wales.
Miner, Colonists of Van Diemen's Land.

Myzantha garrula, Gould, Birds of Australia, fol., vol. iv. pl. 76.

Tasmania, and all parts of the colonies of New South Wales, Victoria, and South Australia, are alike inhabited by this well-known bird. On comparing examples from Tasmania with others killed on the continent of Australia, a difference is found to exist in their relative admeasurements, the Tasmanian birds being more robust and larger in every respect; still as not the slightest difference is observable in the markings of their plumage, I consider them to be merely local varieties and not distinct species.

The natural habits of this bird lead it to frequent the thinly timbered forests of *Eucalypti* clothing the plains and low hills, rather than the dense brushes.

It moves about in small companies of from four to ten in number. In disposition it is restless, inquisitive, bold, and noisy, and frequently performs the most grotesque actions, spreading out the wings and tail, hanging from the branches in every possible variety of position, and keeping up all the time an incessant babbling: were this only momentary or for a short time, their droll attitudes and singular note would be rather amusing than otherwise; but when they follow you through the entire forest, leaping and flying from branch to branch, they become very troublesome and annoying.

The nest is cup-shaped and about the size of that of the European Thrush, very neatly built of fine twigs and coarse grass, and lined either with wool and hair, or fine soft hair-like strips of bark, frequently mixed with feathers; it is usually placed among the small upright branches of a moderately sized tree. The eggs, which vary considerably, are thirteen lines long by nine and a half lines broad, are of a bluish white, marked all over with reddish brown, without any indication of the zone at the larger end so frequently observable in the eggs of other species.

The sexes offer no other external difference than that the female is a trifle smaller than the male.

Face grey; crown of the head dull black; ear-coverts and a crescent-shaped mark inclining upwards to the angle of the bill glossy black; all the upper surface light greyish brown;

the feathers at the back of the neck tipped with silvery grey; primaries dark brown margined externally with grey; secondaries dark brown on their inner webs, the outer webs grey at the tip, and wax-yellow at the base; tail greyish brown, with dark brown shafts, and all but the two centre feathers largely tipped with brownish white; chin grey, a patch of dark brown down the centre; under surface grey; the feathers of the breast with a narrow crescent-shaped mark of brown near the tip of each; irides dark hazel; naked space beneath the eye, bill and feet yellow.

Sp. 354.   MYZANTHA OBSCURA, *Gould*.

SOMBRE HONEY-EATER.

Myzantha obscura, Gould in Proc. of Zool. Soc., part viii. p. 159.
Manorhina obscura, Gray Gen. of Birds, vol. i. p. 127, *Manorhina*, sp. 8.
Bil-ya-goo-rong, Aborigines of the lowland, and
Bil-your-ga, Aborigines of the mountain districts of Western Australia.

Myzantha obscura, Gould, Birds of Australia, fol., vol. iv. pl. 77.

This species inhabits Swan River and the south-western portion of Australia generally, where it beautifully represents the *Myzantha garrula* of New South Wales. In habits, actions, and disposition the two birds closely assimilate.

Gilbert's notes supply me with the following information, which I give in his own words:—

"It inhabits every variety of wooded situation, in all parts of the colony, and is generally met with in small families. In flying the wings are moved very rapidly, but the bird does not make progress in proportion to the apparent exertion; at times, when passing from tree to tree, its flight is graceful in the extreme.

"The stomach is small but tolerably muscular; and the food, which consists of coleopterous and other insects, seeds, and berries, is procured both on the ground and among the branches.

"The nest is built on an upright fork of the topmost branches of the smaller gum-trees, and is formed of small dried sticks lined with soft grasses and feathers. The eggs are eleven and a half lines long by nine lines broad, of a rich orange-buff, obscurely spotted and blotched with a deeper tint, particularly at the larger end."

The sexes offer but little difference in colour, but the female is somewhat smaller in all her admeasurements.

Forehead yellowish olive; lores, line beneath the eye, and ear-coverts black; head and all the upper surface dull grey, with an indistinct line of brown down the centre of each feather, giving the whole a mottled appearance; wings and tail brown, margined at the base of the external webs with wax-yellow, the tail terminating in white; throat and under surface dull grey, becoming lighter on the lower part of the abdomen and under tail-coverts; the feathers of the breast with a crescent-shaped mark of light brown near the extremity, and tipped with light grey; irides dark brown; bare skin round the eye, bill, and bare patch on each side of the throat, bright yellow; legs and feet dull reddish yellow; claws dark brown.

Total length $9\frac{1}{2}$ inches; bill $1\frac{1}{4}$; wing $6\frac{1}{2}$; tail $4\frac{3}{8}$; tarsi $1\frac{1}{4}$.

Sp. 355.   MYZANTHA LUTEA, *Gould.*

LUTEOUS HONEY-EATER.

Myzantha lutea, Gould in Proc. of Zool. Soc., part vii. p. 144.
Manorhina lutea, Gray, Gen. of Birds, vol. i. p. 127. *Manorhina,* sp. 5.

Myzantha lutea, Gould, Birds of Australia, fol., vol. iv. pl. 78.

I consider this to be by far the finest species of the genus yet discovered, exceeding as it does every other both in size and in the brilliancy of its colouring. I am indebted to Messrs. Bynoe and Dring for fine specimens of this beautiful bird, which were obtained by those gentlemen on the north-west coast of Australia, in which part of the country it supplies the place of the *Myzantha garrula* of New South Wales. The law of representation is rarely carried out in a more beautiful manner, than in the members of the present genus; the *Myzantha garrula* being, so far as is yet known, confined to the south-eastern portion of the country, the *M. lutea* to the neighbourhood of the north coast, the *Myzantha obscura* to Swan River and the *M. flavigula* to the north-eastern portion of the country.

Naked space behind the eye, forehead, and the tips of several feathers on the sides of the neck, fine citron-yellow; lores blackish brown with silvery reflexions; upper surface grey, the feathers of the back of the neck and back crossed near the tip with white; rump, upper tail-coverts, and under surface white; throat and chest tinged with grey, each feather crossed by an arrow-shaped mark of brown; wings and tail brown, the external margins of the feathers dull citron-yellow; tail tipped with white; bill fine citron-yellow; feet yellowish brown.

Total length $10\frac{3}{4}$ inches; wing $5\frac{7}{8}$; tail $5\frac{1}{4}$; tarsi $1\frac{1}{4}$.

Sp. 356. MYZANTHA FLAVIGULA, *Gould*.

YELLOW-THOATED MINER.

Myzantha flavigula, Gould in Proc. of Zool. Soc., part vii. p. 143.
Manorhina flavigula, Gray and Mitch. Gen. of Birds, vol. i. p. 127, *Manorhina*, sp. 4.

Myzantha flavigula, Gould, Birds of Australia, fol., vol. iv. pl. 79.

This species is tolerably abundant in the belts of *Eucalypti* bordering the river Namoi and all similar situations in the interior of New South Wales. Although it has many of the habits and actions of its near ally the *Myzantha garrula*, it is much more shy in disposition, less noisy, and more disposed to frequent the tops of the trees; and so exclusively does it replace the common species in the districts alluded to that the latter does not occur therein.

I did not succeed in finding the nest, but the fact of my having shot very young individuals affords indubitable evidence that the bird breeds in the localities above-mentioned.

The sexes are alike in plumage.

Naked space behind the eye, forehead, upper part of the throat, and the tips of several feathers on each side of the neck citron-yellow; rump and upper tail-coverts white; back of the neck and back grey, each feather obscurely barred with white near the tip; lores and ear-coverts black, the latter crossed with silvery grey; throat, cheeks, and all the under surface white, the feathers of the chest crossed by an arrow-shaped mark of brown; wings and tail dark brown, the outer webs of the primaries, many of the secondaries, and the basal portion of the tail-feathers dull citron-yellow; all the tail-feathers tipped with white; bill bright orange-yellow; feet yellow; irides leaden-brown.

Total length $9\frac{3}{4}$ inches; bill 1; wing $5\frac{1}{4}$; tail 5; tarsi $1\frac{5}{16}$.

Genus MANORHINA, *Vieillot.*

The single species of this form is a native of the south-eastern parts of Australia; it is very nearly allied to the *Myzanthæ*, but differs from them in some minor points.

Sp. 357.   MANORHINA MELANOPHRYS.

BELL-BIRD.

Turdus melanophrys, Lath. Ind. Orn., Supp. p. xlii.
Manorhina viridis, Vieill. Gal. des Ois., pl. 149.
Myzantha flavirostris, Vig. and Horsf. in Linn. Trans., vol. xv. p. 319.
Manorina viridis, Bonn. et Vieill. Ency. Méth. Orn., part ii. p. 692.
Dilbong and *Dilring*, Aborigines of New South Wales (Latham).
Bell-bird of the Colonists.

Myzantha melanophrys, Gould, Birds of Australia, fol., vol. iv. pl. 80.

The present bird evinces a decided preference for, and

appears to be strictly confined to dense and thick brushes, particularly such as are of a humid and swampy nature, and with the foliage of which the peculiar tint of its plumage closely assimilates. I frequently met with it in companies of from ten to forty, and occasionally still greater numbers were seen disporting among the leafy branches in search of insects and displaying many varied actions, at one time clinging to and hanging down from the branches by one leg, and at another prying beneath the leaves, or flying with outspread wings and tail from tree to tree, and giving utterance to a peculiar garrulous note totally different in sound from the faint monotonous tinkle usually uttered, which has been justly compared to the sound of distant sheep-bells, and which, when poured forth by a hundred throats from various parts of the forest, has a most singular effect. The same appellation of Bell-bird having been given by the colonists of Swan River to a species inhabiting that part of Australia, I must here warn my readers against considering them identical, by informing them that the two birds are not only specifically but generically distinct.

This bird has not as yet been observed out of New South Wales, where its peculiar province is the brushes; and if it departs from those which stretch along the coast from Port Philip to Moreton Bay, I believe it will only be found in those which clothe the sides of the higher hills, such as the Liverpool range and others of a similar character.

Like the *Myzanthæ* it is of a prying and inquisitive disposition, and the whole troop may be easily brought within the range of observation by uttering any kind of harsh squeaking note, when they will descend to ascertain the cause, and evince the utmost curiosity. Its flight is of the same skimming motionless character as that of the Garrulous Honey-eater; and upon some given signal the whole flock, or the greater portion of it, fly off simultaneously and collect on some neighbouring branch in a cluster.

The sexes are precisely alike in plumage, and the young soon attain the colouring of the adult.

The whole of the plumage, with the exception of the primaries and secondaries, yellowish olive, but the under surface much paler than the upper; forehead, stripe from the angle of the lower mandible, ring encircling and dilated into a spot above the eye, black; ear-coverts olive-brown; primaries and secondaries dark brown, the former margined with grey and the latter with yellowish olive; bill fine yellow; tarsi and toes fine orange yellow; eye dark leaden brown; eyelash leaden grey; bare space below and behind the eye orange-red.

Genus DICÆUM, *Cuvier*.

The continent of India, the Indian Islands, and New Guinea are the countries in which the members of this genus abound; as yet only a single species has been found in Australia.

Sp. 358. DICÆUM HIRUNDINACEUM.

Swallow Dicæum.

Sylvia hirundinacea, Shaw, Nat. Misc., vol. iv. pl. 114.
Swallow Warbler, Lath. Gen. Syn., Supp. vol. ii. p. 250.
Pipra desmaretii, Leach, Zool. Misc., vol. i. p. 94. pl. 41.
Crimson-throated Honey-sucker, Lewin, Birds of New Holl., pl. 7.
Desmaretian Manakin, Shaw, Gen. Zool., vol. x. p. 18.
Dicœum atrogaster, Less. Traité d'Orn., p. 303.
—— *pardalotus*, Cuv. De la Fres. Mag. de Zool., 1833, pl. 14.
Microchelidon hirundinacea, Reich. Handb. der Spec. Orn., p. 243.
Moo-ne-je-teng, Aborigines of the lowland districts of Western Australia.

Dicæum hirundinaceum, Gould, Birds of Australia, fol. vol. ii. pl. 34.

By far the greater number of the Australians are, I believe, unacquainted with this beautiful little bird, yet there is scarcely an estate in either of the colonies in which it may not be found either as a permanent resident or an occasional visitor.

Its natural disposition, leading it to confine itself almost exclusively to the topmost branches of the loftiest trees, is doubtless the cause of its not being more generally known than it is; its rich scarlet breast, not even attracting notice at the distance from the ground at which it generally keeps; and, in obtaining specimens, I was more frequently made aware of its presence by its pretty warbling song than by its movements among the branches; so small an object, in fact, is most difficult of detection among the thick foliage of the lofty *Casuarinæ*, to which trees it is extremely partial, particularly to those growing on the banks of creeks and rivers. It is also frequently to be seen among the clusters of the beautiful parasitic *Loranthus*, which is very common on the *Casuarinæ* in the neighbourhood of the Upper Hunter. Whether the bird is attracted to this misseltoe-like plant for the purpose of feeding upon its sweet and juicy berries I could not ascertain; its chief food is insects, but in all probability it may occasionally vary its food.

The Swallow Dicæum has neither the actions of the Pardalotes nor of the Honey-eaters; it differs from the former in its quick darting flight, and from the latter in its less prying, clinging, and creeping actions among the leaves, &c. When perched on a branch it sits more upright, and is more Swallow-like in its contour than either of the forms alluded to; the structure of its nest and the mode of its nidification are also very dissimilar.

Its song is a very animated and long-continued strain, but is uttered so inwardly, that it is almost necessary to stand beneath the tree upon which the bird is perched, before its notes can be heard.

It would appear that the range of this species extends to all parts of the Australian continent, since I have received specimens from every locality yet explored. I found it breeding on the Lower Namoi, which proves that the interior of the country is inhabited by it as well as those portions between the ranges and the coast.

Mr. White, of the Reed-beds near Adelaide, says—

"This little bird is sometimes rather numerous here. It appears to be wholly frugivorous, for all of those I have dissected had fruit in them; it has no regular stomach, not even an enlargement of the intestine, which averages above five inches and a half in length, and through which the food passes whole. It arrives at Adelaide about February, and stays but a short time. I have met with it very far north."

Its beautiful purse-like nest is composed of the white cotton-like substance found in the seed-vessels of many plants, and among other trees is sometimes suspended on a small branch of a *Casuarina* or an *Acacia pendula*. The ground-colour of the eggs is dull white, with very minute spots of brown scattered over the surface; they are nine lines long by five and a half lines broad.

The male has the head, all the upper surface, wings, and tail black, glossed with steel-blue; primaries black; throat, breast, and under tail-coverts scarlet; flanks dusky; abdomen white, with a broad patch of black down the centre; irides dark brown; bill blackish brown; feet dark brown.

The female is dull black above, glossed with steel-blue on the wings and tail; throat and centre of the abdomen buff; flanks light brown; under tail-coverts pale scarlet.

Genus NECTARINIA, *Illiger*.

It gives me great pleasure to state that at least one species of this genus is found in Australia, a circumstance which might naturally be expected when so many inhabit New Guinea and the adjacent islands. Birds of this form are also spread throughout the Philippines to Malasia, China, and

India; nor are they wanting in Africa. That the *Nectarinidæ* and *Meliphagidæ* are closely allied must be evident to every one who attends to ornithology as a science.

Sp. 359. NECTARINIA AUSTRALIS, *Gould*.

AUSTRALIAN SUN-BIRD.

Nectarinia frenata, Müll. Verb. Nat. Gesch., p. 61. t. 8. f. 1?
——— *australis*, Gould in Proc. of Zool. Soc., part xviii. p. 201.
Terridirri, Aborigines of Cape York.

Nectarinia australis, Gould, Birds of Australia, fol., Supplement, pl.

The *Nectarinia australis* offers a very close alliance to the *N. frænata* of the Celebes; it will be found, however, to differ from that species in its larger size, in the mark above the eye being less conspicuous, and in the straighter form of the bill. For my first knowledge of this bird I am indebted to the researches of the late Commander Ince, R.N., who, while attached to H.M.S. Fly, paid considerable attention to the natural history of the northern parts of Australia. Since then many other specimens have been forwarded to me by Mr. Macgillivray and others.

Mr. Macgillivray informed me that "this pretty Sun-bird appears to be distributed along the whole of the north-east coast of Australia, the adjacent islands, and the whole of the islands in Torres Straits. Although thus generally distributed, it is nowhere numerous, seldom more than a pair being seen together. Its habits resemble those of the *Ptilotes*, with which it often associates, but still more closely to those of *Myzomela obscura*; like those birds, it resorts to the flowering trees to feed upon the insects which frequent the blossoms, especially those of a species of *Sciadophyllum*: this singular tree is furnished with enormous spike-like racemes of small scarlet flowers, which attract numbers of insects, and thus furnish an abundant supply of food to the present bird and many species of the *Meliphagidæ*. Its note, which is a sharp shrill cry, prolonged for about ten seconds, may be represented by '*Tsee-tsee-tsee-tss-ss-ss-ss*.' The male appears to be of a pugnacious disposition, as I have more than once seen it drive away and pursue a visitor to the same tree; perhaps, however, this disposition is only exhibited during the breeding-season. I found its nest on several occasions, as will be seen by the following extracts from my note-book:—

"Nov. 20, 1840.—Cape York. Found two nests of *Nectarinia* to-day: one on the margin of a scrub, the other in a clearing. The nests were pensile, and in both cases were attached to the twig of a prickly bush: one, measuring seven inches in length, was of an elongated shape, with a rather large opening on one side close to the top; it was composed of shreds of *Melaleuca* bark, a few leaves, various fibrous substances, rejectamenta of caterpillars, &c., and lined with the silky cotton of the *Bombax Australis*. The other, which was similar in structure, contained a young bird, and an egg with a chick almost ready for hatching. The female was seen approaching with a mouthful of flies to feed the young. The egg was pear-shaped, generally and equally mottled with obscure dirty brown on a greenish-grey ground.

"Dec. 4th.—Mount Ernest, Torres Straits. A nest of *Nectarinia* found to-day differs from those seen at Cape York in having over the entrance a projecting fringe-like hood composed of the panicles of a delicate grass-like plant. It contained two young birds, and I saw the mother visit them twice with an interval of ten minutes between; she glanced past like an arrow, perched on the nest at once, clinging to the lower side of the entrance, and looked round very watchfully for a few seconds before feeding the young, after which she disappeared as suddenly as she had arrived."

Mr. Ramsay, in his "List of Birds received from Port Denison, Queensland," published in the 'Ibis' for 1865, says:—

"According to Mr. Rainbird, numbers of this beautiful little Sun-bird may be seen, on bright mornings, among the leafy tops of the mangrove-belts near Port Denison, which is nearly in lat. 20° S., long. 148° E. of Greenwich. They are ever darting out to capture some insect on the wing, returning and disappearing again in the thick foliage, or perching upon some topmost twig, to devour their captures, and show their shining purple breasts glittering in the sun. During the hottest part of the day the Sun-birds betake themselves to the thick scrub, which in many places runs down to the water's edge. They breed in the months of November and December. One pair chose a little break in the scrub within a few yards of the water, where, facing the rising sun, they suspended their nest by the top from the dead twig of a small shrub at the foot of a large 'Bottle-tree' (*Sterculia rupestris*). The nest is of an oval form, much resembling, and suspended in the same way as that of *Acanthiza lineata*, with a small hood over the opening, which is near the top. It is composed of fibrous roots and shreds of cotton-tree (*Gomphocarpus fruticosus*) bark, firmly interwoven with webs and cocoons of various spiders, and a few pieces of white seaweed ornamenting the outside. It is lined with feathers and the silky native cotton, and is about five inches long by three and a half in diameter. The eggs, I regret to say, I did not receive, as Mr. Rainbird was obliged to come away before they were laid."

The male has the crown of the head and upper surface olive-green; over and under the eye two inconspicuous marks of yellow; throat and chest steel-blue; remainder of the under surface fine yellow; irides chestnut; bill and feet black.

Total length $4\frac{3}{4}$ inches; bill $\frac{7}{8}$; wing $2\frac{1}{2}$; tail $1\frac{1}{2}$; tarsi $\frac{5}{8}$.

The female differs in having the whole of the under surface yellow, without a trace of the steel-blue gorget so conspicuous in the male.

Family ——?

Genus ZOSTEROPS.

The members of this genus are very widely dispersed over the Old World, except its extreme northern portions; three well-defined species inhabit the continent of Australia and Tasmania; two are found on Norfolk Island.

In placing this group next to the Honey-eaters, I have been influenced by their approximation to those birds both in form and habits, and to which they exhibit a further degree of affinity in the form and structure of their nest, but not in the colouring of their eggs, which are always blue. The sexes are alike in plumage.

Sp. 360. ZOSTEROPS CŒRULESCENS.

Grey-backed Zosterops.

Certhia cærulescens, Lath. Ind. Orn. Supp., p. xxxviii.
Sylvia lateralis, Lath., id., p. lv.
Certhia diluta, Shaw, Gen. Zool., vol. viii. p. 244.
Philedon cæruleus, Cuv.
Meliphaga cærulea, Steph. Cont. of Shaw's Gen. Zool., vol. xiv. p. 264.
Sylvia annulosa, var. β, Swains. Zool. Ill., 1st ser., pl. 16.
Zosterops dorsalis, Vig. and Horsf. in Linn. Trans., vol. xv. p. 235.
White-eye, Colonists of New South Wales.

Zosterops dorsalis, Gould, Birds of Australia, fol., vol. iv. pl. 81.

This bird is stationary in all parts of Tasmania, New South Wales, and South Australia, where it is not only to be met with in the forests and thickets, but also in nearly every garden. It even builds its nest and rears its young in the shrubs and rose-trees bordering the walks. Among the trees of the forest the beautiful *Leptospermum* is the one to which at all times this species evinces a great partiality.

Its flight is quick and darting, and when among the

branches of the trees it is as active as most birds, prying and
searching with scrutinizing care into the leaves and flowers
for the insects, upon which it feeds. It is sometimes seen
singly or in pairs, while at others it is to be observed in great
numbers, on the same or neighbouring trees. It is of a
familiar disposition, and utters a pretty and very lively song.

The breeding-season commences in September and continues to January. The nest is one of the neatest structures possible; it is of a round deep cup-shaped form, composed of fine grasses, moss, and wool, and most carefully lined with fibrous roots and grasses. The eggs are usually three in number, of a beautiful uniform pale blue, eight and a half lines long by six broad.

The sexes present no difference of plumage.

Crown of the head, wings, and tail olive; back dark grey, eyes surrounded by a zone of white feathers, bounded in front and below with black; throat, centre of the abdomen, and under tail-coverts greyish white with a slight tinge of olive; flanks light chestnut brown; upper mandible dark brown, under mandible lighter; irides and feet greyish brown.

In some specimens the throat and sides of the head are wax-yellow, and the flanks are only stained with chestnut brown.

Sp. 301.   ZOSTEROPS GOULDI, *Bonaparte*.
GREEN-BACKED ZOSTEROPS.

Zosterops chloronotus, Gould in Proc. of Zool. Soc., part viii. p. 165.
—— *Gouldi*, Bonap. Consp. Gen. Av., tom. i. p. 308, *Zosterops*, sp. 2.
Jule-we-de-lung, Aborigines of the lowland districts of Western Australia.
Grape- and *Fig-eater*, Colonists of Swan River.

Zosterops chloronotus, Gould, Birds of Australia, fol, vol. iv. pl. 82.

The *Zosterops gouldi* is an inhabitant of the western

coast of Australia, where it constitutes a beautiful representative of the *Z. cærulescens* of the southern and eastern coasts. As might be supposed, the habits, manners, actions, and economy of two species so nearly allied are very similar; hence the settlers of Swan River were not long in discovering that in this species they had found no friend to their gardens during the season when the fruits are ripening, whatever good it may effect by the destruction of insects at other periods.

Gilbert informed me that " This bird is particularly fond of figs and grapes, it consequently abounds in all the gardens where those plants are cultivated; and it is often to be seen as numerous as sparrows in England; besides feeding upon fruits, I have also observed it taking flies while on the wing after the manner of the true Flycatchers.

" Its note is a single plaintive one, several times repeated; and its flight is irregular, and of short duration.

" The breeding-season commences in August and ends in November; those nests that came under my observation during the earlier part of the season, invariably contained two eggs; but in October and November I usually found the number to be increased to three, and upon one occasion to four. The nest is small, compact, and formed of dried wiry grasses, bound together with the hairy tendrils of small plants and wool, the inside being lined with very minute fibrous roots; its breadth is about two inches, and depth one inch; the eggs are greenish blue without spots or markings, eight lines long by six lines broad."

Lores black; crown of the head and all the upper surface olive-green; primaries and tail-feathers brown, margined with olive-green; throat and under tail-coverts light greenish yellow; breast and under surface grey, tinged with brown on the abdomen and flanks; irides wood-brown; bill brown, lighter on the under mandible; legs and feet dark grey.

Total length $4\frac{1}{4}$ inches; bill $\frac{5}{8}$; wing $2\frac{1}{4}$; tail $1\frac{3}{4}$; tarsi $\frac{5}{8}$.

The specific term *chloronotus* having been previously

assigned to another bird of this form, the late Prince Charles Bonaparte was pleased to dedicate the present one to myself.

Sp. 362. ZOSTEROPS LUTEUS, *Gould.*

YELLOW ZOSTEROPS.

Zosterops luteus, Gould, Birds of Australia, fol., vol. iv. pl. 83.

This new species is an inhabitant of the northern portion of Australia. "I first met with it," says Gilbert, "in August, on Greenhill Island, Van Diemen's Gulf, dwelling among the mangroves or the densest thickets. It is much more wild and solitary than *Zosterops cærulescens*, and does not resort like that bird to the gardens and the neighbourhood of the houses of the settlers; its note is also very different, being a pretty canary-like song. When disturbed it usually left the thicket for the higher branches of the gum-trees, where it was effectually hidden from view by the thick foliage. It was generally met with in small families of from three to seven or eight in number."

All the upper surface olive-yellow; primaries and tail-feathers brown, margined with olive-yellow; forehead and throat pure yellow; lores and lines beneath the eye black; eye encircled with a zone of white feathers; abdomen and under tail-coverts dull yellow; irides light reddish brown; upper mandible blackish grey, the basal half rather lighter; apical third of the lower mandible blackish grey; basal two-thirds light ash-grey; legs and feet bluish grey.

Total length 4¼ inches; bill ½; wing 2⅜; tail 1¾; tarsi ⅝.

Family EPIMACHIDÆ.

Many authors place the three following birds in the family *Paradiseidæ*; Cabanis makes them part of the subdivision of the subfamily *Epimachinæ*. Mr. G. R. Gray retains them in the same subfamily, but makes it form a part of the family *Upupidæ*. It has always appeared to me that they bear a strong resemblance to the *Climacteres*, preceding which I shall therefore place them.

Genus PTILORHIS, *Swainson*.

Of this genus two well-defined species inhabit Australia, viz. *P. paradisea* and *P. Victoriæ*.

Sp. 363.   PTILORHIS PARADISEA, *Swains.*

RIFLE-BIRD.

Ptiloris paradiseus, Swains. Zool. Journ., vol. i. p. 481.
Epimachus regius, Less. Zool. de la Coq., pl. 28.
—— *brisbanii*, Wils. Ill. of Zool., pl. 9.
Ptiloris paradisea, G. R. Gray, Gen. of Birds, 2nd edit. p. 15.
Epimachus paradiseus, Gray and Mitch. Gen. of Birds, vol. i. p. 94,
  Epimachus, sp. 4.
Ptilorhis paradiseus, Cab. Mus. Hein., Theil i. p. 214.
—— *paradisea*, Reich. Handb. der Spec. Orn., p. 328.

Ptiloris paradiseus, Gould, Birds of Australia, fol., vol. iv. pl. 100.

Hitherto this magnificent bird has only been discovered in the brushes of the south-eastern portion of Australia; so limited in fact does its range of habitat seem to be, that the river Hunter to the southward, and Moreton Bay to the eastward, may be considered its natural boundaries in either direction. I have been informed by several persons who have seen it in its native wilds that it possesses many habits in common with the *Climacteres*, and that it ascends the upright boles of trees precisely after the manner of those birds. It was a source of regret to me that I had no opportunity of verifying these assertions, but an examination of the structure of the bird induces me to believe that such is the case: that its powers of flight are very limited, is certain from the shortness and peculiarly truncate form of the wing, and this mode of progression is doubtless seldom resorted to further than to transport it from tree to tree, or from one part of the forest to another. That it is stationary, and breeds in South-eastern Australia, is evident from the numerous specimens of all ages that have been sent from thence to Europe.

Since the above remarks were published in the folio edition, the late F. Strange forwarded me the following note, which I give in his own words:—

"The principal resort of the Rifle-Bird is among the large cedar-brushes that skirt the mountains and creeks of the Manning, Hastings, MacLeay, Bellenger, Clarence, and Richmond Rivers, and there, during the pairing-months of November and December, the male bird is easily found. At that time of the year, as soon as the sun's rays gild the tops of the trees, up goes the Rifle Bird from the thickets below to the higher branches of the pines (*Araucaria macleayana*) which there abound. It always affects a situation where three or four of these trees occur about two hundred yards apart, and there the morning is spent in short flights from tree to tree, in sunning and preening its feathers, and in uttering its song each time it leaves one tree for another. The sound emitted resembles a prolonged utterance of the word "Yass," by which the bird is known to the natives of the Richmond River. In passing from tree to tree, it also makes an extraordinary noise resembling the shaking of a piece of new stiff silk. After 10 a.m. it descends lower down, and then mostly resorts to the thick limb of a Cedar-tree (*Cedrela australis*), and there continues to utter its cry of *Yass* at intervals of two minutes' duration; at this time, owing to the thickness of the limb and the closeness with which the bird keeps to it, it is very difficult of detection; wait with patience, however, and you will soon see him, with wings extended, and his head thrown on his back, whirling round and round, first one way and then another."

The adult male has the general plumage rich velvety black,

glossed on the upper surface with brownish lilac; under surface similar to the upper, but all the feathers of the abdomen and flanks broadly margined with rich olive-green; feathers of the head and throat small, scale-like and of a shining metallic blue-green; two centre tail-feathers rich shining metallic green, the remainder deep black; bill and feet black.

The female has the whole of the upper surface greyish brown; the wings and tail edged with ferruginous; the feathers of the head with a narrow line of white down the centre; line passing down the side of the head from behind the eye, chin, and throat buffy white; all the under surface deep buff, each feather with a black arrow-shaped mark near the tip.

Sp. 364.   PTILORHIS VICTORIÆ, *Gould.*

QUEEN VICTORIA'S RIFLE-BIRD.

Ptiloris victoriæ, Gould in Proc. of Zool. Soc., 1849, p. 111, Aves, pl. 12.
Ptilorhis victoriæ, Reich. Handb. der Spec. Orn., p. 329.

Ptiloris victoriæ, Gould, Birds of Australia, fol., Supplement, pl.

This Rifle-Bird is smaller in all its admeasurements than the *Ptilorhis paradiseus*, and may be distinguished by the purple of the breast presenting the appearance of a broad pectoral band, bounded above by the scale-like feathers of the throat, and below by the abdominal band of deep oil-green, and by the broad and much more lengthened flank feathers which show very conspicuously.

"This bird," says Mr. Macgillivray, "was seen by us during the survey of the N.E. coast of Australia on the Barnard Isles, and on the adjacent shores of the mainland at Rockingham Bay, in the immediate vicinity of Kennedy's first camp. On one of the Barnard Isles (No. III. in lat. 17° 43′ S.), which is covered with dense brush, I found Queen Victoria's Rifle Bird in considerable abundance. Females and young males were

common, but rather shy; however, by sitting down and
quietly watching in some favourite locality, one or more
would soon alight on a limb or branch, run along it with
great celerity, stop abruptly every now and then to thrust its
beak under the loose bark in search of insects, and then fly
off as suddenly as it had arrived. Occasionally I have seen
one anxiously watching me from behind a branch, its head
and neck only being visible. At this time (June) the young
males were very pugnacious, and upon one occasion three of
them were so intent upon their quarrel that they allowed me
to approach sufficiently near to kill them all with a single
charge of dust shot. The adult males were comparatively
rare, always solitary and very shy. I never saw them upon
the trees, but only in the thick bushes and masses of climbing
plants beneath them; on detecting the vicinity of man they
immediately shuffled off among the branches towards the
opposite side of the thicket and flew off for a short distance.
I did not observe them to utter any call or cry; this, however, may have arisen from my attention not having been so
much directed to them as to the females and young males,
which I was more anxious to procure, the very different style
of their colouring having led me to believe they were a new
species of *Pomatostomus*."

The male has the general plumage rich deep velvety black,
glossed on the upper surface, sides of the neck, chin, and
breast with plum-colour; feathers of the head and throat
small, scale-like, and of a shining, metallic bronzy green;
feathers of the abdomen very much developed, of the same
hue as the upper surface, but each feather so broadly margined with rich deep olive-green, that the colouring of the
basal portion of the feather is hidden, and the olive-green
forms a broad abdominal band, which is sharply defined
above, but irregular below; two centre tail-feathers rich
shining metallic green, the remainder deep black; bill and
feet black.

The female has all the upper surface greyish brown, tinged with olive; head and sides of the neck dark brown, striated with greyish brown; over each eye a superciliary stripe of buff; wing-feathers edged with ferruginous; chin and throat pale buff; remainder of the under surface, under wing-coverts, and the base of the inner webs of the quills rich deep reddish buff, each feather with an irregular spot of brown near the tip, dilated on the flanks in the form of irregular bars; bill and feet black.

Male, length 10½ inches; bill 1¾; wing 6; tail 3¼; tarsi 1¼.

Genus CRASPEDOPHORA, *G. R. Gray.*

The *Epimachus magnificus* of Cuvier differing from the members of the genus *Ptilorhis* in form and colouring, Mr. G. R. Gray has made it the type of his genus *Craspedophora*.

Sp. 305.   CRASPEDOPHORA MAGNIFICA.

MAGNIFICENT RIFLE-BIRD.

Le Promefil, Levaill. Ois. de Parad., p. 36, pl. 16.
Falcinellus magnificus, Vieill. Nouv. Dict. d'Hist. Nat., tom. xxviii. p. 167, pl. G. 30. No. 8.
Epimachus magnificus, Cuv. Règn. Anim., pl. 4. fig. 2.
——— *paradiseus*, Gray and Mitch. Gen. of Birds, vol. i. pl. xxxii.
——— *splendidus*, Steph. Cont. of Shaw's Gen. Zool., vol. xiv. p. 77.
Craspedophora magnifica, G. R. Gray, List of Gen. of Birds, 2nd edit. p. 15.

Ptiloris magnifica, Gould, Birds of Australia, fol., Supplement, pl.

"It is New Guinea," says Vieillot, "that country in which are found the most beautiful birds in the world, and the most remarkable for the singularity of their plumage, that is the habitat of this species, one of the richest of its family." "It is still," says M. Lesson, writing in 1830, "very rarely met with in collections; the individual in the gallery of the Museum (at Paris) was procured in London, at the sale of

Bullock's collection. During our sojourn at New Guinea with the corvette 'La Coquille,' we only obtained two mutilated skins; and M. Dumont-Durville, commander of the expedition of the 'Astrolabe,' secured only a single skin deprived of its wings and feet, the manner in which they are usually prepared by the natives. It is in the dense and vast forests which surround the harbour of Dorey in New Guinea, that this fine species resides." The researches of Mr. Macgillivray and others enable me to state that it also inhabits the north-eastern portion of Australia, a circumstance of no ordinary interest, since besides adding another fine species to the already exceedingly rich fauna of that country, we now know that our museums will ere long be graced with fine and perfect specimens in lieu of the mutilated skins hitherto procurable. We have abundant evidence of its being frequently met with at Cape York, since nearly every officer of the 'Rattlesnake' procured and brought home specimens from that locality.

The following are Mr. Macgillivray's notes respecting it:—

"This fine Rifle-Bird inhabits the densest of the brushes in the neighbourhood of Cape York. The natives are familiar with it under the name of '*Yagoonya*'; the Darnley Islanders also recognized a skin shown them, and described it to be a native of *Dowde*, or the south coast of New Guinea, near Bristow Island. Its cry is very striking: upon being imitated by man, which may be easily done, the male bird will answer; it consists of a loud whistle resembling *wheeoo* repeated three times and ending abruptly in a note like *who-o-o*. Both sexes utter the same note, but that of the male is much the loudest. The old males were generally seen about the tops of the highest trees, where, if undisturbed, they would remain long enough to utter their loud cry two or three times at intervals of from two to five minutes. If a female be near, the male frequently perches on a conspicuous dead twig in a crouching attitude, rapidly opening and closing his wings, the feathers of which by their peculiar form and texture produce a loud rustling noise, which in the comparative stillness of these solitudes may be heard at the distance of a hundred yards, and may be faintly imitated by moving the feathers of a dried skin. The full-plumaged males are much more shy than the females or immature birds. According to the testimony of several of the Cape York natives whom I questioned upon the subject, the *C. magnifica* breeds in a hollow tree and lays several white eggs. The ovary of a female shot in November, the commencement of the rainy season, contained a very large and nearly completely formed egg.

"From the shyness of this Rifle-Bird, it is difficult to catch more than a passing glimpse of it in the dense brushes which it inhabits; I once, however, saw a female running up the trunk of a tree like a Creeper, and its stomach was afterwards found to be filled with insects only, chiefly ants; while the stomach of a male, shot about the same time, contained merely a few small round berries, the fruit of a tall tree, the botanical name of which is unknown to me."

Let me add that differences too slight to be considered specific are observable in Australian and New Guinea specimens; one of them being the greater length of the black side plumes in the New Guinea examples.

The male has the general plumage deep velvety black, slightly tinged with purple; wings dull purplish black, glossed with a greenish hue on the margins of the feathers; feathers of the head small, scale-like, and of a shining metallic bronzy green; feathers of the throat similar in form, and of a shining metallic oil-green, bounded below by a crescent of velvety black, to which succeeds a narrower crescent of shining yellowish green; under surface purplish black, the flank-feathers prolonged into a filamentous form and reaching beyond the extremity of the tail; two central tail-feathers shining metallic green, the remainder deep black; irides umber-brown; feet lead-colour, the soles ochraceous.

The female has all the upper surface brown; wings reddish brown, margined with bright rufous; tail rufous; over each eye a superciliary stripe of buffy white; throat buffy white; from the lower angle of the bill on each side a narrow streak of brown; breast and under surface buffy, crossed with numerous irregular bars of dark brown.

Family CERTHIADÆ?

Genus CLIMACTERIS, *Temminck*.

Great additions have been made to the species of this well-defined and singular group of Australian birds, two out of the six now known being all that had been described prior to the publication of the folio edition. With the exception of Tasmania, every colony is inhabited by one or other of the following species.

Sp. 300.　CLIMACTERIS SCANDENS, *Temm.*

BROWN TREE-CREEPER.

Buff-winged Honey-eater, Lath. Gen. Hist., vol. iv. p. 178.
Climacteris scandens, Temm. Pl. Col., 281. fig. 2.

Climacteris scandens, Gould, Birds of Australia, fol., vol. iv. pl. 93.

The Brown Tree-Creeper inhabits the whole of the south-eastern portion of the Australian continent, from South Australia to New South Wales. It gives a decided preference to the open thinly-timbered forests of *Eucalypti*, as well as the flats studded with the apple-trees (*Angophoræ*), the bark of which, being rough and uneven, affords numerous retreats for various tribes of insects; its food, however, is not only sought for upon the boles and branches of the trees, but is obtained by penetrating the decayed and hollow parts; and it even dives into the small hollow spouts of the branches in search of spiders, ants, and other insects: although its form would

lead to a contrary supposition, it spends much of its time on
the ground, under the canopy and near the boles of the larger
trees, in a similar pursuit, and also traverses the fallen trunks
with a keen and scrutinizing eye. While on the ground it
has a pert lively action, passing over the surface in a succession
of quick shuffling hops, carrying its head erect with the
feathers puffed out, almost in the form of a crest. Among
the trees it assumes all the actions of the true Creeper,
ascending the upright boles, and traversing with the greatest
facility both the upper and under sides of the branches. It
never descends with the head downwards, like the members
of the genera *Sitta* and *Sittella*; still I have seen it descend
an upright bole for a short distance, by hopping or shuffling
backwards, as it were, generally making a spiral course.

It flies with a skimming motion of the wings, during which
the brown marking of the primaries is very conspicuous.

Like many other insectivorous birds in Australia it seldom,
if ever, resorts to the water for the purpose of drinking. It
has a sharp piercing cry, which is frequently uttered, especially if the tree upon which it is climbing be approached.

The breeding-season commences in August and continues
until January. The nest is generally placed deep down in a
hollow branch: those I found were entirely composed of
the hair of the Opossum, which, judging from its brightness
and freshness, had doubtless been plucked from the living
animal while reposing in the hollow trees. The eggs in all
the nests I took were two in number, of a reddish flesh-colour, thickly blotched all over with reddish brown; they
are ten and a half lines long by eight lines broad.

The male has the crown of the head blackish brown; lores
black; line over the eye and the throat dull buff; at the base
of the throat a few indistinct blackish-brown spots; all the
upper surface rufous brown; primaries blackish brown at the
base and light brown at the tip, all but the first crossed in
the centre by a broad band of buff, to which succeeds another

broad band of blackish brown; tail brown, all but the two centre feathers crossed by a broad band of blackish brown; all the under surface greyish brown, each feather of the chest and abdomen having a stripe of dull white, bounded on either side with black, running down the centre; under tail-coverts reddish buff, crossed by irregular bars of black; irides, bill, and feet blackish brown.

Little difference is observable either in the colour or size of the sexes; the female may, nevertheless, be at once distinguished from the male by the spots at the base of the throat being rufous instead of blackish brown as in the male.

Sp. 307.   CLIMACTERIS RUFA, *Gould*.

Rufous Tree-Creeper.

Climacteris rufa, Gould in Proc. of Zool. Soc., part viii. p. 149.
Jin-ner, Aborigines of the mountain districts of Western Australia.

Climacteris rufa, Gould, Birds of Australia, fol, vol. iv. pl. 94.

In its robust form and general contour this species closely resembles the *Climacteris scandens*, but from which it is readily distinguished by the rufous colouring of its plumage.

It is a common bird at Swan River, where Gilbert states it is most abundant in the gum forests abounding with the white ant: it ascends the smooth bark of the *Eucalypti*, and traverses round the larger branches with the greatest facility, feeding, like the other members of the genus, upon insects of various kinds; but is frequently to be seen on the ground, searching for ants and their larvæ, and in this situation presents a most grotesque appearance, from its waddling gait.

Its note is a single piercing cry, uttered more rapidly and loudly when the bird is disturbed, and having a very singular and striking effect amidst the silence and solitude of the forest.

It makes a very warm nest of soft grasses, the down of flowers and feathers, in the hollow part of a dead branch, generally so far down that it is almost impossible to reach it, and it is, therefore, very difficult to find. I discovered one by seeing the old birds beating away a Wattle-bird that tried to perch near their hole; the nest, in this instance, was fortunately within arm's length; it contained three eggs of a pale salmon-colour, thickly blotched all over with reddish brown, eleven lines long by eight and a half lines broad: this occurred during the first week in October.

The stomach is large and tolerably muscular.

The male has the crown of the head, all the upper surface, and wings dark brown; rump and upper tail-coverts tinged with rufous; primaries brown, all but the first crossed by a broad band of rufous, to which succeeds a second broad band of dark brown; two centre tail-feathers brown, indistinctly barred with a darker hue; the remainder pale rufous, crossed by a broad band of blackish brown, and tipped with pale brown; line over the eye, lores, ear-coverts, throat, and under surface of the shoulder rust-brown; chest crossed by an indistinct band of rufous brown, each feather with a stripe of buffy white, bounded on each side with a line of black down the centre; the remainder of the under surface deep rust-red, with a faint line of buffy white down the centre of each feather, the white line being lost on the flanks and vent; under tail-coverts light rufous, with a double spot of blackish brown at intervals along the stem; irides dark reddish brown; bill and feet blackish brown.

The female is rather less in size; is of the same colour as the male, but much lighter, without the bounding line of black on each side of the buff stripes on the breast, and having only an indication of the double spots on the under tail-coverts.

Total length 6 inches; bill $\frac{7}{8}$; wing $3\frac{1}{2}$; tail $2\frac{3}{4}$; tarsi $\frac{7}{8}$.

Sp. 308. CLIMACTERIS ERYTHROPS, *Gould*.

RED-EYEBROWED TREE-CREEPER.

Climacteris erythrops, Gould in Proc. of Zool. Soc., part viii. p. 148.

Climacteris erythrops, Gould, Birds of Australia, fol., vol. iv. pl. 95.

I obtained this interesting species while encamped on the low grassy hills under the Liverpool range; but whether it is generally distributed over the colony, or merely confined to districts of a similar character to those in which I found it, I had no opportunity of ascertaining. So far as I could observe, its habits and manners bore a striking resemblance to those of the *Climacteris leucophæa*.

One singular feature connected with this species is the circumstance of the female alone being adorned with the beautiful radiated rufous markings on the throat, the male having this part quite plain; a fact which I ascertained beyond a doubt by the dissection of numerous specimens of both sexes; it is true that a faint trace of this character is observable both in *Climacteris scandens* and *C. rufa*, but the present is the only species of the genus in which this reversion of a general law of nature is so strikingly apparent.

The male has the crown of the head blackish brown, each feather margined with greyish brown; lores and a circle surrounding the eye reddish chestnut; back brown; sides of the neck, lower part of the back, and upper tail-coverts grey; primaries blackish brown at the base and light brown at the tip, all but the first crossed in the centre by a broad band of buff, to which succeeds another broad band of blackish brown; two centre tail-feathers grey, the remainder blackish brown, largely tipped with light grey; chin dull white, passing into greyish brown on the chest; the remainder of the under surface greyish brown, each feather having a broad stripe of dull white, bounded on either side with black running down the centre, the lines becoming blended, indistinct, and tinged with buff on the centre of the abdomen; under tail-coverts buffy white, crossed by irregular bars of black; irides brown; bill and feet black.

Total length 5 inches; bill ⅞; wing 3¼; tail 2⅜; tarsi ¾.

The female differs in having the chestnut marking round the eye much richer, and in having, in place of the greyish brown on the breast, a series of feathers of a rusty red colour, with a broad stripe of dull white down their middles, the stripes appearing to radiate from a common centre; in all other particulars her plumage resembles that of the male.

Sp. 309. CLIMACTERIS MELANONOTA, *Gould*.

BLACK-BACKED TREE-CREEPER.

Climacteris melanotus, Gould in Proc. of Zool. Soc., part xiv. p. 106.

Climacteris melanotus, Gould, Birds of Australia, fol., vol. iv. pl. 96.

For this additional species of the limited genus *Climacteris*, a form confined to Australia, we are indebted to Dr. Leichardt's Expedition from Moreton Bay to Port Essington. It was killed in latitude 15° 57′ south, on the eastern side of the Gulf of Carpentaria, and is rendered particularly interesting to me as being one of the birds procured by poor Gilbert on the day of his lamented death, the 28th of June 1845, which untoward event prevented him from recording any particulars respecting it: all, therefore, that I can do, is to point out the differences by which it may be distinguished from the other members of the genus, and recommend to future observers the investigation of its habits.

In the dark colouring and thick velvety plumage of the upper surface it is most nearly allied to the *Climacteris melanura*, but differs from that species in being destitute of the lanceolate marks on the throat, and from all others in the dark colouring of the back.

The usual distinction of the sexes—the finer colouring of

the female—exists in this as in the other species of the genus; they may be thus described:—

Superciliary line and throat buffy white; line before and behind the eye, all the upper surface, wings, and tail dark brownish black; the base of the primaries, secondaries, and tertiaries, and the under surface of the shoulder buff; under surface pale vinous brown; the feathers of the abdomen with two stripes of black running parallel to and near the stem, the space between dull white; at the base of the throat several irregular spots of black; under tail-coverts buffy white, crossed by broad bars of black; irides brown.

Total length $5\frac{1}{2}$ inches; bill $\frac{3}{4}$; wing $3\frac{1}{2}$; tail $2\frac{1}{4}$; tarsi $\frac{7}{8}$.

The female differs in having the markings of the abdomen larger and more conspicuous, and in having the spots at the base of the throat chestnut instead of black.

Sp. 370. CLIMACTERIS MELANURA, *Gould.*

BLACK-TAILED TREE-CREEPER.

Climacteris melanura, Gould in Proc. of Zool. Soc., part x. p. 138.

Climacteris melanura, Gould, Birds of Australia, fol., vol. iv. pl. 97.

I formerly believed that all the members of this genus were confined to the southern portions of Australia, but that such is not the case is proved by the circumstance of Mr. Bynoe having killed this bird on the northern coast. It exceeds all the other species in size, and also differs from them in its colouring, particularly in the lanceolate feathers on the throat and in the black colour of the tail. Nothing whatever is known of its habits or general economy, but, judging from its structure, it doubtless closely assimilates to its congeners in all these particulars.

Forehead, all the upper surface, and the tail-feathers velvety brownish black; the occiput and back of the neck stained with ferruginous brown; primaries and secondaries dark brown at the base and at the tip, the intermediate space buff, forming a conspicuous band across the wing when expanded; feathers of the throat white, edged all round with black, giving the throat a striated appearance; abdomen and flanks ferruginous brown; under tail-coverts black, irregularly crossed with bars of buff; bill and feet blackish brown.

Total length $6\frac{3}{4}$ inches; bill $\frac{7}{8}$; wing 4; tail 3; tarsi 1.

Sp. 371.   CLIMACTERIS LEUCOPHŒA.

WHITE-THROATED TREE-CREEPER.

Certhia leucophæa, Lath. Ind. Orn., Supp. p. xxxvi.
—— *picumnus*, Ill. Mus. Berol.
Climacteris picumnus, Temm. Pl. Col. 281. fig. 1.
New Holland Nuthatch, Lath. Gen. Hist., vol. iv. p. 78.
Certhia leucoptera, Lath. Gen. Hist., vol. iv. p. 182.
The Common Creeper, Lewin, Birds of New Holl., pl. 25.
Climacteris leucophæa, Strickl. Ann. and Mag. Nat. Hist., vol. xi. 1843, p. 330.

Climacteris picumnus, Gould, Birds of Australia, fol., vol. iv. pl. 98.

The range of this species is as widely extended as that of the *Climacteris scandens*, being a common bird in New South Wales and the intervening country, as far as South Australia: the precise limits of its habitat northward have not been ascertained; but it does not form part of the Fauna of Western Australia.

The whole structure of this species is much more slender and Creeper-like than any other member of its genus, and I observed that this difference of form has a corresponding influence over its habits, for they are more strictly arboreal than those of its congeners; indeed so much so, that it is questionable whether the bird ever descends to the ground. It also differs from the *C. scandens* in the character of country and kind of trees it inhabits, being rarely seen on the large *Eucalypti* of the open forest lands, but resorting to trees border-

ing creeks, as well as those on the mountains and the brushes.
I have frequently seen it in the brushes of Illawarra and Maitland, in which localities the *C. scandens* is seldom if ever
found. While traversing the trunks of trees in search of
insects, which it does with great facility, it utters a shrill
piping cry: in this cry, and indeed in the whole of its actions,
it strikingly reminded me of the Common Creeper of Europe
(*Certhia familiaris*), particularly in its manner of ascending
the upright trunks of the trees, commencing at the bottom
and gradually creeping up the bole to the top, generally
in a spiral direction. It is so partial to the *Casuarinæ*, that I
have seldom seen a group of those trees without at the same
time observing the White-throated Tree-Creeper, their rough
bark affording numerous receptacles for various kinds of
insects, which constitute its sole diet. I have never observed
this species near the water-holes, and I feel assured it has the
power of subsisting without drinking.

The breeding-season is September and the three following
months. The nest is built of grasses, is warmly lined with
feathers, and is placed in the hollow branch or bole of a tree.
The eggs are three in number, of a dull white thinly speckled
with fine spots of rich brown, and a few larger blotches of
the same colour; they are ten lines long by eight lines broad.

Crown of the head and back of the neck sooty black; back
olive-brown; wings dark brown, all the primaries and secondaries crossed in the centre by a dull buff-coloured band;
throat and centre of the abdomen white, the latter tinged
with buff; feathers of the flanks brownish black, with a broad
stripe of dull white down the centre; rump and upper tail-
coverts dark grey; under tail-coverts white, crossed by several
bands of black, each of which being separated on the stem
appear like a double spot; tail greyish brown, crossed by a
broad band of black near the tip; bill black; the under
mandible horn-colour at the base; feet blackish brown.

The female is precisely the same in colour, with the exception of having a small orange-coloured spot just below the ear-coverts, and by which she is at once distinguished from the male.

Genus ORTHONYX.

Much difference of opinion has arisen among ornithologists respecting the situation of this bird in systematic arrangements, and as to what genus it is most nearly allied; I regret to say that not having seen much of the bird in a state of nature, I am unable to clear up these disputed points. The form is strictly Australian, and the single species known is confined to the south-eastern part of the country.

Sp. 372. ORTHONYX SPINICAUDUS, *Temm*.

SPINE-TAILED ORTHONYX.

Orthonyx spinicaudus, Temm. Pl. Col., 428 male, 429 female.
—— *temminckii*, Vig. and Horsf. in Linn. Trans., vol. xv. p. 294.
—— *maculatus*, Steph. Cont. of Shaw's Gen. Zool., vol. xiv. p. 186.

Orthonyx spinicaudus, Gould, Birds of Australia, fol., vol. iv. pl. 99.

The Spine-tailed Orthonyx is very local in its habitat, being entirely confined, so far as I have been enabled to ascertain, to the brushes which skirt the southern and eastern coasts of Australia, such as those at Illawarra, and in the neighbourhood of the rivers Manning, Clarence, and MacLeay. It is usually found in the most retired situations running over the prostrate logs of trees, large moss-covered stones, &c. I ascertained by an examination of the stomach that the food consists of insects, principally of the order Coleoptera, and that the white throat distinguishes the male and the rufous throat the female.

M. Jules Verreaux, who has written a highly interesting account of this bird, states that it is strictly terrestrial, and scratches among the detritus and fallen leaves for its food, throwing back the earth like the *Gallinaceæ*. It never climbs, as was formerly supposed, but runs over fallen trunks of trees;—is rather a solitary bird, seldom more than two being seen together. Its often-repeated cry of *cri-cri-cri-crite* betrays its presence, when its native haunts, the most retired parts of the forest, are visited. Its chief food consists of insects, their larvæ, and wood-bugs. It builds a large domed nest, of slender mosses; the entrance being by a lateral hole near the bottom. The eggs are white and disproportionately large. The situation of the nest is the side of a slanting rock or large stone, the entrance-hole being level with the surface. —*Revue Zoologique*, July 1847.

The male has the crown of the head and upper part of the back reddish brown, with a large mark of black on each feather; lower part of the back and upper tail-coverts rich rufous brown; wings black; coverts largely tipped with grey; primaries crossed with grey at the base; apical half of the primaries and the tips of the secondaries dark brownish grey; tail dark brown; sides of the head and neck dark grey; throat and chest white, separated from the grey of the sides of the neck by a lunar-shaped mark of deep black; flanks and under tail-coverts grey, stained with reddish brown; bill and feet black; irides very dark hazel.

The female only differs in colour in having the throat rich rust-red.

Genus SITTELLA, *Swainson*.

During the progress of the "Birds of Australia" I had the pleasure of characterizing several new species of this form; one from Southern and Western Australia, another from Moreton Bay, and a third from the north coast. The *Sittella chrysoptera* was the only one previously known.

No species of this genus exists in Tasmania.

These birds build singular, upright nests on the branches of trees, and do not incubate in the holes of trees like the Nuthatches of Europe and India.

Sp. 373. SITTELLA CHRYSOPTERA, *Swains.*
ORANGE-WINGED SITTELLA.

Sitta chrysoptera, Lath. Gen. Syn. Supp., p. xxxii.
Orange-winged Nuthatch, Lath. Gen. Syn. Supp., vol. ii. p. 146, pl. 227.
Sitta ? chrysoptera, Steph. Cont. Shaw's Gen. Zool., vol. xiv. p. 189.
Neops chrysoptera, Vieill. 2nde édit. du Dict. d'Hist. Nat., tom. xxxi. p. 831.
Sittella chrysoptera, Swains. Class. of Birds, vol. ii. p. 317.
Mur-ri-gang, Aborigines of New South Wales.

Sittella chrysoptera, Gould, Birds of Australia, fol., vol. iv. pl. 101.

New South Wales is the true habitat of this species, over nearly every part of which it is rather plentifully distributed. I generally observed it in small companies of from four to eight in number, running over the branches of the trees with the greatest facility, and assuming every possible variety of position; often, like the Nuthatch, traversing the boles of the trees with its head downwards.

During its flight, which is quick and darting, the red mark on the wing shows very conspicuously; its powers of flight are, however, seldom employed, further than to enable it to pass from one tree to another.

The colouring of this bird is more sombre, and has the markings of the head less decided, than any other species of the genus. The darker colouring of the head and ear-coverts of the female, however, at once points out to the ornithologist the sex of any specimen he may possess of this genus.

The male has the head dark brown; all the upper surface grey, with a broad streak of dark brown down the centre of each feather; wings dark brown, with a broad patch of rich rufous crossing the primaries and secondaries; upper tail-coverts white; tail black, the outer feathers tipped with white; all the under surface grey, with a faint streak of

2 K

brown down each feather; under tail-coverts white, crossed near the tip with a spot of brown; bill horn-colour at the base; irides cream-colour; eyelash light buff; feet yellow.

The female differs only in having the head of a darker tint of brown.

I possess a somewhat mutilated specimen of a *Sittella*, which was given to me by Captain Sturt, but I am unaware of the locality in which he obtained it. This bird, which I feel assured is a new species, is very nearly allied to *S. chrysoptera*, but differs from it in having a longer and more upturned bill, the base of which is yellow, and a uniformly coloured back and breast without apparently any trace of the brown striae seen on the feathers of those parts in *S. chrysoptera*; in other respects, particularly in the chestnut coloured band across the wings, it is very similar to that species. If it should hereafter prove to be new, I would propose for it the specific name of *tenuirostris*.

Sp. 374. SITTELLA LEUCOCEPHALA, *Gould*.

WHITE-HEADED SITTELLA.

Sittella leucocephala, Gould in Proc. of Zool. Soc., part v. p. 152.

Sittella leucocephala, Gould, Birds of Australia, fol., vol. iv. pl. 102.

My collection contains several specimens of this species of *Sittella*, two of which were received from the neighbourhood of Moreton Bay; another was procured during Dr. Leichardt's overland expedition to Port Essington, Gilbert having killed it near Peak-Range Camp on the 27th of January 1845: the latter differs from the former in the greater purity of the white colouring of the head, and in the darker tint of the striae, which run down the centre of each of the feathers on the breast; and it is possible that it may hereafter prove to be distinct.

Head and neck pure white; upper surface greyish brown with darker centres; under surface greyish white, with a stripe of brownish black down the centre of each feather; wings dark brown, crossed by a band of pale rusty red; tail brownish black, the middle feathers slightly, and the outer ones largely tipped with white; upper tail-coverts white, the lateral feathers with a patch of dark brown in the centre; under tail-coverts brown, tipped with white; irides greenish yellow; base of the bill, nostrils, and eyelash orange-yellow.

Total length 4½ inches; bill ⅜; wing 2⅔; tail 1½; tarsi ½.

Sp. 375.   SITTELLA LEUCOPTERA, *Gould*.

WHITE-WINGED SITTELLA.

Sittella leucoptera, Gould in Proc. of Zool. Soc., part vii. p. 144.

Sittella leucoptera, Gould, Birds of Australia, fol., vol. iv. pl. 103.

The present bird, which is a native of the northern parts of Australia, is a perfect representative of the *Sittella chrysoptera* of the south coast, to which species it is most nearly allied. The contrasted style of its plumage, together with the white spot in the wings, sufficiently distinguish it from every other species of the genus yet discovered. It is found in the Cobourg Peninsula, but is nowhere very abundant; it moves about in small families of from four to twelve in number. Its note, actions, and general habits are precisely similar to those of the other members of the genus.

The sexes differ from each other in the markings of the head; the male has the summit only black, while the female has the whole of the head and ear-coverts of that colour.

The male has the forehead, crown of the head, and occiput deep black; wings black, with a broad band of white crossing the primaries near the base; tail black, the lateral feathers tipped with white; throat, under surface, and upper tail-coverts white; under tail-coverts white, with a spot of black near the tip of each feather; back greyish brown, the centre of each feather streaked with blackish brown; irides ochre-yellow; eyelash straw-yellow; bill straw-yellow, tipped with black; legs and feet lemon-yellow.

Total length 4 inches; bill 1½; wing 3; tail 1½; tarsi 1½.

Sp. 370.   SITTELLA PILEATA, *Gould.*

BLACK-CAPPED SITTELLA.

Sittella pileata, Gould in Proc. of Zool. Soc., part v. p. 151, male.
—— *melanocephala*, Gould in Ibid., p. 152, female.
Goo-mal-de-dite, Aborigines of Western Australia.

Sittella pileata, Gould, Birds of Australia, fol., vol. iv. pl. 104.

This species of *Sittella* enjoys a range extending over several degrees of longitude. I killed several examples during my excursion into the interior of South Australia, and I transcribe from my journal the following notes on the subject :—
"I met with a flock of these birds on the hills near the source of the River Torrens, about forty miles northward of Adelaide: they were about thirty in number and were extremely shy, keeping on the topmost branches of the trees, and the whole company flying from tree to tree so quickly, that I and my companion were kept at a full run to get shots at them."

The following is from Gilbert's notes made in Western Australia :—

"An extremely active bird, running up and down the trunks and branches of the trees with the utmost rapidity, always in families of from ten to twenty in number. It utters a weak piping note while on the wing, and occasionally while running up and down the trees. Its flight, which is generally performed in rather rapid undulating starts, is of short duration."

Gilbert subsequently informed me, on the authority of Mr. Johnson Drummond, that this species "makes a nest of short strips of bark attached together and fastened to the branch with cobwebs, and so covered over with them as to be very nearly smooth; the cobweb is laid or felted on, not wound round the pieces; and portions of lichen are frequently attached. The nest is generally placed in the highest and most slender fork of an Acacia, and is most difficult to detect, from its very diminutive size and from its resembling a slight excrescence of the wood; the eggs are three in number, of a whitish colour, with circular green spots regularly distributed over the whole surface. The bird breeds in September.

On reference to the synonyms given above, it will be seen that, prior to my visit to Australia, I regarded, described, and named the two sexes of this bird as distinct species, an error which the opportunity I subsequently had of observing the bird in a state of nature and of dissecting recent specimens has enabled me to correct; the black-headed specimens proving to be females, and those with a black cap only, males.

The male has the forehead, stripe over the eye, throat, breast, and centre of the abdomen white; crown of the head black; ear-coverts, back of the neck and back greyish brown, with a small stripe of dark brown down the centre of each feather of the latter; rump white; upper and under tail-coverts greyish brown, crossed with an arrow-shaped mark of dark brown, and tipped with white; tail black, the centre feathers slightly and the outer ones largely tipped with white; wings blackish brown, with a large patch of rufous in the centre, interrupted by the blackish brown margins of some of the secondaries; all the feathers slightly tipped with greyish brown; flanks and vent greyish brown; bill yellow at the base, black at the tip; feet beautiful king's-yellow; irides buffy hazel; eyelash buff.

The female differs in being somewhat darker on the upper surface, and in having the whole of the upper part of the head including the orbits deep black.

Total length 4¾ inches; bill ⅜; wing 3½; tail 1⅞; tarsi ⅝.

Family CUCULIDÆ.

The species of this extensive family, many of which are rendered remarkable by their parasitic habits, are universally dispersed over the surface of the globe; they abound in the old world, but are much less numerous in the new. In Africa,

Asia, the Indian and Polynesian islands they are about equally abundant; generally speaking the range of the various species is somewhat limited, while the genera are more widely spread. All the Australian species, with the exception of the members of the genus *Centropus* are parasitic, the huge *Scythrops* and the diminutive *Chrysococcyx* alike depositing their eggs in the nests, and entrusting their young to the fostering care of other birds.

Genus CUCULUS, *Linnæus*.

Müller, Bonaparte, Cabanis, and other writers having separated the Cuckoos of the southern portion of Australia from the genus *Cuculus*, only one species of that form, as now restricted, finds a place in the avifauna of the country.

Sp. 377.   CUCULUS CANOROIDES, *Müller?*

AUSTRALIAN CUCKOO.

Cuculus canoroides, Müll. Verh. Nat. Gesch. &c., Land-en Volk., p. 235.
——— *horsfieldi*, Moore's Cat. Birds E. I. Comp., vol. xi. p. 703.
——— *optatus*, Gould in Proc. of Zool. Soc., part xiii. p. 18.
Nicoclarius optatus, Bonap. Consp. Vol. Zygod., p. 6.

Cuculus optatus, Gould, Birds of Australia, fol., vol. iv. pl. 84.

The northern part of Australia is the only locality in which this bird has been found; the specimens I have seen were killed in the month of January: whether it utters the word 'Cuckoo' or not I am unable to say, but it is most likely that in its voice as in its form and general appearance it closely assimilates to its European relative.

In the Australian bird the black bands on the breast are broader and more defined than in the European *C. canorus*; the claws of the Australian bird are also smaller and more delicate than those of its European ally.

All the upper surface slaty grey; inner webs of the primaries broadly barred with white; tail-feathers dark violet-brown, with a row of oblong spots of white, placed alternately on either side of the stem, and slightly tipped with white; the lateral feathers have also a row of white spots on the margin of their inner webs; chin and breast light grey; all the under surface buffy white, crossed by bands of black; irides, bill, and feet orange.

Total length 13 inches; bill 1¼; wing 7¾; tail 6½; tarsi ¾.

Genus CACOMANTIS, *Müller*.

This genus was founded for the *Cuculus flavus* of Gmelin, a form which is freely represented in Australia, where there are at least four species, some inhabiting the southern, and others the western and northern parts of the country.

Sp. 378.   CACOMANTIS PALLIDUS.

PALLID CUCKOO.

Columba pallida, Lath. Gen. Syn., Supp. vol. ii. p. 270.
Cuculus inornatus, Vig. and Horsf. in Linn. Trans., vol. xv. p. 207.
—— *albostrigatus*, Vig. and Horsf. Ib., p. 208, young.
—— *variegatus et cinereus*, Vieill. Nouv. Dict. d'Hist. Nat., tom. viii. pp. 224, 226.
Cacomantis inornatus, Bonap. Consp. Gen. Av., tom. i. p. 103, Cacomantis, sp. 1.
—— *cinereus*, Bonap. Consp. Vol. Zygod., p. 6.
Heteroscenes pallidus, Cab. et Hein. Mus. Hein., Theil iv. Heft 1. p. 26
—— *occidentalis*, Cab. et Hein. Ib., Theil iv. Heft. 1. p. 27, note.
Djü-dŭ̄-rŭi, Aborigines of Western Australia.
Greater Cuckoo of the Colonists.

Cuculus inornatus, Gould, Birds of Australia, fol., vol. iv. pl. 85.

The southern portion of Australia generally, and the island of Tasmania, are inhabited by this species of Cuckoo; to the latter country, however, it is only a summer visitant, and a partial migration also takes place in the adjacent portion of the continent, as is shown by its numbers being much fewer during winter. It arrives in Tasmania in the month of September, and departs northward in February. During the vernal season it is an animated and querulous bird, and may then be seen either singly, or two or more males engaged in chasing each other from tree to tree. Its ringing whistling call, which consists of a succession of running notes, the last and highest of which are several times rapidly repeated, is often uttered while the bird is at rest among the branches, and also occasionally while on the wing. Its food consists of caterpillars, *Phasmidæ*, and coleopterous insects, which are generally procured among the leafy branches of the trees, and in searching for which it displays considerable activity, and great power of traversing the smaller limbs. When desirous of repose after feeding, it perches on the topmost dead branches of the trees, on the posts and rails of the fences, or any other prominent site whence it can survey all around. Its flight is straight and rapid, and not unlike that of the *Cuculus canorus*.

In respect to its reproduction it is strictly parasitic, devolving the task of incubation on the smaller birds, many species of which are known to be the foster-parents; among them may be enumerated the various *Melithrepti*, *Ptilotes*, *Maluri*, *Acanthizæ*, &c. After the young Cuckoo has left the nest, it selects some low dead branch in an open glade of the forest as a convenient situation for its various foster-parents to supply it with food, for the procuring and supplying of which the smaller birds appear to have entered into a mutual compact.

The specimens of this bird from Western Australia are somewhat smaller, and have the white marks of the tail less distinct than specimens from Tasmania, but these differences are, in my opinion, too trivial to be regarded as other than mere local variations; but M.M. Cabanis and Heine think otherwise, and have assigned to them the specific appellation of *occidentalis*. When fully adult the plumage is nearly of a uniform brown, with the inner webs of the wing and tail-feathers relieved by bars and markings of white; the immature colouring, on the contrary, presents a variegated and very diversified character, which, owing to the constant change taking place, cannot be described so as to render it clear to my readers. When the young leaves the nest, the throat, face, and shoulders are black, the feathers of the remainder of the body crossed and spotted with buff; the black colouring gradually gives place to the grey of the under surface, while the buffy marks of the upper surface are retained even after the second or third moult; it breeds in this state, and it is doubtful whether in the female it is ever entirely cast off.

The stomachs of those dissected were found to be capacious, membranous, and thickly lined with hair.

The egg is about seven-eights of an inch long by five-eighths broad, and is of a cream-colour, speckled all over with markings of brown.

The adult male has the head, neck, and all the under surface brownish grey, with a streak of dark brown down the sides of the neck; all the upper surface olive-brown, becoming much darker on the wings and tail; basal portion of the inner webs of the primaries broadly barred with white; tail-feathers barred on the margins of both webs with white, slightly on the outer and deeply on the inner; all the feathers tipped with white, and with a mark of white on the stem near the tip, this mark being very small on the central tail-feather, and gradually increasing on the lateral feathers until on the outer it forms a band; under-irides very dark brown; eyelash yellow; gape and inside of the mouth rich deep orange; feet olive.

The female differs in having the upper surface mottled with buff and rufous, in having a triangular spot of reddish buff at the extremity of each of the wing-coverts, and the markings of the tail buff instead of white; all which markings may in very old birds give place to a style of colouring similar to the male.

The young, independently of the differences pointed out above, has the feet yellowish olive, the soles of the feet yellow; the bill yellowish olive, the corner of the mouth and the tip of the bill being more yellow than the rest of that organ; irides greyish brown.

Sp. 379. CACOMANTIS FLABELLIFORMIS.

Fan-tailed Cuckoo.

Cuculus flabelliformis, Lath. Ind. Orn., Supp. vol. ii. p. 30.
—— *rufulus*, Vieill. Nouv. Dict. d'Hist. Nat., tom. viii. p. 284.
—— *pyrrhophaenus*, Vieill. Ib., tom. viii. p. 234.
—— *prionurus*, Licht. Verz. Doubl., p. 9.
—— *cineraceus*, Vig. and Horsf. in Linn. Trans., vol. xv. p. 298.
—— *incertus*, Vig. and Horsf. Ib., vol. xv. p. 299.
—— *cariulosus*, Vig. and Horsf. Ib., vol. xv. p. 300.
—— *flavus*, pt., Less. Traité d'Orn., p. 152.
—— *cinerascens*, Gray and Mitch. Gen. of Birds, vol. ii. p. 463, Cuculus, sp. 41.
—— *pyrrhophanes*, Gray and Mitch. Ib., vol. ii. p. 463, Cuculus, sp. 46.
Cacomantis flabelliformis, Bonap. Consp. Gen. Av., tom. i. 104, Cacomantis, sp. 7.
—— *incertus*, Bonap. Consp. Vol. Zygod., p. 6.
Du-laar, Aborigines of the lowland districts of Western Australia.
Lesser Cuckoo of the Colonists.

Cuculus cineraceus, Gould, Birds of Australia, fol., vol. iv. pl. 86.

This is a migratory species, arriving in Tasmania in September, and, after spending the summer months therein, departing to the northward in January and February. In the southern parts of the continent of Australia solitary individuals remain throughout the winter, as evidenced by my having observed it round Adelaide in July: I have never seen individuals from the north coast; I therefore infer that its migratory movements are somewhat restricted; in all probability the 28th degree of latitude may be the extent of its range to the northward. During the summer months, its distribution over the southern portion of the continent may be said to be universal, but withal it is rather a solitary bird and loves to dwell in secluded situations, where, but for its loud ringing call, which much resembles its aboriginal name, it would easily escape detection.

It flies rather heavily, and on alighting moves the tail up and down for some time; a similar movement of the tail also invariably precedes its taking flight.

Like the other species of Cuckoo, it deposits its single egg in the nest of some one or other of the smaller kinds of birds: it is of a perfectly oval form, of a flesh-white sprinkled all over with fine spots of purplish brown, nine or ten lines long by seven and a half lines broad.

The stomach is capacious, membranous, and lined with hairs; and the food consists of the larvæ of insects of various kinds.

The sexes are alike in plumage, but the female is a trifle smaller than the male.

Head and all the upper surface dark slate-grey; wings brown, glossed with green; tail dark glossy greenish brown, each feather toothed on the edge with white, the extent of which gradually increases until on the lateral feathers they assume the form of irregular interrupted bars; on the edge of the shoulder a short narrow stripe of white; on the under surface of the wing an oblique band of white; chin grey; under surface ferruginous; bill black, except at the base of the lower mandible, where it is fleshy orange; irides dark brown; eyelash beautiful citron-yellow; feet yellowish olive.

Sp. 380. CACOMANTIS INSPERATUS, *Gould.*

BRUSH-CUCKOO.

Cuculus insperatus, Gould in Proc. of Zool. Soc., part xiii. p. 19.
Cacomantis insperatus, Bonap. Consp. Gen. Av., tom. i. p. 104, *Cacomantis*, sp. 2.

Cuculus insperatus, Gould, Birds of Australia, fol., vol. iv. pl. 87.

While traversing the cedar brushes of the Liverpool range

on the 20th of October, 1839, my attention was attracted by the appearance of a Cuckoo, which I at first mistook for the *Cacomantis flabelliformis*, but which on examination proved to be a new species; this example was the only one I ever saw living, and a single skin is all that has since been sent to me from New South Wales; it must therefore be very rare in the south-eastern portion of the continent. More recently I have received examples from Western Australia.

On comparison, this species will be found to differ from *C. flabelliformis*, for which it might be readily mistaken, in its smaller size, in the more square form of the tail, and in that organ being destitute of white markings on the outer webs of the feathers. In its structure and colouring it will be found to approximate to the members of the genus *Mesocalius*, and in fact to form one of the links which unite the two groups.

Head, throat, and all the upper surface dark slate-grey; back and wings glossed with green; tail glossy brownish green, each feather tipped with white, and with a row of triangular-shaped white markings on the margins of the inner webs; primaries and secondaries with a patch of white on their inner webs near the base; edge of the shoulder white; under surface of the shoulder, vent, and under tail-coverts rufous; remainder of the under surface grey, washed with rufous; bill black; feet olive.

Total length 9¼ inches; bill 1; wing 6½; tail 5; tarsi ⅜.

Sp. 381. CACOMANTIS DUMETORUM, *Gould*.

SQUARE-TAILED CUCKOO.

Cuculus dumetorum, Gould in Proc. of Zool. Soc., part xiii. p. 19.
Cacomantis dumetorum, Bonap. Consp. Gen. Av., tom. i. p. 104, Caco- mantis, sp. 8.

This species, which inhabits the north-western coast, differs from *C. insperatus* in being of a much smaller size and in the whole of the plumage being browner.

Head, neck, and rump dark slate-grey; back, wings, and tail bronzy brown; tail-feathers slightly tipped with white, and with a row of small triangular-shaped spots on the margins of their inner webs; breast grey, washed with rufous; under surface of the shoulder, flanks, vent, and under tail-coverts deep rufous; irides brown.

Total length 8½ inches; bill ⅞; wing 5; tail 4½; tarsi ¼.

Genus MESOCALIUS, *Cabanis et Heine.*

MM. Cabanis and Heine have established the above genus for the bird I had called *Chalcites osculans*, and as I have adopted many of the new genera into which the *Cuculidæ* are now divided, I have no alternative but to adopt this one also. The only species of the form yet discovered is a larger or more robust bird than the little Bronze Cuckoos, and it also differs from them in its colouring.

Sp. 392.   MESOCALIUS OSCULANS, *Gould.*

BLACK-EARED CUCKOO.

Chalcites osculans, Gould in Proc. of Zool. Soc., part xv. p. 32.
Cuculus osculans, Gray and Mitch. Gen. of Birds, vol. ii. p. 468
    Cuculus, sp. 29.
Chrysococcyx osculans, Gould, Birds of Australia, vol. i. Introd. p. lxi.
Misocalius palliolatus, Cab. et Hein. Mus. Hein. Theil iv. Heft i.
    p. 16, note.
Black-eared Cuckoo, Colonists of Swan River.

Chalcites osculans, Gould, Birds of Australia, fol., vol. iv. pl. 89.

Four examples of this species came under my notice during the time I was engaged on the folio edition of the Birds of Australia—one from Swan River, two killed by myself in New South Wales, and one in the collection of the late H. E. Strickland, Esq.; since its completion a fine example has been sent to me by G. French Angas, Esq., from South Australia.

Judging from the little I saw of this species in a state of nature, its habits were those of the members of the genus *Lamprococcyx*; thick shrubby trees of moderate height appeared to be its favourite resort, and its food to consist of insects obtained among the branches and from off the leaves, in search of which it hops about with stealthiness and quietude; further than this, little is known respecting it. One of my specimens was killed near Gundermein on the Lower Namoi, on the 24th of December, 1839; but the true habitat of the species has not yet been discovered. That it is confined to Australia is almost certain, but this can only be verified by future research.

Gilbert, who observed this bird in Western Australia, states that it is very shy, and that he only met with it in the interior of the country. It utters a feeble, lengthened, and plaintive note at long intervals. It flies slowly and heavily, and but short distances at a time. The stomach is thin and capacious, and slightly lined with hairs of caterpillars.

Head, all the upper surface, and wings glossy olive-brown, becoming darker on the shoulders and primaries, and fading into white on the upper tail-coverts; tail dark olive-brown, each feather tipped with white, and the lateral one on each side crossed on the inner web with five bars of white; ear-coverts black, encircled with white; under surface of the wing, throat, breast, and abdomen pale cinnamon-brown, fading into white on the under tail-coverts; bill very dark brown; irides dark blackish brown; tarsi and upper surface of the feet greenish grey; under surface of the feet and the back of the tarsi mealy fleshy grey.

Total length $7\frac{1}{2}$ inches; bill $\frac{7}{8}$; wing $4\frac{3}{8}$; tail $3\frac{7}{8}$; tarsi $\frac{3}{4}$.

That this bird is not identical with the *Cuculus palliolatus* of Latham as supposed by MM. Cabanis and Heine is, in my opinion, quite certain; Latham's description does not agree with it in any particular; besides which it is not likely that the bird, which is strictly confined to the interior of the country, could have been sent to England at the period at which he wrote; it is even now extremely rare in our collections.

Genus LAMPROCOCCYX, *Cabanis et Heine*.

The members of this genus are widely dispersed, being found in New Zealand, Australia, Java, and Africa. At least three inhabit Australia, of which, the two frequenting the southern portions of that country have been considered identical, but with a little care they may be easily distinguished.

Sp. 363.   LAMPROCOCCYX PLAGOSUS.

BRONZE-CUCKOO.

Cuculus plagosus, Lath. Ind. Orn., Sup. p. xxxi.
—— *metallicus*, Vig. and Horsf. in Linn. Trans., vol. xv. p. 802.
—— *versicolor*, Gray, Gen. of Birds, vol. ii. p. 464, *Cuculus* sp. 80.
Golden or Bronze-Cuckoo of the Colonists.

Chrysococcyx lucidus, Gould, Birds of Australia, fol., vol. iv. pl. 89, centre figure.

The New Zealand *Lamprococcyx lucidus* being now considered distinct from the species found in New South Wales, it becomes necessary to determine which specific appellation was first applied to the latter; this I believe to be *C. plagosus* of Latham, which I therefore adopt, and reduce the *C. metallicus* of Vigors and Horsfield, and the *C. versicolor* of Mr. G. R. Gray's Genera of Birds to the rank of synonyms.

The *Lamprococcyx plagosus* is very widely dispersed over every part of the Australian continent, and if it be not migratory in New South Wales, the greater number certainly retire in winter to the northward, where insect food is more abundant. I have, however, seen it in the Botanic Garden at Sydney in the month of March. Its food consists of insects of various orders, the stomachs of those examined containing the remains of *Hymenoptera*, *Coleoptera*, and caterpillars. While searching for food, its motions, although very active, are characterized by a remarkable degree of quietude, the bird leaping from branch to branch in the gentlest manner possible, picking an insect here and there, and prying for others among the leaves and the crevices of the bark with the most scrutinizing care. Its flight is quick and undulating, and when passing from one tree to another on a sunny day, the brilliant green colouring of the male shows very beautifully. Like the true Cuckoos, it always deposits its single egg in the nest of another bird, those of the *Maluri* and *Acanthizæ* being generally selected; in New South Wales the *Malurus cyaneus* and the *Geobasileus chysorrhous* are among others the foster-parents; in Western Australia the nests of the *Malurus splendens* are resorted to; and it is a remarkable fact, that the egg is mostly deposited in nests of a domed form, with a very small hole for an entrance.

The stomach is capacious, membranous, and slightly lined with hair.

Its note is a mournful whistle, very like that usually employed to call a dog.

The egg is of a clear olive-brown, somewhat paler at the smaller end, about eleven-sixteenths of an inch long by half an inch in breadth.

The adult male has the head, all the upper surface and wings of a rich coppery bronze; primaries brown with a bronzy lustre; tail bronzy brown, crossed near the tip with a dull black band; the two lateral feathers on each side with a series of large oval spots of white across the inner web, and a series of smaller ones opposite the interspaces on the outer web; third and fourth feathers on each side with a small oval spot of white at the tip of the inner web; all the under surface white, crossed by numerous broad conspicuous bars of rich deep bronze; irides brownish yellow; feet dark brown, the interspaces of the scales mealy.

The female is similarly marked, but has only a wash of the bronzy colouring on the upper surface, and the bars of the under surface much less distinct.

The young, which are brown, with a still fainter wash of bronze, have the throat and under surface grey, without any trace of the bars, except on the under surface of the shoulder; the base of the tail-feathers deep rusty-red, the irides bright grey, and the corners of the mouth yellow.

Sp. 384. LAMPROCOCCYX MINUTILLUS, Gould.

Little Bronze Cuckoo.

Chrysococcyx minutillus, Gould in Proc. of Zool. Soc., part xxvii. p. 128.
Lamprococcyx minutillus, Cab. et Hein. Mus. Hein., Theil iv. Heft i. p. 15, note.

Chrysococcyx minutillus, Gould, Birds of Australia, fol., Supplement, pl.

Nothing further is known respecting this little bronze Cuckoo than that it is a native of Port Essington, whence the only specimen I have yet seen was sent. The example alluded to is fully adult, and differs very considerably from every other species with which I am acquainted. It is one of the smallest species of the genus, yet it has as stout a bill as some of the larger kinds.

Head, all the upper surface, and wings shining bronzy green; all the under surface white, barred with bronzy green, the bars being most distinct on the flanks; primaries and secondaries white on the basal portion of their inner webs; two centre tail-feathers bronzy green, the next on each side bronzy green on the outer web, rufous on the inner web, crossed by a broad band of black near the tip, and with an oval spot of white across the tip of the inner web; the two next on each side bronzy green on their outer webs, their inner webs rufous with large spots of black near the shaft, most conspicuous on the outermost of the two feathers; their inner webs are also crossed near the tip with a very broad band of black, and have an oval spot of white at the tip; the outer feather on each side is barred alternately on the outer web with dull bronzy green and dull white, and on the inner one with broad decided bars of black and white, and tipped with white; bill black; feet olive.

Total length 5¼ inches; bill ⅜; wing 3¼; tail 2½; tarsi ¼.

Sp. 385. LAMPROCOCCYX BASALIS.

NARROW-BILLED BRONZE CUCKOO.

Cuculus auratus, var., Vieill. Ency. Meth. Orn., part iii. p. 1338.
—— *basalis*, Horsf. in Linn. Trans., vol. xiii. p. 179.
—— *malayanus*, Raff. in Linn. Trans., vol. xiii. p. 286.
—— *chalcites*, Blyth, Journ. Asiat. Soc. Beng., 1842, p. 919.
Chrysococcyx basalis, Blyth, id., 1846, p. 64.
—— *chalcites*, Bonap. Consp. Gen. Av., tom. i. p. 106, *Chrysococcyx*, sp. 8.
Chalcites basalis, Bonap. Consp. Vol. Zygod., p. 7.
Chrysococcyx malayanus, Horsf. and Moore, Cat. of Birds in Mus. East Ind. Comp., vol. ii. p. 707.
Lamprococcyx basalis, Cab. et Hein. Mus. Hein., Theil iv. Heft 1. p. 12.
Chrysococcyx pœcilurus, G. R. Gray, Proc. of Zool. Soc., 1861, p. 431 ?

Chrysococcyx lucidus, Gould, Birds of Australia, fol., vol. iv. pl. 89; lower figure adult, upper figure young.

If the residents in the southern portion of Australia will examine the little Bronze Cuckoos which annually visit them in summer, they will find that they are of two distinct species. They bear a general resemblance; but one will be found to have a stouter bill than the other, and a nearly uniformly coloured tail, the outer feather on each side only being barred. This bird may be observed in all the southern parts of Australia from east to west, and I believe in Tasmania. The other species is about the same size, but has a narrower bill, a lighter-brown head, a paler-coloured back; the outer feathers of the tail strongly barred, as in the last, and the basal portion of the next three feathers on each side rufous-chestnut, which colour must, I presume, show very conspicuously when the bird is flying, or when sitting on a tree with its tail spread. I have specimens of this species from South Australia and Moreton Bay, and I believe I may state that it is the Common Bronze Cuckoo of Tasmania, but of this I am not certain; the chances are that both it and the *L. lucidus* is found there. After a careful examination I have come to the conclusion that the stout-billed bird is the *C. plagosus* of Latham, and that the narrow-billed one is identical with the Javan species to which Horsfield gave the appellation of *C. basalis*. Having the type specimen of *C. basalis*, New Zealand skins to which the specific term *lucidus* was originally applied, and examples of *C. plagosus*, wherewith to compare it, I am the more certain of being correct in these conclusions.

Having said thus much about these little parasitic Cuckoos, I leave to the rising ornithologists of Australia the task of investigating the subject, and of informing the scientific world whether there be any differences in the eggs of the two birds, and the character of the plumage of their nestlings. If their first dress be not nearly uniform, and destitute of any bars on the throat and under surface, then there is another species yet to be described.

Crown of the head and nape bronzy brown; over the eye a stripe of dull white; feathers of the back, wing-coverts, upper tail-coverts, and two centre tail-feathers dark shining green, edged with grey; wings brown, glossed with green and margined with grey; outer tail-feather on each side alternately and broadly barred with blackish brown and white; the three rest on each side rufous chestnut at the base, passing into green towards the extremity, and ending in blackish brown spotted with white; ear-coverts and sides of the neck brown; under surface buffy white, mottled with

bars of pale brown on the throat, and strongly barred on the
flanks with bronzy brown; under surface of the shoulder
similarly but not so strongly barred.

Genus SCYTHROPS, *Latham.*

The only known species of this remarkable form inhabits
the eastern parts of Australia, and according to the information gained from the notes made by Gilbert during Dr.
Leichardt's Expedition, extends it range northward from
thence to Torres Straits.

Sp. 380. SCYTHROPS NOVÆ-HOLLANDIÆ, *Lath.*

CHANNEL-BILL.

Scythrops novæ-hollandiæ, Lath. Ind. Orn., vol. i. p. 141.
Psittaceous Hornbill, Phil. Bot. Bay, pl. in p. 165.
Anomalous Hornbill, White's Journ., pl. in p. 142.
Channel-Bill, Lath. Gen. Syn. Supp., vol. ii. p. 90, pl. 124.
Australasian Channel-Bill, Shaw, Gen. Zool., vol. viii. p. 378, pl. 50.
Scythrops australasiæ, Shaw, id., p. 378.
—— *goerang*, Vieill. Nouv. Dict. d'Hist. Nat., tom. xxx. p. 450.
—— *australis*, Swains. Class. of Birds, vol. ii. p. 299.
Curriay-gun, Aborigines of New South Wales.

Scythrops novæ-hollandiæ, Gould, Birds of Australia, fol., vol. iv.
pl. 90.

This remarkable bird, which has been considered a Hornbill by some authors, and as nearly allied to the Toucans
by others, is in reality a member of the family *Cuculidæ*.
An examination of its structure and a comparison of it
with that of the other species of the family will render
this very apparent, and I may add, that the little I saw of it
in a state of nature fully confirms the opinion here given; its
habits, actions, and mode of flight are precisely the same, as
is also the kind of food upon which it subsists, except that it
devours the larger kinds of *Phasmiæ* and *Coleoptera* instead

of the smaller kinds of insects eaten by the other members of the family, and that it occasionally feeds upon fruits; the changes too which it undergoes from youth to maturity are very similar.

The Channel-Bill is a migratory bird in New South Wales, arriving in October and departing again in January; whither it proceeds is not known. As I had but few opportunities of observing it myself, I cannot do better than transcribe the particulars recorded by Latham, who, in the second volume of his 'General History of Birds,' says, "It is chiefly seen in the morning and evening, sometimes in small parties of seven or eight, but more often in pairs; both on the wing and when perched it makes a loud screaming noise when a hawk or other bird of prey is in sight. In the crop and gizzard the seeds of the red gum- and peppermint-trees have been found; it is supposed that they are swallowed whole, as the pericarp or capsule has been found in the stomach; exuviæ of beetles have also been seen, but not in any quantity. The tail, which is nearly the length of the body, is occasionally displayed like a fan, and gives the bird a majestic appearance. The natives appear to know but little of its habits or haunts; they consider its appearance as an indication of blowing weather, and that its frightful scream is through fear. It is not easily tamed, for Mr. White observes, that he kept a wounded one alive for two days, during which it would eat nothing, but bit everything that approached it very severely."

In some notes by the late Mr. Elsey on the birds observed by him during Mr. Gregory's Expedition, and which were kindly made for my use, he says, "This bird appeared on the northern side of the ranges. It settled in a tree close to our camp, and for five minutes at a time pumped out its awful notes. Sometimes it was quite indifferent to our presence, but generally it was very shy. I have never seen it on the ground, but always at the tops of large trees. One, shot by Mr. Gregory and preserved, proved to be an incubating female;

it contained several eggs, one nearly matured, and from the state of the oviduct another must have been recently extruded. Its habits seem to indicate that it is parasitic," and this view is confirmed by Lady Dowling informing me that a young specimen, kindly presented to me by her Ladyship, was one of two taken from a branch of a tree while being fed by birds not of its own species.

In some notes on the habits of the *Scythrops* forwarded to me by my friend Dr. Bennett, of Sydney, in June 1858, he says:—"I have much pleasure in telling you that when the young *Scythrops* was introduced into Mr. Denison's aviary it was placed in a compartment already occupied by a *Dacelo gigas*, and doubtless feeling hungry after its journey, immediately opened its mouth to be fed, and its wants were readily attended to by the *Dacelo*, who, with great kindness, took a piece of meat, and, after sufficiently preparing it by beating it about until it was in a tender and pappy state, placed it carefully in the gaping mouth of the young *Scythrops*; this feeding process continued until the bird was capable of attending to its own wants, which it now does, feeding in company with the *Dacelo* in the usual manner. When I saw it in the morning it was perched upon the most elevated resting-place in the aviary, occasionally raising itself, flapping its wings, and then quietly settling down again after the manner of Hawks in confinement, and presenting much the appearance of a member of that tribe of birds. It comes down for food every morning, and immediately returns to its elevated perch. Judging from what I saw of this specimen, I should imagine that the bird might be readily tamed, and would bear confinement very well. In the young state it is destitute of the scarlet orbits so conspicuous in the adult."

I once possessed an egg sent me by Strange, which he informed me was taken by himself from the ovarium after he had shot the bird. It was of a light stone-colour, marked all over, but particularly at the larger end, with irregular blotches of reddish brown, many of which were of a darker hue and appeared as if beneath the surface of the shell; it was one inch and eleven-sixteenths long by one inch and a quarter broad; without wishing to cast a doubt upon Strange's veracity, I should much like to see an authenticated mature egg of this bird, as it may differ in colour from the one described.

The sexes are alike in plumage, but the female is somewhat smaller than the male.

Head, neck, and breast grey; all the upper surface, wings, and tail greenish olive-grey, each feather largely tipped with blackish brown; tail crossed near the extremity by a broad band of black and tipped with white, which gradually increases in extent as the feathers recede from the centre; the inner webs are also largely toothed with white, which is bounded posteriorly with a broad streak of black; under surface of the wing and body buffy white crossed with indistinct bars of greyish brown, which gradually deepen in colour on the flanks and thighs; orbits and lores scarlet; bill light yellowish horn-colour; feet olive-brown.

Genus EUDYNAMIS, *Vigors and Horsfield.*

One species only of this form inhabits Australia; others are found in the Indian Islands and on the continent of India; in which latter country the trivial name of Koel has been applied to them. They are all parasitic, depositing their egg in the nests of Crows, and doubtless in those of other birds. The sexes differ considerably in size, the female being the larger; moreover, her plumage is spotted, while that of the male is of a uniform colour.

Mr. Blyth states that the Indian bird of this genus, which is very nearly allied to the Australian species, ejects from its mouth the seeds of the fruits upon which it feeds.

Sp. 387.   EUDYNAMIS FLINDERSI.

Australian Koel.

Cuculus cyanocephalus, Lath. Ind. Orn., Supp. vol. ii. p. xxx.
—— *flindersii*, Lath., Vig. and Horsf. in Linn. Trans., vol. xv. p. 305.
Eudynamys orientalis, Vig. and Horsf. Ib., p. 304.
—— *flindersii*, Vig. and Horsf. Ib., p. 305.
Eudynamis australis, Swains. Anim. in Menag., &c., p. 344.
Eudynamys australis, Gray, Gen. of Birds, vol. ii. p. 464, *Eudynamys*, sp. 6.
Eudynamis flindersi, Reich. Vög. Neuholl., tom. ii. p. 216.

Eudynamys flindersii, Gould, Birds of Australia, fol, vol. iv. pl. 91.

It will be seen by the list of synonyms quoted above, that the young and the adult have been considered as distinct species, and that the specific name *Flindersi*, which I have retained from its priority, has been applied to the bird in one of the earliest stages of its existence after leaving the nest, when the prevailing tints of its plumage are rufous brown, with transverse markings of dark brown; from this state until the bird attains maturity, many parti-coloured changes of plumage occur; but whether the sexes when fully adult are alike in colouring, I have not been able to ascertain; I am inclined to think they are not, and that the specimens having the upper surface regularly spotted with white on a bronzed olive ground, and with zigzag marks or bars on the buffy white of the under surface, are adult females.

This bird is very abundant in all the brushes of the east coast, from the river Hunter to Moreton Bay, and thence round to Torres Straits; it was also found in considerable abundance by Sir George Grey on the north-west coast. I did not meet with it myself, and I regret to say that no information has yet been obtained respecting its habits and manners. I should be glad to know if it be parasitic or not, and also the size and colour of its egg.

The adult male has the entire plumage deep glossy greenish

blue-black, the green tint predominating on the back and wings; irides red; bill yellowish olive; feet purplish black.

The adult female has the head and neck glossy greenish black; back, wings, and tail bronzy brown, with numerous oblong spots of white on the back and wing-coverts, the remainder of the wing crossed by irregular bars of white, stained with rufous; tail regularly barred with white stained with rufous, and slightly tipped with white; line from the angle of the mouth and all the under surface white stained with buff, spotted with black on the sides of the throat, and crossed on the abdomen and under tail-coverts with narrow irregular lines of blackish brown.

The young has the head and upper surface mingled bronze and buff, disposed in large patches; wing-coverts reddish buff, crossed by narrow bands of brown; remainder of wings and tail bronzy brown, crossed by bands of rufous; under surface rufous, crossed by narrow bars of blackish brown; tail-feathers longer and more pointed than in the adult.

Genus CENTROPUS, *Illiger*.

On reference to my account of the *Centropus phasianus*, it will be seen I have stated that some difference occurs in specimens from different localities, intimated a belief of there being more than one species, and remarked that should such prove to be the case, the term *macrourus* might be applied to the Port Essington birds, and *melanurus* to those from the north-west coast; and these names are provisionally retained until future research has proved that they are the same species. The birds of this genus have a harsh and spiny kind of plumage, and one of their hind toes armed with a lengthened spur-like claw. The old and young differ considerably in colour, the prevailing hue of the former being black or blue, while the latter are brown. Species of this form are found in Asia, Africa, and Australia.

Sp. 389.   CENTROPUS PHASIANUS
           Pheasant-Coucal.

Cuculus phasianus, Lath. Ind. Orn., Supp. vol. ii. p. 30.
Polophilus phasianus, Leach, Zool. Misc., vol. i. p. 116. pl. 46.
—— *variegatus*, Leach, Ib., pl. 51.
—— *leucogaster*, Ib., p. 117. pl. 52.
—— *gigas*, Steph. Cont. of Shaw's Gen. Zool., vol. ix. p. 45.
Cuculus gigas, Cuv. Règn. Anim., tom. i. p. 420, not. 1.
Corydonix phasianus, Vieill. N. Dt. d'Hist. Nat., tom. xxxiv. p. 293.
—— *giganteus*, Vieill. Ib., p. 295.
—— *variegatus*, Vieill. Ib., p. 298.
—— *leucogaster*, Vieill. Ib., p. 299.
Centropus gigas, Steph. Cont. of Shaw's Gen. Zool., vol. xiv. part i. p. 214.
—— *variegatus*, Steph. Ib., p. 214.
—— *phasianus*, Steph. Ib., p. 214.
—— *leucogaster*, Steph. Ib., p. 214.
—— *phasianinus*, Blyth, Cat. of Birds Mus. Asiat. Soc., Calc. p. 78.

Centropus phasianus, Gould, Birds of Australia, fol., vol. iv. pl. 92.

The *Centropus* inhabiting New South Wales differs from that found at Port Essington in having a much shorter and more arched bill, and in being somewhat smaller in size; specimens from the western coast again differ in being smaller than the bird of New South Wales, in having a more attenuated bill and a more uniform colouring of the tail. The greater part of the coast-line of New South Wales, the eastern, northern, and north-western portions of Australia generally are tenanted by *Centropi*, but only in such situations as are favourable to their habits, namely swampy places among the brushes abounding with tall grasses and dense herbage, among which they run with facility, and when necessity prompts, fly to the lower branches of the trees, from which they ascend in a succession of leaps from branch to branch until they nearly reach the top, and then they fly off to a neighbouring tree. The most westerly part of New South Wales in which I have heard of their existence is Illawarra, where they are rare, and from whence to Moreton Bay they gradually increase in numbers.

The nest, which is placed in the midst of a tuft of grass, is of a large size, composed of dried grasses, and is of a domed form with two openings, through one of which the head of the female protrudes while sitting, and her tail through the other. At Port Essington the nest is sometimes placed among the lower leaves of the *Pandanus*, but this occurrence seems to be rare; a large tuft of long grass being most frequently selected, as affording a better shelter. The eggs are from three to five in number, nearly round, and of a dirty white, in some instances stained with brown, and with a rather rough surface, somewhat like that of the eggs of the Cormorant; they are about one inch and four lines long by one inch and two lines broad.

By dissection I learn that the males are always smaller than the females; it also appears that when fully adult both sexes are alike in plumage, and have the bill, head, neck, and abdomen black, whereas the young has the bill horn-colour, and the same parts which are black in the adult, of a deep brown with a tawny stripe down the centre of each feather.

The adults of the present species have all the feathers of the upper and under surface dull black with glossy black shafts; wing-coverts mottled tawny brown and black, each feather with a conspicuous tawny shaft; remainder of the wing rich reddish chestnut crossed with irregular double bars of black, the interstices between which fade into tawny on the outer webs of the primaries; lower part of the back and upper tail-coverts deep green freckled with black; tail dark brown glossed with green, and minutely freckled with rufous and pale tawny, the latter hue assuming the form of irregular and interrupted bars, all but the two centre feathers tipped with white; bill black; feet leaden black, the scales lighter.

The young have all the upper surface reddish brown with glossy conspicuous tawny shafts; the throat and breast tawny with lighter-coloured shafts; in other respects the colouring is similar to the adult, except that the markings of the tail are more distinct.

The eyes of the birds at Port Essington are said to be red.

Sp. 389. CENTROPUS MACROURUS, *Gould.*
GREAT-TAILED COUCAL.

Centropus macrourus, Gould, Birds of Australia, fol., vol. 1. Introd. p. 68; and vol. iv. text to pl. 92.

Mr. Gregory informs me that "this bird is almost invariably found in thickets or cane-brakes near water, and appears to live principally upon seeds and insects; but one was observed devouring a Cockatoo that had been recently shot and fallen in a jungle. In flight it somewhat resembles the Common Hawk, but with an irregular and uncertain movement; it runs over the ground with great speed, and then resembles a hen Pheasant; in the trees their actions are much like those of Jays and Magpies. The eye has a rich golden tint, and is remarkably keen and Hawk-like. The body is slender, and by no means fleshy. We observed it generally in small companies, and sometimes only in pairs."

Sp. 390. CENTROPUS MELANURUS, *Gould.*
BLACK-TAILED COUCAL.

Centropus melanurus, Gould, Birds of Australia, fol., vol. 1. Introd. p. 68; and vol. iv. text to pl. 92.

See remarks in *Centropus phasianus.*

END OF VOL. I.

A HANDBOOK
TO THE
BIRDS OF AUSTRALIA.

BY

JOHN GOULD, F.R.S., &c.

This day is published, Vol. I. of the above work, in royal octavo, containing 630 pages, price £1 5s.; and on the 2nd of December will be issued the second and concluding volume, at the same price.

Extract from Preface, vol. i.

"Nearly twenty years have elapsed since my folio work on the Birds of Australia was completed. During that period many new species have been discovered, and much additional information acquired respecting those comprised therein; it therefore appeared to me that a careful *résumé* of the entire subject" [in an octavo form] "would be acceptable to the possessors of the former edition, as well as to the many persons in Australia who are now turning their attention to the ornithology of the country in which they are resident. Indeed I have been assured that such a work is greatly needed to enable the explorer during his journeyings, or the student in his quiet home, to identify the species that may come under his notice, and as a means by which the curators of the museums now established in the various colonies may arrange and name the collections entrusted to their charge."

PUBLISHED BY THE AUTHOR,

AT 26 CHARLOTTE STREET, BEDFORD SQUARE, W.C.

September 1st, 1865.

www.ingramcontent.com/pod-product-compliance
Lightning Source LLC
Chambersburg PA
CBHW021222300426
44111CB00007B/399